The Jonglei Canal: impact and opportunity

Frontispiece. Dinka cows resting in a cattle camp in the early morning. They are tethered to pegs to which they return each night. The white colour and the long horns are characteristic; both animals have their ears cut as marks of ownership. Photo: Alison Cobb.

THE JONGLEI CANAL

Impact and Opportunity

Edited by

PAUL HOWELL, MICHAEL LOCK AND STEPHEN COBB

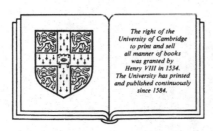

The right of the
University of Cambridge
to print and sell
all manner of books
was granted by
Henry VIII in 1534.
The University has printed
and published continuously
since 1584.

CAMBRIDGE UNIVERSITY PRESS

Cambridge

New York   New Rochelle   Melbourne   Sydney

CAMBRIDGE UNIVERSITY PRESS
Cambridge, New York, Melbourne, Madrid, Cape Town, Singapore, São Paulo, Delhi

Cambridge University Press
The Edinburgh Building, Cambridge CB2 8RU, UK

Published in the United States of America by Cambridge University Press, New York

www.cambridge.org
Information on this title: www.cambridge.org/9780521105491

First published 1988
This digitally printed version 2009

A catalogue record for this publication is available from the British Library

Library of Congress Cataloguing in Publication data

The Jonglei Canal: impact and opportunity / edited by P. Howell, M. Lock, S. Cobb.
p.   cm. – (Cambridge studies in applied ecology and resource management)
Bibliography: p.
Includes index.
ISBN 0 521 30286 2
1. Jonglei Canal (Sudan) – Environmental aspects.   2. Ecology–Sudan–Jonglei Canal
Region.   3. Agriculture–Economic aspects–Sudan–Jonglei Canal Region.   4. Jonglei
Canal Region (Sudan)–Economic conditions.   5. Jonglei Canal Region (Sudan)–Social
conditions.   I. Howell, P. P. (Paul)   II. Lock, J. M. (Michael)   III. Cobb, S. M.
(Stephen)   IV. Series.
HE509.3.Z5J6645 1988
330.9269'3–dc 19

ISBN 978-0-521-30286-9 hardback
ISBN 978-0-521-10549-1 paperback

CONTENTS

Contents

Contents   xv

PREFACE

A diversion canal to carry water past the swamps of the *Sudd* region in the Sudan has been under consideration since the beginning of the century. Preparations for the first phase of the present scheme began in earnest in 1974 and construction in 1978. Six years and 260 kilometres later this was brought to an abrupt halt by dissident southern Sudanese forces. The canal and its impact is neither central to the dispute that underlies the rebellion nor a primary reason for the present civil war. It has, nonetheless, been the cause of suspicion and controversy, partly because of misunderstanding of the nature and extent of its local impact, and partly because of fears that promises of measures to alleviate difficulties it might cause may not be fulfilled. The book has therefore been written in a sensitive situation and in circumstances in which gaps in available knowledge could not be filled by further investigation or enquiry on the spot.

The book may be critical. For example, it draws attention to the restriction of recent research and development activities to the most easily accessible parts of the area and the disregard of the interests of those living farther afield who will also be affected by the operation of the canal. It is, however, in no sense partisan. It is a purely factual presentation of the nature of the canal and its local impact and an assessment of some of the disadvantages it will bring as well as the new opportunities for progress it will offer.

From 1955 until the Addis Ababa agreement of 1972, civil disturbance made research and development virtually impossible in the area. Between then and 1983, when hostilities were renewed, the time for such endeavours was all too short, especially since investigation did not begin in earnest until 1976. In the short period available more knowledge of the area was acquired and progress towards solutions to the very difficult problems was achieved, but those concerned inevitably found that the constraints of an intractable and highly unstable environment made short-term answers elusive. Despite frustrations and failures however, promising results were just beginning to emerge when work was halted – another lost opportunity to add to the tragedy of the present civil war.

Even if there were adequate quantitative data to do so, which is not the case, no attempt has been made to weigh in the balance the benefits of conveying much needed water for irrigation downstream against the much less clearly evident advantages and disadvantages of the project to the people of the area through which the canal passes. The reader must form an independent opinion on whether the canal, if ever completed, is likely to result in an unqualified ecological disaster as some would claim – though that is not a view to which the editors subscribe – or a development measure designed for the area (which it is not) that will bring immense local benefits with few, if any, undesirable effects. That is not a view shared by the editors either. It is simply to be hoped that the book, by presenting a summary of the best evidence so far available, supported by an extensive bibliography, will provide a starting point and some guidance for those, Sudanese and expatriate alike, who may be concerned with the rehabilitation and subsequent development of a totally war-devastated land, as well as for those involved with the negotiation and planning for the revival of the canal project when peace returns.

The book is part of a series established by the Cambridge University Press of studies in Applied Ecology and Resource Management. In considering the *Sudd* region, the need for reconciliation between conservation and development, one of the main themes of the series, is especially apparent. In a sense the complexity of the issues presents a whole range of case studies rather than a single one. These issues begin with hydrological effects, though it must be borne in mind that hydrological changes due to natural climatic fluctuations over the centuries have been far, far greater than any that could ever be caused by a canal of the present proposed capacity. Hydrological changes are followed by ecological modifications, and, given the way in which the local economy is finely adapted to the environment, these will require adjustments in the present mode of livelihood of the people as the book shows.

The scale of the area, and the intricate interrelationships of environmental and human factors within it, carry the questions posed across the boundaries of an unusually wide range of disciplines. Some attempt has therefore been made to order the diverse contents in such a way as to present a coherent picture that crosses quite naturally the confines of the biological sciences at one end of the spectrum to those of the social sciences at the other. Sources of information have been by no means equal in extent, depth and reliability. In particular, reliable quantitative economic data are almost entirely lacking.

Many of the authors had the advantage of working together in the field, although the investigation of the Jonglei area in more recent years has manifestly lacked the full co-ordination and inter-disciplinary control that the complexities of the problems demand. There have, however, been the opportunities for dialogue and consultation between contributors, and in some cases the editors have put together chapters from the contributions of a number of people. It is the hope of the editors that a measure of continuity has thus been achieved, resulting in an integrated and cohesive presentation rather than a succession of loosely connected essays which is often a feature of multi-authored works of this kind.

ACKNOWLEDGEMENTS

Our thanks are due to the Royal Geographical Society and its Director, Dr John Hemming, who arranged the conference in 1982 which gave rise to the idea of this book, and under whose auspices it is presented. Our special thanks go to Dr B. H. Farmer, Fellow of St John's College, Cambridge and Editor of the *Geographical Journal*, who has been a constant source of advice and encouragement in the four and a half years of its gestation. We are especially grateful to St. John Armitage, CBE, as the promoter of material support without which the enterprise might well have foundered. We extend thanks to the following for advice on the text: G. L. Ackers, MA, MEng, FICE, R. L. Raikes, CEng, FICE, FSA, and Dr Godfrey Lienhardt, Fellow of Wolfson College, Oxford. Special thanks go to the Ford Motor Company, whose generosity has helped with the costs of producing this book.

We are particularly grateful to D. M. Jones, BVM, MRCVS, Director of Zoos, Zoological Society of London, for constructive advice on the chapter on livestock, and to A. T. Grove, Fellow of Downing College, Cambridge, for his contribution to Chapter 4 and advice on Appendix 2. We are also grateful to Jonathan Kingdon not only for advice on the chapter on Wildlife [but also for the design for the cover], and to Dr Gordon Wells of the Lunar and Planetary Institute, Houston, for arranging for the supply of photographs from Space Shuttle missions from the National Aeronautics and Space Administration (NASA). Most importantly, we wish to thank all those workers in the field in the southern Sudan, both Sudanese and expatriate, who have provided material of great importance in the production of this book, especially Ian Aikman, John Goldsworthy (who also helped in the preparation of the photographs), Dr M. R. Litterick, B. M. McWeeny, Giuseppe Pezzotta, Mrs B. M. Trottier and Paul Waring.

Finally, we must thank the successive Commissioners of the Executive Organ of the National Council for the Development Projects for the Jonglei Canal Area, and their staffs, at whose request much of the investigational and research work described in this book was undertaken.

Errata (1988 printing)

p.xxii, line 15   *insert* Author of Chapter 9; contributor to
                   Chapters 19 and 20. *after* 1980-3.

p.xxii, line 16   *for* Nuer, 1975-6 *read* Dinka, 1982-4

EDITORS AND CONTRIBUTORS

The terms 'author', 'co-author' and 'contributor' are not used rigidly in this list; as explained in the Preface, there has been extensive collaboration and editorial re-writing throughout.

EDITORS

Dr Paul Howell, social anthropologist. Chairman of the Jonglei and Southern Development Investigation Teams, Sudan, 1948–55; Chairman of the East African Nile Waters Co-ordinating Committee,, 1955–61; Head of the Middle East Development Division, 1961–9; Fellow of Wolfson College and Director of Development Studies Courses, University of Cambridge, 1969–83. Author of Chapters 1, 2 and 20; co-author of Chapters 10, 11 and 18.

Dr Michael Lock, consultant botanist and ecologist. Member of the Range and Swamp Ecology Surveys of the Jonglei Area, 1979–83 Currently Research Fellow, the Royal Botanic Gardens, Kew. Author of Chapter 7; co-author of Chapters 8, 12, 13, 15, 16 and 17.

Dr Stephen Cobb, zoologist. Study Director of the Range Ecology Survey of the Jonglei Area, 1979–83. Director for IUCN of research team in the Niger Inland Delta, 1984–7. Co-author of Chapters 12, 15 and 19; contributor to Chapters 2 and 20.

CONTRIBUTORS

Dr Roland Bailey, fish and fisheries biologist. Study Director of the Swamp Ecology Survey of the Jonglei Area, 1981–3. Lecturer, Department of Human Environmental Science, King's College, London. Author of Chapter 14; co-author of Chapter 17.

Dr Tjark Struif Bontkes, agronomist. Agricultural adviser, Ilaco, concerned with agricultural field trials in Bor District, 1980–2. Lecturer, agricultural University, Wageningen, The Netherlands. Contributor to Chapters 13 and 19.

Timothy Fison, veterinary adviser, Range Ecology Survey of the Jonglei Area, 1979–83. Currently Veterinarian, Tanzania. Contributor to Chapters 12 and 15.

Dr Douglas Johnson, historian. Field research among the Nuer, 1975–6. Assistant Director for Archives in the Regional Ministry of Culture and Information, Juba, Sudan, 1980–3.

Andrew Mawson, social anthropologist. Fieldwork among the Nuer, 1975–6. Ioma Evans-Pritchard Junior Research Fellow, St Anne's College, Oxford, 1985–7. Co-author of Chapters 10 and 11.

Yvonne Parks, hydrologist, Institute of Hydrology, Wallingford, Oxon. Co-author of Chapters 5 and 16.

Michael Prosser, consultant botanist, limnochemist and algologist. Member of Swamp Ecology Survey, 1981–2. Author of Chapter 6.

Michael Rae, livestock production adviser, Range Ecology Survey, 1979–83. Livestock Project Officer, Wau, Sudan and Botswana, 1984–8. Contributor to Chapters 12 and 17.

Dr John Sutcliffe, hydrologist. Member of Jonglei and Southern Development Investigation Teams, 1950–55. Institute of Hydrology, Wallingford, Oxon., 1961–85. Currently hydrological consultant. Author of Chapter 4; co-author of Chapters 5, 8 and 16.

Professor John Waterbury, William Stewart Tod Professor of Politics and International Affairs, Woodrow Wilson School of Public and International Affairs, Princeton University, USA. Author of Chapter 3.

Sjöerd Zanen, social anthropologist, Ministry of Foreign Affairs, The Netherlands Government. Researcher in socio-economic aspects of development projects in Bor District under Ilaco and later BADA (1978–84). Co-author of Chapter 11.

ORTHOGRAPHY

The spelling of place and tribal names, as well as vernacular terms, present particular difficulties. The rendering of Arabic names has been wholly inconsistent for a long period of time. Take the Bahr al Jabal (the main channel of the White Nile running through the *Sudd*) as an example. Early travellers usually spoke of this simply as the White Nile, though they did refer to its western tributary, as, e.g. the *Bahr el Gasall* (Werne, 1849), *Bahr el Gazal* (Baker, 1863), *Bahr el Ghazal*, (Petherick, 1869), while the *Bahr al Zaraf* sometimes *Bahr ez Zeraf*, was simply the Giraffe River. Since then there have been various versions – e.g. Egyptian: Behr el Jebel, Behr el Gebel. The correct version is undoubtedly *Bahr al Jabal* and is now commonly used. However, maps produced during this century and the Gazetteer of the Sudan Survey Department consistently refer to the *Bahr el Jebel, Bahr el Zeraf, Bahr el Ghazal* spellings (apart from the Egyptian variant the hard G for the soft J) adopted in most of the engineering and hydrological literature and reports to which reference is made. We retain these spellings. In Arabic the definite article *al* is assimilated to certain consonants. *Bahr el Zeraf*, for example, is pronounced *Bahr ez Zeraf (Bahr az Zaraf)*, but it is incorrect to spell it so. We do not adopt phonetic symbols for Nilotic languages (e.g. ŋ for *ng* or ɣ for *gn*), except the fairly standard *c* for ch. Thus: *cic* for chich Dinka, *toic* for toich.

GLOSSARY

aying	Dinka word for rain-flooded or 'intermediate' grassland; notably the plains east of Bor/Kongor Districts – part of the Eastern Plain (q.v.)
Bahr	Arabic for sea or river – hence Bahr el Jebel, 'the river of the mountain', etc.
Creeping flow	Overland or sheet flooding (See Appendix 3).
Eastern Plain	Originally defined by the JIT as 'the country north of a line from Bor to Pibor and east of the Bahr el Jebel'. The northern boundary inferred from JIT maps is the River Sobat. The Eastern Plain is associated with rain-flooded grasslands, mainly *Hyparrhenia rufa* and *Setaria* spp. The SDIT extended the use of this geographical term to include four sections, stretching from the Imatong mountains and adjacent hills in the south, west to the Zeraf Island, east to the Pibor-Kengen catchment and north to the Sobat. The term is very loosely applied, but more recently has come to refer more specifically, to the grasslands east of Bor and Kongor Districts and east of the Duk ridge.
Gilgai	An Australian term to describe a regular pattern of low mounds and depressions occurring on fairly level ground (see Appendix 1).
hafir	Arabic word used to describe an excavated water tank fed by rainfall.
heglig	Arabic for the tree *Balanites aegyptiaca*
high land	term used to describe higher, better-drained land where

	people have their more permanent homesteads and cultivations.
intermediate land	land lying in level between high land and *toic*, usually associated with rain-flooded grasslands.
jebel	Arabic for hill or mountain.
khor	Arabic for watercourse or drainage channel.
luak	Common Nilotic word for cattle-byre – a large conical structure (see Plate 19) of timber, mud and wattle walls and a thatched roof.
Machar Marshes	The marshland and swamp flooded by inflow from torrents from the Ethiopian uplands and overspill from the rivers Sobat and Baro, and lying north of the Sobat and east of the White Nile.
Monythany	Nilotic word describing people who have taken to fishing as their principal mode of livelihood and living mainly in the heart of the *Sudd*.
Sudd, sudd	from the Arabic root ـسد meaning 'to block', as in a pipe. This has come to mean a 'dam' (e.g. *Sadd al a'ali*: the high dam at Aswan), 'barrier', 'barrage', etc. The word is used loosely to describe the whole area of swamps through which the White Nile system (here the Bahr el Jebel and Bahr el Zeraf) flows. Sometimes referred to as the 'White Nile floodplain', but incorrectly since that could apply to the whole length of that river where flooding occurs. *sudd* is used to refer more specifically to permanent swamp and the vegetation associated with it (e.g. Papyrus, *Typha*, *Vossia* etc), particularly in the context of blockages in river channels – i.e. *sudd* blocks. We refer in this book to the *Sudd* region when the context is the whole geographical area of the floodplain, including seasonally inundated grasslands, of the Bahr el Jebel and its distributary, the Bahr el Zeraf. While similar blockages occur in the Bahr el Ghazal and might be referred to as *sudd* blocks, the Bahr el Ghazal basin, enclosed in the west by the Nile–Congo Divide, is not the *Sudd* or *Sudd* region, except that there is a variable and very indeterminate line where runoff from the Bahr el Ghazal basin meet overspill from the Bahr el Jebel. The JIT and SDIT referred to the whole of this hydrologically and ecologically similar region, including also

the Machar Marshes, as the 'Flood Region'. This description has not been current recently and is not used in this book, but could prove relevant in the future when describing the whole area likely to be subject to the drainage effects of the future schemes described in Chapter 20.

thung   (Nuer), *apuk* (Dinka), 'bloodwealth' or compensation, mainly in cattle, for homicide: Arabic: *dia*.

toic   (pronounced *toich*) common Nilotic word describing seasonally inundated floodplain or river-flooded grass-lands to which people and livestock migrate in the dry season.

tukl   Arabic for conical dwelling hut of timber structure, mud and wattle walls, and plain or ridged thatched roof (see Plate 25).

Western Plain   Not used in this book. Sometimes used very loosely and imprecisely to describe grasslands (comparable but not hydrologically identical to the Eastern Plain) west of the Bahr el Jebel.

Zeraf island   (sometimes Zeraf Island) the area bounded by the Bahr el Jebel in the west, Bahr el Zeraf in the east, and the Bahr el Abiad (White Nile) in the north.

ABBREVIATIONS

Institutional and general

ARE	Arabic Republic of Egypt
BADA	Bor Area Development Activities
CCI	Compagnie de Constructions Internationales
DRE	Democratic Republic of the Sudan
EDF	European Development Fund
EWMP	Egyptian Water Master Plan
FAO	Food and Agriculture Organisation of the UN
Hydromet	Hydro-meteorological survey of the catchments of Lakes Victoria, Kioga and Albert (WMO)
IBRD	International Bank for Reconstruction and Development (World Bank)
ILACO (Ilaco)	International Landuse Company Arnhem, The Netherlands
ILCA	International Livestock Commission for Africa
IUCN	International Union for the Conservation of Nature and Natural Resources
JEO	Jonglei Executive Organ (National Council for the Development Projects for the Jonglei Area)
JIT	Jonglei Investigation Team (1948–53)
JSERT	Jonglei Socio-Economic Research Team
PCIWE	Public Corporation for Irrigation Works and Earthmoving
PDU	Project Development Unit (of the Regional Ministry of Agriculture)
PJTC	Permanent Joint Technical Commission for Nile Waters
RES	Range Ecology Survey (1979–83)
RWDD	Rural Water Development Department

SDIT	Southern Development Investigation Team (1953–54)
SES	Swamp Ecology Survey (1979–83)
SPLA	Sudan Peoples Liberation Army
UNCDF	United Nations Capital Development Fund
UNDP	United Nations Development Programme
UNICEF	United (International) Nations Childrens (Emergency) Fund
UNOPE	United Nations Office of Project Execution
WHO	World Health Organization
WMO	World Meteorological Organization

Scientific

CBPP	Contagious Bovine Pleuro-Pneumonia
CF	Crude Fibre
CP/DCP	Crude Protein/Digestible Crude Protein
DCP	Digestible Crude Protein
ITCZ	Inter-Tropical Convergence Zone
IV	*In Vitro*
ME	Metabolisable Energy

UNITS AND CONVERSION FACTORS

Units of flow

(In engineering reports on the Nile, agreements between Egypt and the Sudan, etc, annual discharges have usually been expressed in 'milliards', that is 1,000 million cubic metres. We have preferred to use cubic kilometres, km^3, which is gaining in international usage.)

$km^3 = m^3 \times 10^9 = $ milliard

million cubic metres per day = million m^3/day = $m^3 \times 10^6$ per day (= 11.6 cubic metres per second = 408 cusecs)

Units of area

1 feddan	$= 4,200$ m^2 $= 0.42$ ha $= 1.037$ acres
1 km^2	$= 100$ ha $= 1,000,000$ m^2

Units of energy

cal m^{-2}	= calories per square metre
MJ kg^{-1}	= megajoules per kilogram (4.1868 cal = 1 joule)

Units of concentration

g dl^{-1}	= grams per decilitre
l l^{-1}	= litres per litre

Units of standing crop biomass (of plants and animals)

g m^{-2} = grams per square metre
tonnes ha^{-1} = tonnes per hectare = g m^{-2} ÷ 100
kg ha^{-1} = kilograms per hectare = tonnes ha^{-1} × 1,000
kg km^{-2} = kilograms per square kilometre = kg ha^{-1} × 100

Units of production

g m^{-2} y^{-1} = grams per square metre per year

Units of animal density

Numbers
km^{-2} = numbers per square kilometre

Introduction

Background to the book

The White Nile, in its long course to the junction with the Blue Nile and onwards to the sea, passes through a great variety of country. In places it descends precipitously over falls and rapids; in others it meanders through fringes of marshland and swamp. In the southern Sudan it discharges its waters into the great wetland of the *Sudd*, a maze of channels, lakes and swamps in which as much as half the inflowing water is lost by evaporation.

The two Niles and their tributaries derive their waters from no less than eight different countries: Ethiopia, Uganda, Kenya, Tanzania, Zaïre, Rwanda, Burundi and the Sudan. The ninth country of the Nile basin, Egypt, adds nothing to the flow but the lack of rainfall there makes the contributions from upstream an absolute necessity for the livelihood of her people. Egypt, dependent on the waters of the Nile for millennia, has not only been much concerned to preserve what she regards as her natural and historic rights, but has perforce taken the lead in studying the hydrology of the Nile Basin and preparing projects for its control to suit her own irrigation interests. These endeavours, in partnership with the Sudan, have led to one such project, the Jonglei Canal, designed to bypass the *Sudd* and direct downstream a proportion of the water that is lost from the river each year by spill and evaporation in the swamps. That these 'losses' in fact create resources in pasture and fisheries for local use will be apparent later.

The scale of this undertaking is large: a canal with a width from bank to bank of about 75 metres, a channel bed-width averaging 38 metres, a varying depth of between 4 and 8 metres, and a length of 360 kilometres, over twice the length of the Suez canal. The measure of additional water thus made available will be substantial in terms of increased irrigation and agricultural production of importance to downstream users. Yet its total contribution in water saved, 'new' water as it is sometimes called, will be relatively small when expressed as a proportion of the total average Nile flow – around 5% of the mean at Aswan.

On 5th October 1982 a conference was convened by the Royal Geographical Society in London to discuss the Jonglei Canal Project, its potential benefits, and particularly its effects on the livelihood of the people through whose territory it passes in the southern Sudan. Speakers from the Sudan, both north and south, and people from other countries who had been working in the area, introduced papers on a variety of relevant topics. The meeting reached no conclusions and passed no resolutions, but gave the opportunity for debate which revealed a number of explosively controversial points of view as well as some obvious misconceptions. A brief summary of the papers presented was published in the *Geographical Journal* (Howell, 1983).

The canal must be seen in the perspective of its long history, for it has been under consideration since the beginning of the century. It must be viewed not only in the context of its downstream irrigation benefits but of its impact on the area through which it passes, and the complexity of the human problems that this raises. These problems have been under investigation, first between 1946 and 1954 when a diversion canal was one essential component of the proposed Equatorial Nile Project, and since 1974 when the decision to embark on the current canal was taken. Yet there are still unanswered questions which require further research. Moreover some effects will not be immediately apparent and continuous monitoring after the canal is operational will be essential (see Chapter 20).

A book to present the scientific facts, the product of a number of different professional disciplines, together with objective interpretation, seems desirable, as a general contribution to knowledge as well as to provide greater comprehension of the local issues and the way in which they might be resolved. The fact that the canal has come to an abrupt halt owing to violent confrontation between south and north in the Sudan seems to the authors no good reason to withhold publication. Indeed, since the canal itself became an early target in the conflict, it is to be hoped that what we have to say may lead to greater mutual understanding. If the book has any message, it is that those concerned with technical and policy decisions in connection with the project and its ultimate aim – more water for irrigation downstream – should avoid complacency about any possible adverse effects, for there are obvious local disadvantages which are often ignored as well as benefits that are sometimes overstressed. Equally, those whose interests lie in the southern Sudan and in the Jonglei area in particular must not allow their judgement to be distorted

by over-anxiety or conclusions based on misconceptions. All·must strive for
the maximisation of benefits and the minimisation of drawbacks.

The key here is closer consultation between those who seek to benefit from
increased water downstream, and those who live in the area and whose
livelihood depends also on the optimum use of the same water resources. Such
objectives can only be achieved by continued investigation, and in particular
at this stage by programmes of longer-term research, experiment and trial,
such as were started again in recent years but have now once more been
abandoned. In an environment which is beset with problems of unpredictability and in which there are physical obstacles to improved production in both
rain-fed agriculture and animal husbandry, the bases of the local economy,
these may take many years to resolve.

There is a second reason why publication seems desirable. The Jonglei
project and the *Sudd* region have been the subject of scientific investigation
and observation in time and scope rarely paralleled in Africa. Such results as
are available are of great importance not only in the solution of development
problems in that area, irrespective of whether the canal is completed or not,
but relevant to the examination of similar problems elsewhere. The book,
which contains what must necessarily be summary presentations, is also
intended to provide a guide to more detailed accounts which can be found by
reference to the bibliography.

The current project unfortunately started in a highly charged political
atmosphere. The sudden announcement in 1974 of its impending implementation caused unfavourable reactions among southern Sudanese, partly due to
general unease over the lack of consultation on a matter that was clearly of
deep concern to them, but also owing to confusion between the present
scheme and that proposed as part of the former Equatorial Nile Project, the
predicted effects of which were a cause of some anxiety in the mid-1950s. As
will be made clear, however, any adverse effects caused by the operation of the
present canal are not likely to reach the magnitude of those foretold under the
earlier plan. The most alarming feature of that scheme was the manipulation
of discharges in the upper White Nile in timing and amounts to serve the
irrigation interests of downstream users in a way which would virtually have
reversed the natural seasonal rise and fall of the river in the *Sudd* region. This
alteration of the natural river regime would in a number of ways have
deprived the people of the Jonglei area of a high proportion of the resources
essential for the all-important pastoral component of their economy. This no
longer applies since seasonal irrigation requirements are now controlled at the
Aswan High Dam. All subsequent references in this book to the Jonglei Canal
refer to the present project, construction of which began in 1978. Previous

schemes and projected expansions of the current project will be discussed in their turn.

Since the early 1950s there have been very marked changes in environmental conditions. Massive increases in flow into the *Sudd* from the East African catchment beginning in 1961 have caused a spectacular expansion of the swamps (see Figures 8.1, 8.2 and 8.3). Hydrological measurements have shown that the greater the flow of water into the *Sudd*, the greater the percentage losses from evaporation (see Table 1.1). This has led to ecological conditions in many ways less favourable to the local economy, and in some years and some areas there have been severe and very damaging floods. A harsh and unreliable environment has become more so. By carrying part of the flow down the canal the consequent reduction in discharges down the natural channels at times of peak levels or in peak years may have the effect of moderating exceptional floods, an important local benefit. However, if the river system returns naturally to a period of low years, which in the course of time seems likely, this advantage would not apply; physical conditions may change too much in the opposite direction, reducing the availability of seasonally river-flooded grassland at critical periods in the land-use cycle which is the basis of the rural economy.

A better understanding of the differences between the likely effects of the present Jonglei Canal and those of the earlier Equatorial Nile Project, and also of the change in natural conditions in the area since then, might have served to moderate unfavourable local and international reactions to the project. In 1974, the Sudanese Government were wise enough to establish the National Council for the Development Projects for the Jonglei Canal Area and its so-called Executive Organ. The work of the Jonglei Executive Organ (JEO), in particular a number of studies commissioned on its initiative, has been a primary source of information.

The role of applied environmental studies in Africa

Inadequate attention to factors other than the technical engineering and projected economic implications of large-scale irrigation or drainage schemes in Africa has all too frequently led to great difficulties. Decisions to embark on these costly ventures have often been made in the absence of sound objective assessments of their environmental and social implications.

This book forms a part of a series that aims to demonstrate that applied ecological research is not just valuable, but essential; political and economic decisions about managing natural resources are made without a true assessment of their environmental implications at the peril of the environment itself and of succeeding generations who might otherwise have depended on it. The book does not provide a systematically complete or mathematically precise cost–benefit analysis or suggest in quantitative economic terms a comparison between the natural use of water locally and its controlled use downstream; it does not attempt to model, or even to describe, all aspects of an ecosystem. We have, however, endeavoured to assemble such material as is available which has some bearing on the problems likely to be caused by the canal, its operation and local effects, believing – like the Jonglei Investigation Team (JIT) before – that in the examination of a project of this magnitude there needs to be the closest possible co-ordination of applied research in all relevant disciplines.

Research in the Bahr el Jebel floodplain

As we shall show in Chapter 1, the *Sudd* has been subjected to over a century of scrutiny, in the nineteenth century in the form of observations by early travellers, opportunistic yet remarkable for their perspicacity, such as Werne, Baker, Petherick, Poncet, Schweinfurth, Casati, Gordon, the dauntless Mademoiselle Tinné, and many others. At the turn of the century more systematic hydrological enquiry began, starting with the journeys up and down the river by engineers of the Egyptian Ministry of Works, the foundation of one of the most comprehensive hydrological survey and gauging services on any international river. Following the emergence of successive plans to direct water past the *Sudd* in order to reduce the losses and the eventual realisation that this would have substantial local repercussions, there have been two periods of intensive research in the area. It is these investigations that have made the writing of this book possible.

The Jonglei Investigation Team and something of its work is described in the next chapter. Between 1946 and 1954, the date of publication of its final report, this body was concerned with *The Equatorial Nile Project and its Effects in the Anglo-Egyptian Sudan*. That project included plans for annual and over-year storage in the Great Lakes of East Africa and an artificially controlled river regime the effects of which would have been felt in the Sudan all the way from Nimule, on the Uganda border, to Kosti some 1,100 km below. Of necessity this ambitious project also included a diversion Canal – the Jonglei Canal – to bypass the *Sudd*. This book leans heavily on the findings of the team, many of them as pertinent today as they were 30 years ago, and the work is referred to, throughout the text, as JIT (1954).

After 20 years in abeyance brought about both by the construction of the Aswan High Dam in Egypt and by the protracted civil war in the southern Sudan, the Jonglei Canal once more became an imminent reality and a new phase of serious enquiry began after the cessation of north–south hostilities in

1972. The Dutch Government undertook to finance a long-term investigation of the feasibility of large-scale mechanised irrigated crop production in the vicinity of the Jonglei Canal as then planned, for at that time the area was thought to have vast, untapped agricultural potential, which it was supposed could be exploited for cereal production on a massive scale. In addition to the agronomic studies that this undertaking involved, there were a number of other enquiries in the vicinity of the canal, including the socio-economic circumstances of people in Bor and Kongor districts. Much of this work was undertaken by Ilaco, a Dutch agricultural and rural consultant development company. Meanwhile, their sister company, Euroconsult, was engaged in the design of the Jonglei Canal itself, which association enabled them to contribute a number of short reports on certain aspects of its potential effects. The University of Khartoum, too, had research interests in the *Sudd* region, notably through the Hydrobiological Research Unit and through the Social and Economic Research Unit. The latter, in particular, was engaged by the Sudanese government to undertake basic socio-economic research in the area; some field-work was accomplished and a number of reports published between 1975 and 1979.

In 1978, the European Development Fund undertook to assist the Sudanese government by financing a further series of studies that were felt necessary in the light of the imminent start of construction work on the Canal. The purpose of these was never to determine whether or not it was appropriate, locally, to proceed with the construction of the Canal, for that was a decision that had long since been made. Rather, it was to investigate aspects of the local natural environment and its traditional exploitation, to comment on the likely effects of the Canal on these and, in the light of those effects, to make development proposals for the future.

The work was divided into three distinct parts; the first of these, a soil survey and suitability classification of the land 30 km either side of the line of the Canal, was awarded jointly to the Sudanese Soil Survey Administration and to Euroconsult. A number of technical problems delayed this work. The second part was a study of the terrestrial ecology of the area to be immediately affected by the operation of the canal. This study was directed towards the pastoral system generally and livestock health and production in particular, but also concerned the wildlife and the availability of water supplies, both for people and for animals. The results of this work, carried out in the field between 1979 and 1982, is referred to in the text as the Range Ecology Survey, RES (1983). The third part of the work was a study of the ecology of the river and the associated swamps. Referred to as the Swamp Ecology Survey, the work in the field took place in 1982 and 1983 (SES, 1983).

Enormous practical difficulties face anyone wishing to conduct field research in and around the *Sudd*. Not the least of these is that throughout the wet season, that is between early May and late November in a normal year, the entire area is impassable to motor vehicles, the concrete-hard black soil of the dry season having turned to the most slippery kind of mud. During the whole period of 1979 to 1983 the research teams, as did everyone else in southern Sudan, suffered from acute shortages of fuel (which was not, at the time, legally for sale anywhere in southern Sudan without a special signed purchase authority) and from the other chronic problems of communication and supply that beset what is, without doubt, one of the most remote and least-developed corners of the globe. Research methods and objectives had, as a consequence, constantly to be amended in the light of the general dearth of resources essential for an undertaking of this kind.

The methods used by the Range and Swamp Ecology teams are discussed later in the book, but it is worth outlining the overall survey strategy adopted. Each part of the programme tried to combine a mixture of intensive sequential sampling at one or two key sites, with more extensive sampling to give an overall picture. Thus most of the work of the Swamp Ecology Survey was concentrated on a series of lagoons and channels a few kilometres south of the village of Jonglei; this detailed work, which spanned all seasons of a complete year, was complemented by river trips throughout the length of river channels in the *Sudd*, and interpretation of satellite photographs, aerial photographs and low-level aerial survey information.

Exactly the same sequence of extensive data was available to the Range Ecology Survey, who based their work at Nyany, an old cattle-camp site some ten kilometres east of Jonglei. From this base, it was feasible to set up a transect of semi-permanent plots in each of the main grassland types, in which grass production, quality and consumption were measured over a period of 16 months, to follow monthly the life histories of several hundred cattle, sheep and goats, and to study the migratory wildlife. One feature of the work, drawn on repeatedly throughout the book, was a series of three low-level systematic aerial surveys, which were used to estimate the livestock and wildlife populations, to map their density and distribution, the major vegetation types and the distribution of people, their wet season habitations, and the features of their main economic activities. The aerial survey techniques involved are described more fully in Appendix 5.

The Jonglei area

Why Jonglei? Jonglei has never been more than a small, obscure Dinka village close to the Atem channel at a point where the earlier canal alignment was planned to begin (Plate 15); but although the offtake will now be farther south at Bor, the canal is still so named and Jonglei has given its name to a province as well.

We refer to the Jonglei area (see p. 201) throughout this book; by this we mean that area that will be directly affected by the physical presence or operation of the Jonglei Canal between Bor, the offtake point, and Malakal downstream of the tail of the Canal. This area is more or less equivalent to the 67,900 km² that were covered in each of the three aerial surveys conducted during the Range Ecology Survey, and is shown in Figure 10.1. Two areas have been neglected in the more recent field studies and hence in this narrative. These are the west bank of the Bahr el Jebel, between Shambe and Lake No, territory of western Nuer tribes, with some Dinka to the south and north of them; and the White Nile itself downstream of Tonga, home of the majority of the Shilluk, and on the opposite bank the most northerly Dinka tribes. We do not know precisely to what extent the livelihood of the people on the west bank was affected by the rise in the river in the early 1960s, nor are we in a position to predict the effect on that area of the operation of the Jonglei Canal. That requires further investigation as, indeed, is the case with the area enclosed by the Bahr el Jebel and Bahr el Zeraf, the 'Zeraf Island', where aerial survey was not accompanied by field observation. The Canal will cause the White Nile discharges to increase north of Malakal, with consequent effects on the river-flooded pastures that are so important in the dry season to Shilluk livestock and those of the northern Dinka opposite them; we have all too little information about this. Here again more research is needed.

The Jonglei Canal will have its effects far wider afield than just the Jonglei area; the improved communications it will provide may help to unite the Sudan; controversy over its impact may continue to divide it. Downstream the flow of additional water will help to meet the needs of ever-increasing populations in the northern Sudan and Egypt. This book is, however, primarily concerned with the local effects, though we have tried, where appropriate, to put them in the wider perspective of the Nile Basin as a whole.

The physical environment

The most remarkable topographic feature of the Jonglei Area is its flatness: for four hundred kilometres, from south to north, the slope of the land averages a mere 10 cm per km, and much of it is flatter even than that (see Figure 1.4). In only one place is this interrupted, by four small rocky outcrops near the confluence of the White Nile and the Bahr el Zeraf. The flatness of the landscape belies a fact well known to every car traveller and, more important, every livestock herder: that the ground is extremely rough and uneven. Paradoxically, in a landscape with so flat a gradient, the ground may rise and fall by 70 cm in only ten metres. The explanation of this roughness is partly to be found in *gilgai* formation, an Australian term that describes micro-topographical features in clay soils which expand on wetting and contract on drying. The soils of the whole area are generally rich in clay and poor in nutrients. They are fine-grained and well weathered, typical of an alluvial floodplain; material coarser than sand is non-existent. A more detailed description of the soils is to be found in Appendix 1.

Differences in height of just a few centimetres, undetectable to the eye, are all-important to plants and, in turn, to people. These small differences determine whether or not a place will receive floodwater from the river or from accumulations of rainfall, or a combination of both, moving slowly across the plains; and it is on this type of distinction that classifications of the landscapes have been based. The first of these was that of the JIT, who classified the land for descriptive purposes into 'high land', 'inter-mediate land', *toic*, or seasonally inundated floodplain, and permanent swamp. A slightly different approach was used by those involved in the Range Ecology Survey, who preferred to classify both the grasslands and the woodlands on the basis of the dominant species. Six such grasslands were identified, but these in turn, for the final analysis, were regrouped into the fundamentally distinct rain-flooded grasslands (equivalent to the JIT's high and intermediate lands) and river-flooded grasslands (equivalent to the JIT's *toic*).

These are by no means the only terms that have been used to describe the landscape of the Jonglei Area: we have already introduced the general term *Sudd*, which is used to describe the whole floodplain of the Bahr el Jebel between Bor and the mouth of the Zeraf river, but also more specifically the permanent swamp and the vegetation associated with it (see Glossary). Other authors have used the term 'Sudan Plains', which has relevance when one is thinking of the Nile basin as a whole, and in describing the descent of the Bahr el Jebel from the highlands of East Africa onto the flat plain north of Juba in the Sudan; but this is a term that we do not use.

Climate is the other feature of the physical environment that is crucial in the Jonglei area. Typically for its latitude, the rain falls in a single season, lasting from late April to November and varying from about 900 mm in the south to 800 mm in the north of the area. The rainy season coincides with, though is slightly shorter than, the flood season of the river, but together they produce an environment of remarkable extremes: a land of water and mud for half the year, and, away from the rivers, a land of desert-like dryness for the other half. Dry season temperatures in March and April regularly rise to

45 °C. This, and other features of the climate, are discussed in more detail in Appendix 2.

The inhabitants of the Jonglei area

The occupants of the Jonglei area are almost exclusively Nilotes, apart from a few people from other regions of the Sudan to be found in the main towns, such as Malakal, and some of the smaller administrative and trading centres. Nilotic peoples make up the majority of the inhabitants of the southern Sudan, living within, or adjacent to, the *Sudd* region. A standard but somewhat simplistic classification is shown in the diagram.

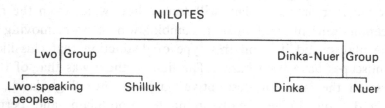

The main factor in this classification, but not the only one, is linguistic, for all these people have language characteristics indicating some elements of a common origin, though, except in the case of the Lwo-speakers, their languages are not mutually intelligible.

Nuer and Dinka are a group apart and have no close ethnic connections with other peoples in this or any other part of Africa. The Shilluk, who call themselves *Collo*, belong to a much larger, widely dispersed ethnic group (Crazzolara, 1950). These include the Anuak living on both sides of the Sudan–Ethiopian frontier east of the Jonglei area, and a number of small fragmented groups in the western Ghazal region, as well as the Pari of Jebel Lafon to the far south of the Sudan, and the Acholi clans that straddle the Sudan–Uganda border. Farther south in Uganda are the Lango, Teso, and Alur, and the Luo tribe, the second largest ethnic group in Kenya, living round the Kavirondo Gulf in Lake Victoria.

The definition of 'tribe' in this context is somewhat indeterminate and its use is necessarily relative. The Shilluk or Anuak, for example, are often spoken of as tribes, though we would refer to them as 'people', but the Nuer and Dinka, using the definition adopted by Evans-Pritchard (1940), are divided into a number of 'tribes'. This definition is that within the framework of the Dinka and Nuer peoples a tribe is the largest political unit occupying a common territory, and having a common name, whose members consider that they should combine in defence or aggression against other such groups, and consider that violence between their members should be composed by peaceful arbitration (Evans-Pritchard, 1940). Though thus divided into tribes the Nuer recognise themselves as a distinct people, somewhat arrogantly calling themselves *nei te naadh*, the 'people of the People', to distinguish themselves collectively from other groups, and Evans-Pritchard went as far as to refer to the 'Nuer nation'. Dinka, who call themselves *jieng*, would likewise recognise their ethnic affinity, though they are much more widely scattered geographically, generally speaking are divided into smaller units, and in many cases their territories separated by the Nuer whose country forms a central contiguous strip from west to east of the Nile (see Figure 10.2).

Nilotic societies are segmentary, that is tribes are further divided into recognised segments or 'sections', these into smaller units each having similar structural features to the tribe as a whole. This can lead to further difficulties of definition. Among the Nuer, tribal units and the sections within them have fairly clear lines of distinction as well as genealogical explanations of their origins, however far removed and fictitious such relationships may have become. The bases of Dinka units in the Jonglei area are often less easy to disentangle, many groupings having been scattered over the last two centuries by Nuer incursions into their territory. In many cases tribes or their segments were adopted as administrative and judicial units by the Condominium Government, tending to stabilise Nilotic groupings as they were found at that time. Such distinctions continue to be reflected in the organisation of courts and local government divisions today. They are significant also in the context of the Canal since seasonal migration routes across its alignment are determined by the territorial boundaries of tribes or sections thereof.

Presentation

Local problems likely to arise as the result of the Jonglei Canal are complex. Its operation and effects on the hydrological regime of the natural channels of the river, the repercussions of these in turn on the ecology of the area, and the impact of consequent changes in natural resources upon human interests, in some instances considerable distances away from the canal, are matters requiring the attention of a whole range of professional disciplines. For those living in its vicinity the physical presence of the canal will be a handicap to the seasonal movement of livestock on the one hand; on the other it will provide a source of water in an area that is waterless for at least four months in the year. Its advantages as a new major line of communication have other economic and social implications, and it may provide relief from the damaging effects of excessive flooding, at least in years of peak flows.

Most of the relevant disciplines are represented among our authors. A series of separate essays on the many different aspects would not, however, be appropriate even if it were possible to isolate the salient interrelated factors one from the other. An attempt has therefore been made to present the background and the many facets of the local impact the canal may have as well as the opportunities it may offer – the reason for the sub-title of the book – in some kind of sequential order. The book is itself addressed to a readership drawn from the same range of disciplines, and others too. It is written in a way which we hope will be intelligible to all. The book is intended to present the facts and their analyses; but while these in some instances do not fully substantiate the local benefits sometimes claimed for the project, we are concerned with facts, not with criticisms. We therefore hope that it may provide some guidance towards the solution of the problems, and there are many, as well as towards the most beneficial use of such advantages the project may bring to those who live in the Jonglei area.

Space precludes any detailed assessment of the merits of the project in connection with those downstream who will receive and make use of the water saved from the swamps. It is taken for granted that ways and means must be found to increase the amount of water required for the extension of irrigated agriculture in the Sudan and Egypt, particularly to meet the needs of expanding population. But while increased efficiency in water control suggests that some measure of drainage of the swamps of the upper White Nile is a necessary target, full account must continue to be taken of local human interests as well as of the reasonable requirements of environmental conservation such as the world now demands.

Part I outlines, in Chapter 1, the topographical and hydrological features of the Nile basin, the background history of plans for optimal control of its waters, and early debates on different possibilities for reducing losses in the *Sudd* region – the genesis of the Jonglei canal. Mention is also made of the earlier Equatorial Nile Project, the extensive investigation of the local effects of that scheme, and the differences from the present Canal plan. This is important because confusion between the effects of these two projects has led to many misconceptions. The nature of the present Canal project is then described in outline in Chapter 2: its inception, objectives, alignment, construction and plans for its management. The book is not, however, concerned with the detailed engineering aspects and does not attempt a review of the canal structures or morphological effects. In Chapter 3, Professor John Waterbury discusses the present and projected Nile water balance, the potential conflict of national interests in what is essentially an international river system, and makes his forecast of future supply and demand. This is to present the canal project in the perspective of the Nile basin as a whole and to gauge its relative importance.

One of the themes of this book is the interdependence of the people and their physical environment. This is described later in Part III, but Part II provides a more detailed introduction to the main characteristics of that environment.

Climatic change and variation in river flows of the White Nile in prehistoric and historic times are described in Chapter 4. These demonstrate the uncertainties which underlie predictions of the effects of the canal's operation in the future, as well as the fact that they are not likely to be anything near as great as the massive natural changes in the flooding regime that have taken place in the past. This does not mean, however, that they will not be substantial and in certain circumstances damaging. Chapter 5 gives a brief outline of the hydrology of the *Sudd* and describes the simple model that has been developed to quantify the hydrological processes in order to assess, at least in broad terms, areas of swamp and seasonally inundated land, as they were in the 1950s and early 1980s, as a basis for estimating (in Chapter 16) what they will be in the future under different natural river regimes modified by the operation of the canal. Chapter 6 describes the way in which the waters are modified in quality by their passage through the complex of lakes, swamps and channels of the *Sudd*. Chapter 7 is concerned with the ecology of plants in swamps and floodplains, and, together with Chapter 8, explains how the grassland types alter in value as grazing for domestic livestock and wild animals, both seasonally or annually. In addition large-scale changes in vegetation in response to periods of major variation in river discharges are demonstrated. Two other essential components of the environment, soils and climate, are described in Appendices 1 and 2.

The overall picture is one of instability. There is evidence of striking fluctuations in White Nile discharges in the past, but the massive increases in the 1960s have demonstrated the huge changes in flooding patterns and vegetation that result. While this instability may well be one of the factors that have given rise to the floristic poverty of the region, it is also a phenomenon which means that the effects of the canal, if completed, may be only one more disturbance in an historically changeable environment. Ecological changes caused by the operation of the canal may prove damaging to the pastoral component of the economy and fisheries resources, but must be viewed in this perspective.

Part III concerns the inhabitants of the Jonglei area, the nature of their society, and the way in which their predominantly rural economy is adapted to the risks and imbalances of the environment described in the previous chapters. Chapter 9 introduces the main features of human reaction to physical environment in historical perspective, demonstrating how fluctuations in natural conditions, including major environmental disturbances as well as external intervention, have affected movement, settlement patterns and the configuration of social and political units. In introducing Chapter 10 it is necessary to stress that there is no intention of implying that the people of the area have themselves remained unchanged in the last 50 years or more. There are evident features of modernisation, especially in the growth of the educated sector of the population, which includes many persons of intellectual distinction, administrative ability and political talent. Nevertheless, despite incipient economic diversification and some urbanisation, mainly concentrated in the towns of Malakal and Bor, the great majority of the people are still scattered throughout the rural areas and largely dependent on a mixed agricultural economy. Chapter 10 therefore begins with a description of the ways in which they endeavour to regulate their form of livelihood by varying their reliance on the exploitation of different natural resources, not only from season to season within one year, but from year to year according to fluctuations in climatic and ecological circumstances.

However, climate and ecological conditions, fundamental though they may be, are not the only factors contributing to economic and social change. Chapters 9 and 11 touch on the effect of the introduction of systematic administration by the government of the Anglo-Egyptian Condominium during the first half of the century. The imbalance in education and economic development during that period between the southern and northern parts of the Sudan, and a policy of separation rather than national integration except in the very last years before independence, was followed almost immediately after by grave public disturbance and civil war. The war spread to the Jonglei area in the mid-1960s, adding its destructive effects to the damaging consequences of the excessive flooding described in previous chapters. Chapter 11 includes an assessment of Dinka and Nuer responses to the economic and social problems people had to face during this period of great hardship and political upheaval. Peace was restored under the Addis Ababa Agreement of 1972 which provided regional autonomy for the southern Sudan. Compared with the previous 17 years the decade that followed gave the opportunity for growth and advancement, supported as it was by extensive national and international efforts in rehabilitation and development, which are described later in Chapter 19. These were not rewarded with immediate success; local disillusionment and political tensions generated by other events combined to create the Southern Peoples Liberation Movement, very much an expression of Nilotic revolt, and the outbreak of a new civil war. It is at this point that any evaluation of the processes of change which are identified and described in Chapter 11 comes to an abrupt end. The scale of violence, hunger and suffering is clearly much greater now than ever before, and much of the interpretation of research in the social sciences as was carried out before the new hostilities began could be a misleading indicator of what may happen in the future.

Part IV provides a more detailed account of the technical aspects of the different components of the local economy. Chapter 12, derived from the Range Ecology Survey, describes the significance of livestock, animal populations, types of stock, herd structures, productivity, growth rates, disease and animal nutrition, as well as the annual cycle of management and day-to-day care of the herds. Neither the Range nor the Swamp Ecology Survey were concerned with crop production, but the book would not be complete without an outline of this aspect of traditional agricultural practice and a rather more detailed explanation of the constraints mentioned briefly in Chapter 10 – difficulties of labour availability, erratic rainfall, lack of fertility in the soils, pests and plant diseases.

Much of this account derives from the prolonged and wide ranging survey carried out by the JIT 35 years ago, for recent experience does not extend much beyond the boundaries of Bor and Kongor Districts, but conditions prevailing on the higher, better drained land where cultivation takes place, except that it has become more restricted owing to the increased flooding, have not altered; customary methods of production have changed not at all. Mention is made, however, of the rather discouraging results of more recent painstaking research and trials undertaken by agencies of the Netherlands Government and investigations funded by the UNDP/FAO.

This underlines the special difficulties that face the cultivator and is necessary to dispel the illusion of any easy way to large-scale increase in agricultural productivity and production. Chapter 14, based on the findings of the Swamp Ecology Survey, describes the incidence and distribution of fish species in the area and the potential for the expansion of commercial fisheries, one of the most promising fields of development in the region. Chapter 15, again based on observations made during the Range Ecology Survey, is concerned with wildlife, the numbers and distribution of different species, wild animals as a possible reservoir of disease and risk to domestic livestock, and the relative importance of wildlife as a source of food.

Part V turns to the estimated effects of the canal and its operation in the Jonglei area. Chapter 16 suggests that climatic change is not to be expected, but demonstrates how reduced discharges in the natural channels of the Bahr el Jebel will result in a decrease in permanent swamp and, more importantly, seasonally inundated floodplain which provides the only source of grazing at the height of the dry season. The hydrological model described earlier in Chapter 5 is adapted to predict the extent of these changes in areas of vegetation as well as types of grass species. The offtake of water through the canal will not only reduce the area of the river-flooded grasslands which provide essential dry season pasture; it will also serve to reduce somewhat the peak flows that sometimes cause very serious damage. Little appears to have been done to calculate in quantitative terms the measure of this benefit. In Chapter 16 also, a necessarily brief attempt is made to estimate, again with the help of the model, the extent to which this may serve as a form of flood protection.

In Chapter 17, the consequences of these changes are examined in connection with the pastoral element of the economy, the effects on livestock productivity, existing and potential, and on fisheries as a seasonal component of the livelihood of the people as a whole and on the development of specialised commercial fishing industry. The effects on wildlife are also considered both in respect of migration patterns and increased vulnerability.

Chapter 18 is concerned with the more direct impact of the canal itself on those who live in its vicinity – the benefits of better communications and water supply; changes in settlement patterns and the need for controls to avoid over-concentration and consequent environmental degradation; the build-up of overland flooding against the east bank of the canal – which will have both beneficial and harmful effects; and above all the significance of the need for crossings of the canal by massive numbers of livestock and people in pursuance of the migratory pastoral cycle.

In Part VI, Chapter 19 is concerned with a summary and by no means comprehensive account of development endeavours in the Jonglei area during the brief period of peace and comparative stability between 1972 and 1984. In response to local aspirations and anxieties, the National Council for Development Projects for the Jonglei area and its Executive Organ were responsible for a wide-ranging and ambitious, if somewhat disjointed, programme of investigation, research and trial, followed in many cases by active attempts to follow up promising leads. There were formidable logistical problems and, as well, the environment proved more intractable than had been expected, despite the cautions embodied in earlier reports. As is so often the case in remote and neglected parts of the world, positive results were not immediately evident, a cause for local concern and criticism. The tragedy is, perhaps, that progress was halted by further civil disturbance just at the stage when solutions to the more troublesome problems were likely to emerge.

The final chapter, Chapter 20, turns to the future, the effect which local disadvantages, on the one hand, and undoubted benefits, on the other, may have on those indigenous economic strategies and social mechanisms that had thus far assured a measure of food security and distribution even in times of scarcity, bearing in mind that neither canal-induced nor natural environmental change, nor the effect of improved communications and all that goes with it, could have anything approaching the devastating impact of the current civil war. Attention is then drawn to the eventual need for the establishment of an overall co-ordinating and controlling body (something in the nature of a Canal Authority has sometimes been mentioned) responsible for all local management, mindful that it cannot also be concerned with the wider aspects of Nile control which is the function of the Permanent Joint Technical Commission. In this connection some of the more important matters are suggested with which such a body should concern itself. These include the all-important need to examine the possibility of the regulation of discharges between the canal and the natural channels of the river in such a way as to ensure optimum downstream benefits while minimising damaging local ecological effects.

Reference is also made to the effect on the local population if in fact the canal is never completed; and then, by contrast, attention is drawn to further much more widespread water conservation schemes that are also in mind for the future: Jonglei Stage II, an enlarged canal, probably with upstream control and storage in the Great Lakes of East Africa; and the drainage by canalisation of the Machar Marshes and virtually the whole of the Bahr el Ghazal basin. This book is not concerned with present political circumstances that relate, almost coincidentally, to the Jonglei canal which in any event is only one part of the political scene. But political factors cannot be wholly ignored, and in the final paragraphs we review, very briefly, some of the more general issues at stake.

Part I

Controlling the Nile

1

The genesis of the Jonglei canal

Introduction

The huge area of swamp and marshland of the southern Sudan known as the *Sudd* is a massive 'blockage' or 'dam' of vegetation (see Glossary) which can absorb and dissipate half or more of the water it receives. Speaking before the Royal Geographical Society in 1908 Sir William Garstin, the first great engineer to suggest ways and means of reducing these water losses, said 'Through this region the White Nile wanders for nearly 400 miles. It enters these marshes a fine river, with a considerable discharge. It issues from them a comparatively insignificant stream, having lost, by evaporation in the swamps and by absorption of the water-plants, from fifty to eighty per cent of the supply that it brought down from the lakes' (1909). Although subsequent records have revealed that the losses have not reached the upper limit suggested by Garstin (see Table 1.1), he was the first man to grasp the significance of the problem in the context of Nile control, a problem that has exercised the minds of engineers ever since.

We are here concerned with the present Jonglei canal, designed to by-pass these swamps, thus to reduce these losses and deliver more water for irrigation use downstream. In concept this is the culmination of Garstin's early ideas for diversion suggested in his *Report upon the Basin of the Upper Nile* (Garstin, 1904). Garstin, who must have surveyed this area largely from the deck of a steamer, his perspective confined to uninhabited swamp, was no lover of the *Sudd*. Speaking on the same occasion he said 'No-one who has not seen this country can have any real idea of its supreme dreariness and its utter desolation. To my mind, the most barren desert that I have ever crossed is a bright and cheerful locality compared with the White Nile marshland' (1909). He can have had little knowledge in those early days of the country bordering the papyrus swamp or the importance of the seasonal variations in river level to the livelihood of the inhabitants who, after all, believe theirs to be the finest land in the world.

Losses in the *Sudd* are a natural process, and from one point of view their reduction can be regarded as the saving of a resource which could be put to better economic use farther downstream; from another, the water involved can be regarded as an asset of fundamental significance in the present local economy as well as a potential for development in the future. Yet extremes – too much flooding in some years, too little in others – could be avoided and control could bring local benefits as well as disadvantages.

To understand plans to reduce these swamp losses and the way in which they relate to other and broader schemes to control the waters of the Nile, a brief description of the topography and hydrology of the Nile Basin and its two main catchments is an essential introduction (see Figures 1.1 and 1.2).

Plate 1. Aerial view of part of the southern end of the *Sudd*. The turbid waters of the main channel of the Bahr el Jebel meander through the middle of the picture. Beyond the riverain fringe of papyrus swamp lie numerous lagoons which may either contain turbid river water, or dark clear water which has reached them only after percolating slowly through the swamps. Note the line of floating water hyacinth in the river. Photo: Alison Cobb.

Hydrological and topographical background

The source of the Nile most distant from the sea is the upper
catchment of the Luvironza river in Burundi, a tributary of the Kagera river

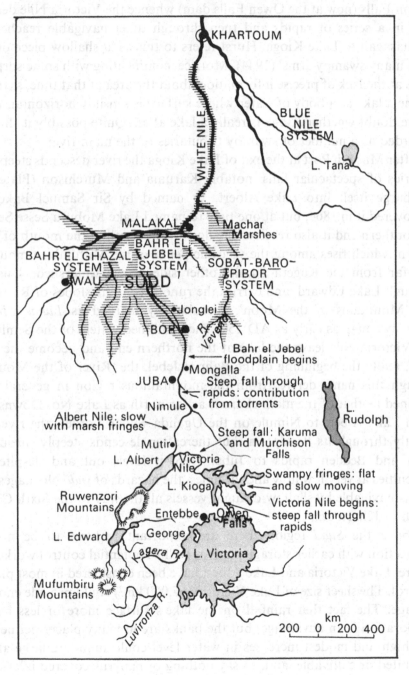

Figure 1.1 The Nile Basin south of Khartoum.

where there are plans for substantial hydro-electric and irrigation development. The Kagera flows into Lake Victoria, the second largest freshwater lake in the world. From this, the only outlet, discovered by Speke in 1862, is at the Ripon Falls (now at the Owen Falls dam) whence the Victoria Nile descends, first in a series of rapids and then through quiet navigable reaches from Namasagali to Lake Kioga. Hurst refers to this as 'a shallow piece of water with many swampy arms' (1944). Morrice, commenting with some surprise in 1958 at the lack of precise information about the area at that time, says 'If we define a lake as a body of water whose surface is sensibly horizontal, there is grave doubt whether Kioga is really a lake at all. Quite possibly it should be regarded as a number of swampy tributaries of the main river'.[1]

After Masindi Port at the exit of Lake Kioga the river descends steeply over a series of spectacular falls, notably Karuma and Murchison (Plate 8), to discharge itself into Lake Albert, so named by Sir Samuel Baker who discovered it in 1864, but at one time re-named Lake Mobutu Sesse Seko. At its southern end it also receives the flow of the Semliki, the mouth of a river system which rises among the volcanic peaks of the Mufumbiro mountains, not far from the Kagera on the other side of the watershed. This flows through Lake Edward, and carries the runoff from the slopes of Ruwenzori, the 'Mountains of the Moon', speculatively marked as *Lunae Mons* on Ptolemy's map as early as AD 150. The combined waters of the Semliki and the Victoria Nile leave the lake at the northern end and become the Albert Nile, really the beginning of the Bahr el Jebel, the 'River of the Mountain', though this name derives from the mountainous region in general and is retained as that of the main channel as far north as Lake No. Downstream, from Lake Albert to Nimule on the Uganda–Sudan border, the river flows quietly through its floodplain, but thereafter descends steeply through the Fola and Bedden rapids to Juba. Here it levels out and despite some difficulties at very low flow and, at times, the hazards of *sudd* blockages (Plate 17), is navigable by shallow-draught vessels all the way to the Sixth Cataract north of Khartoum.

Above the *Sudd* region there are important features to be noted in connection with earlier storage plans as well as potential control works in the future. Lake Victoria and Lake Albert have been considered in most plans for control. The sheer size of Lake Victoria (67,000 km^2) could provide enormous storage. The fact that rainfall on the lake's surface more or less balances evaporation is an advantage, but the banks are in many places neither steep nor high and modest increases in water level could inundate large areas of inhabited or cultivable land, to say nothing of papyrus-covered inlets, many of which have been drained and put to good agricultural use. By contrast,

Lake Albert is much smaller in area, the surface averaging about 6,000 km², but the sides are steep and rise to considerable heights. Substantial rises in level would be required to give comparable storage, but relatively less land would be flooded and the area is less densely populated.[2] Garstin describes Lake Albert as the 'true reservoir of the (White) Nile inasmuch as it receives the waters of both systems of supply (Victoria Nile and Semliki). It is at the same time the regulator of the river, and the level of its waters largely dominates the volume passing down the Bahr el Jebel to the north' (1904).

Another important factor is the contribution of the so-called 'torrents', notably the Aswa and Kit, which enter the Bahr el Jebel between its exit from Lake Albert and Juba to the north. These rivers, virtually dry for part of the year, come down in flash floods at times of heavy rainfall, add significant amounts of water (on average about 4 km³ per year) to the mean discharge and are at present uncontrolled.

We then come to the most important feature of the White Nile system in the context of this book, the *Sudd* swamps (Plate 1), where not only does the area covered by spill water expand and contract seasonally with the rise and fall of the river, but varies enormously in extent over periods of years, well illustrated by the huge increase in swamp and marshland since the early 1960s (see Table 1.2). In effect, any substantial increase in water intake at the head (or southern end) of the *Sudd* does not result in a directly proportional increase in discharge at the tail of the swamps; with increased spill the area exposed to evapotranspiration is enlarged. Table 1.1 shows the percentage lost in this way during periods of increased mean annual discharge; indeed it indicates that the greater the inflow, the greater the percentage loss. This demonstrates the supreme importance of some form of diversionary channel to carry any increase in mean discharge provided by upstream control works and storage, quite apart from any plan to reduce losses in natural flow on a run of the river regime.

The main natural channels of the river meander along a series of serpentine courses through the floodplain, flanked first by permanent papyrus and

Table 1.1. *Mean annual discharges in km³*

Period	at Mongalla	at tail of swamps	% loss
1905–60	26.8	14.2	47.0
1905–80	33.0	16.1	51.2
1961–80	50.3	21.4	57.5

bulrush swamp waterlogged throughout the year, and then paralleled by grasslands flooded at high river and exposed when the river drops. This is *toic*, a common Nilotic word for these all-important, seasonal pasture resources. Beyond that, particularly to the east, are vast expanses of marginally higher ground covered with rain-grown perennial grasses. The soils here are cracking clays, which at the height of the dry season are waterless and hard as iron, but in the rainy season become tacky and impermeable, subject to accumulations of rainwater and floods which creep across the plains. An added and new phenomenon, mainly along the fringes of the rivers, is water hyacinth (*Eichhornia crassipes*),[3] introduced from the Zaïre basin and unknown in this region before 1957, when it spread downstream with remarkable rapidity.

Numerous side streams run parallel to the main channel of the Bahr el Jebel through the *Sudd*, especially in the upper reaches, notably the Aliab and Atem, and then the Bahr el Zeraf, now for many years linked at the upper end by artificial cuts excavated by the Egyptian Irrigation Department in 1910 and 1913 as one of the schemes to improve the flow through the *Sudd*. The Zeraf is thus a major distributory, taking water from the Bahr el Jebel system and discharging it into the White Nile not far above the mouth of the Sobat. Though never used as the main commercial channel, the Zeraf was navigable throughout its length after the cuts were made, though it was blocked again for a while in recent years.

From time immemorial the *Sudd* had been not only a barrier in the hydrological sense but a physical constraint to overland communications between north and south, so that trade and exploration, which only began in the last century, were largely confined to navigation along the main channel. Along this there are natural hazards when large masses of papyrus and other vegetation break away as the river rises, lodging in corners of the serpentine course and accumulating in often dangerous blockages. There is historical evidence to suggest that, as might be expected, difficulties in navigation have varied a great deal according to periods of high and low Nile flows (JIT, 1948, pp. 49–51; Newhouse, 1939) and consequent variations in the amount of breakaway vegetation (see Table 8.1). Extreme examples of blockages are recorded at various times in the last century until the main channel was cleared by Major Peake in the first years of this century and subsequently kept free of serious obstructions by regular stern-wheeler steamer services.

Sir Samuel Baker, for example, encountered few obstructions in the main channel of the Bahr el Jebel in 1863, but returning after his discovery of Lake Albert in 1864 describes his encounter with a huge blockage somewhere downstream of Lake No, there being 'considerable danger in the descent upon nearing this peculiar dam, as the stream plunged below it by a subterranean channel with a rush like a cataract. A large Diahbiah (native craft) laden with ivory had been carried beneath the dam and had never been seen afterwards' (1866). Returning in 1870 with a substantial fleet Baker found the Bahr el Jebel totally blocked, 'entirely neglected by the Egyptian authorities'. 'In fact', he says, 'the White Nile had disappeared'. He therefore had to take the alternative route up the Bahr el Zeraf. Forced back on the first attempt by blockages and falling river levels, he observed that 'no dependence can ever be placed on this accursed river. The fabulous Styx must be a sweet rippling brook compared to this horrible creation'. Having spent the rains near what is now Malakal, he was successful in breaking through from the upper Zeraf to the main channel of the river in March the next year (Plate 2). Speaking of the creation of *sudd* blocks and the changing course of the channels, he observes 'violent wind acting upon the high waving mass of sugar-cane grass may create a sudden change; sometimes large masses are detached by the gambols of a herd of hippopotami, whose rude rambles during the night break narrow lanes through floating plains of water-grass, through which the action of the stream may tear large masses from the main body. The water being pent up by

HAULING THE NO. 10 STEAMER THROUGH THE CANALS IN THE MIDST OF THE VEGETABLE OBSTRUCTIONS.
Vol. i. p. 75.

Plate 2. Baker's steamer being hauled through the *Sudd*-blocked channels of the upper Zeraf (from Baker, 1874).

enormous dams of vegetation, mixed with mud and half decayed matter, forms a chain of lakes at slightly varying levels. The sudden breaking of one dam would thus cause an impetuous rush of the stream that might tear away miles of country, and entirely change the equilibrium of the floating masses' (1874).

At Lake No is the inconspicuous exit of the whole of the Bahr el Ghazal system, which drains the very extensive western catchment but with no significant contribution to the Nile (see Figure 1.1). Some 50 kilometres below the Zeraf mouth the river is joined by the Sobat, itself fed by the Baro as well as water from 'torrents' from the Ethiopian foothills and Pibor system whose catchment is largely in the Ethiopian foothills. The significance of the Sobat is that its contribution to the mean discharge of the White Nile, over most of this century, has until recently been roughly equal to the mean annual losses in the *Sudd*.

Below the Sobat mouth the river becomes the Bahr el Abiad, the White Nile, a name which in literature applies to the whole system but has more precise local application. Before the two Niles join at Khartoum there are no further significant tributaries, though the Khor Adar, at any rate in some years, drains water in measurable quantities from another large area of wetland, the Machar Marshes which take water not only from local precipitation but overspill as well as water from 'torrents' from the Ethiopian foothills and from the Baro, and expand and contract in swamp area over periods of years (JIT, 1954). These marshes, like the *Sudd*, are of importance in future plans for drainage and the provision of 'new water' for economic use farther north, plans that will be mentioned later (see Chapter 20).

The Blue Nile and its main tributaries, the Dinder and Rahad, rise in the Ethiopian mountains around Lake Tana, though the bulk of the water comes from a comparatively small area south and east of it. To the north is the Atbara river which provides a significant contribution, again deriving from the Ethiopian plateau north-east of Lake Tana. Neither river is of direct physical consequence to the Jonglei Canal or the *Sudd*, but their annual contributions, together with the Sobat which, though technically part of the White Nile system also receives most of its water from Ethiopia, amount on average to about 86% of the total Nile flow. By contrast at the tail of the swamps the White Nile water, derived from the East African catchment and reduced by losses in the *Sudd*, provides only about 14%. This distinction between the relative contributions of the Blue and White Niles is important. Moreover, there is a conspicuous difference between the seasonal variations in flow in the two systems, the rivers of the Ethiopian catchment marked by the extreme range in discharges between the peak and low periods,

	Sobat	14%
Ethiopian catchment	Blue Nile	59%
	Atbara	13%
East African catchment	Bahr el Jebel	14%

while the flow from the East African catchment is relatively uniform. At its peak the former provides nearly 95% of all water reaching Egypt, the latter only 5%. At low Nile the contributions are nearer 40% and 60%. The high peak discharge of the Blue Nile at its junction with the White Nile, both in amount and velocity, also has a ponding effect on the latter river and in natural circumstances water backs up the White Nile for some 300 km, forming for a while a very large natural storage reservoir which only releases its water when the Blue Nile falls rapidly in September. The White Nile water, supplemented marginally by water held in the same manner by the Sobat upstream, then begins to fill the gap left by the falling levels of the Blue Nile and Atbara systems.

   Garstin described vividly the nature and significance of this annual phenomenon. 'To those who do not know the Nile', he said 'and have not studied its discharges, it is difficult to explain this wonderful arrangement, whereby these rivers automatically compensate one another, so that, at the time when one system is passing on a large volume of water, the other is storing up its discharge, and when the former begins to decrease in volume, the stored water takes its place and makes good the deficiency. The

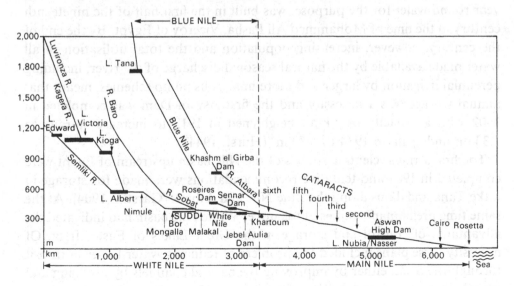

Figure 1.2. Profile of the Nile Basin.

comprehension of these facts is, I consider, one of the most important results of our studies of the Nile since 1898' (1909). It would be more accurate to say that the water so stored *in part* compensates for the seasonal shortfall, but this natural feature of the two Niles is the premise upon which all overall plans for control have been based. In simple terms, the objectives have been to improve upon advantages already inherent in the natural river regimes by artificial storage and control works in various parts of the Nile basin so as to deliver the maximum amount of water, with minimum wastage in transmission and at the right time, for irrigation purposes in Egypt and in the northern Sudan.

Plans for Nile control and irrigation development in Egypt and the Sudan (see Figure 1.3)

Until the middle of the nineteenth century agricultural production in Egypt had for the most part been limited to basin irrigation crops grown in the depressions which flank the river and contained by banks which enabled the people to enclose and control water from the annual flood. The population was very largely dependent on the natural hydrological regime of the Nile as they had been for millennia before, though some perennial irrigation was possible by the use of rudimentary techniques of lifting water; the water-wheel or *saqia* operated by animals, the *shaduf* and Archimedean screw by manpower. Longitudinal canals to provide gravity irrigation had also been dug parallel to the river carrying water from upper reaches to lower, and the first Delta barrage, enabling the extension of the irrigable area as well as providing year round water for the purpose, was built in the first half of the nineteenth century in the time of Mohammed Ali Pasha, Viceroy of Egypt. By the end of the century, however, increasing population and the total utilisation of all water made available by the natural seasonal discharge of the river, including perennial irrigation by larger and more numerous pump schemes, meant that annual storage was a necessity and the first Aswan Dam was completed in 1902 with a capacity of $1 \, km^3$, heightened in 1912 to increase storage to $2.3 \, km^3$ and again in 1934 to $5.3 \, km^3$ (Hurst, 1944).

The first serious scientific studies of the Nile basin upstream of Egypt were completed in 1904 and tentative recommendations were made for storage in Lake Tana and dams along the Blue Nile and Atbara (Garstin, 1904). At the same time preliminary studies of the upper White Nile basin had indicated the advantages of control and storage in the Great Lakes of East Africa. Of necessity these plans included the problem of reducing water losses in transit through the *Sudd*, either by improving the natural channels by dredging and straightening them and raising their banks, Garstin's first idea, or by a diversion canal (suggested by Beresford) somewhere to the east of the swamps, or a combination of all these methods (Newhouse, 1939). This was the genesis of the whole series of schemes outlined in Appendix 4 and the present Jonglei Canal project.

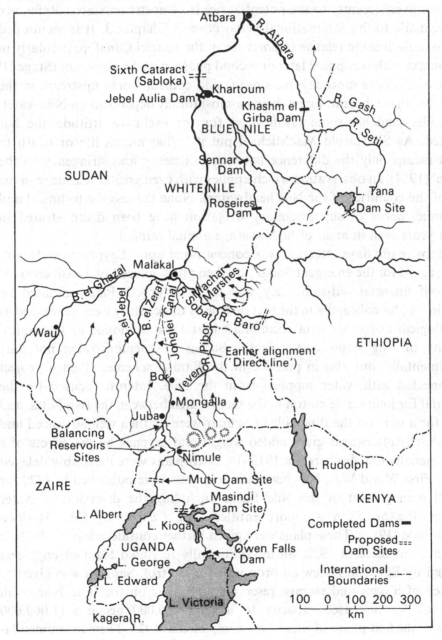

Figure 1.3. Nile Basin south of Atbara. Development works proposed or completed – including the Equatorial Nile Project.

So far as Egypt was concerned further studies and negotiations with upstream users proceeded on the principle of her historic or 'acquired' rights in the waters of the Nile. Subsequent planning has, it is true, given appropriate consideration to the needs of the Sudan, but not much regard was given until relatively recently to the potential needs of upstream states. Reference will be made to this international perspective in Chapter 3. It is mentioned here to underline the relative importance of the Jonglei Canal, particularly in the context of the proposed later or second phase in its development (Stage II) which would once more involve storage and control works upstream in the East African catchment. Egypt, being almost totally dependent on Nile water, has understandably some justification for the exclusive attitude she has adopted. As Sir Harold MacMichael put it 'What means life or death to Egypt means only the difference between sufficiency and stringency to the Sudan' (1954), an observation which applies with even greater relevance in the case of the countries of the Nile headwaters. None the less the technical and economic merits of supplementary irrigation have been demonstrated in recent years even in areas of high average annual rainfall.

In those early days attention was concentrated upon Egyptian needs with some regard for the emergent Sudan, by then much in need of a cash crop for export if financial self-sufficiency was to be achieved. Plans initiated by Garstin and his colleagues in the first decade of this century were developed as hydrological and other data accumulated, and the economic and technical viability of long staple cotton in the Sudan was demonstrated not only experimentally, but also in practice in the Baraka scheme, albeit a project unconnected with water supplied from the Nile. Interest focused on the potential for long staple cotton in the Gezira area between the two Niles, and plans for a dam on the Blue Nile at Sennar were by then well advanced and were shortly afterwards given added weight and urgency by the effects of a phenomenally low discharge in 1913–14. Such plans were inevitably delayed by the First World War, but thereafter proposals were published in 1920 for overall management of the Nile Basin, including the diversion of water through the *Sudd*, in a report entitled *Nile Control* by Sir Murdoch MacDonald (1921). These plans were given further consideration by the Nile Project Committee in 1920, an internationally recruited team of engineers engaged by Egypt to review all proposals for control. Support was given to the idea of a dam and storage reservoir at Sennar on the Blue Nile, with proposals for increased capacity to irrigate 420,000 hectares (1,000,000 feddans), the first phase of which was completed in 1925. Their recommendations also included a proposed dam at Jebel Aulia, just south of Khartoum, to improve upon the natural seasonal storage of White Nile waters created by the annual Blue Nile flood. This dam was eventually built in 1937 with a capacity of $3 \, km^3$. These two dams, Sennar and Jebel Aulia, were the only major works to be completed within the Sudan during the period of the Condominium Government. They were followed later by the Roseires Dam on the Blue Nile, completed in 1966 and designed to increase irrigated agriculture and power generation in the northern Sudan. The Khashm el Girba dam on the Atbara was completed in 1965, initially to provide alternative livelihood for some 70,000 inhabitants of the Wadi Halfa reach of the main Nile, displaced by the rise of water level behind the Aswan High Dam.

In 1929 the first Nile Waters Agreement was signed, whereby the Sudan was allocated $4 \, km^3$ while Egypt would utilise the net remainder, $48 \, km^3$; net because transmission losses by evaporation were inevitably high, and, because there was no overall control of about one-third of the total discharge that ran unused into the Mediterranean at the height of the Nile flood.[4] Negotiations with Ethiopia at this period made little progress and were in any case interrupted first by the Italian invasion of that country in 1936, and then by the Second World War. Plans for wider control of both Nile systems were not at any point forgotten and proposals were detailed in the *Nile Basin*, Vol. VII, by Hurst, Black and Simaika[5] in 1947. These envisaged storage in both the Great Lakes of East Africa and in Lake Tana in Ethiopia, and a series of dams and control works downstream on both branches of the Nile, and, with subsequent versions and embellishments, were later sometimes referred to as the 'Unified Nile Valley Scheme'. This, as will be recounted later, was abandoned in favour of the Aswan High Dam project, which has given a substantial measure of control within Egyptian territory, and has allowed, at least in theory, for the total elimination of all wastage into the sea, albeit at the cost of heavy losses by evaporation and seepage in the reservoir behind the new dam. But this is to anticipate.

The genesis of the Jonglei Canal Project

In the 1920s interest continued in the potential advantages of control and storage in the upper White Nile catchment, later made more urgent by the lack of progress in negotiations with the Ethiopian Government over the development of the Blue Nile. Comparatively massive resources in river and road transport for engineers of the Egyptian Irrigation Department were assembled and a permanent station, the Southern Nile Inspectorate, established at Malakal just below the swamps.

Proposals for diversionary works through the *Sudd*, and there were many, continued to receive close attention (see Appendix 4). One idea was a canal beginning at Jonglei, north of Bor on the edge of the Atem channel. This was to run close to the natural channels of the river and discharge into the Bahr el Zeraf some 100 kilometres above its mouth, its lower course to be straightened and in places banked. This scheme was approved by the Egyptian Council of Ministers in 1925, to the surprise of the Sudan Government of the time who had not been consulted and who opposed the proposal until more detailed information was available on the possible effects of such a canal on the inhabitants of the area through which it would pass. This was the first occasion on which the inhabitants had been considered, and the Sudan Government enlisted the assistance of W. D. Roberts, an engineer from the Egyptian Irrigation Department. His report on *Irrigation Projects on the Upper Nile and Their Effects on Tribal and Other Local Interests* was submitted in 1928 (Collins, 1983). Roberts, who spent nearly a year in touring the area and seeking local opinion, recommended that the Sudan should reserve its position on the canal issue, on the assumption that a diversion channel alone could nowhere near meet Egyptian irrigation needs. Only by storage in Lake Albert could adequate water be made available. This being so, only additional water, it was alleged, would be passed through the canal, and the average discharge in the natural channels would be maintained with no local repercussions.

   Consideration was then given to the possibilities of diverting water from the Bahr el Jebel near Bor through the Veveno (see Figure 1.3) to the River Pibor, with appropriate improvement works to the channel of that river, and thence down the Sobat. This would circumvent the *Sudd* by a wide margin and incidentally allay fears of adverse effects on the Nuer and Dinka of the region. The effects on the interests of the Murle and Anuak, or indeed those Nuer and Dinka living along the Sobat, through whose territory the water would be passed, do not seem to have been regarded as relevant. The Veveno–Pibor scheme (see Appendix 4) was, however, abandoned in 1932 for technical reasons and because of the high costs in relation to limited benefits in water saved (Newhouse, 1939). A. D. Butcher, an engineer of The Egyptian Irrigation Department, who had studied the area for many years, presented his report, *Sadd Hydraulics*, in 1938. This, adding greatly to the understanding of hydrological data on the area, had once more examined the feasibility of a canal starting from the upper Atem at Jonglei and joining the Zeraf lower downstream, this time with a smaller canal flanking it. Despite earlier demands by the Condominium Government that the interests of the Nilotic inhabitants should be taken into account, Butcher's earlier report, entitled *The Jonglei Canal Scheme*, contained little information on local effects, which in any event he regarded as likely to be negligible (Butcher, 1936).

It was John Winder,[6] then District Commissioner of the Zeraf Valley District, who took up the challenge. That was in 1939 after he had made two long excursions on foot, first on the eastern side of the Zeraf River and then westwards round the island formed by the Zeraf and the Bahr el Jebel, plotting cattle camps on the map and movements from rainy season settlements to dry season grazing grounds.

The proposals put forward by the Egyptian Irrigation Department at the time included a canal to be dug in two distinct stages, the later extension of the second stage to coincide with the provision of storage and control upstream in the Great Lakes. Having located the areas of seasonal land use, Winder set out to analyse such hydrological data as were available to him, and to predict the effects on local resources essential for the maintenance of the existing economy to which at the time no alternative could be seen. While stressing the amateur nature of his hydrological analyses, though backed by then unparalleled personal knowledge of the region, he forecast in broad terms the likely nature of these effects with remarkable accuracy.

Winder himself has pointed out that in presenting his report to the Condominium Government he aimed not only to reveal the possible damaging local effects on the resources available to the people in the pursuit of their transhumant pastoral economy, but also to challenge the Egyptian Irrigation Department to provide more detailed and precise information on which predictions could more accurately be based.[7] In this he was successful, but his report was submitted in May, 1940, at the time of the declaration of war by Italy and the grave threat to the Sudan by almost overwhelming Italian military forces in Ethiopia and Eritrea; in consequence it was shelved until 1946.

The Jonglei Investigation, 1946–8

In that year the local impact of a canal in the *Sudd* region once more became a matter of concern to the Sudan Government. A Jonglei Committee, chaired by the Financial Secretary, with the Directors of Irrigation, Agriculture, Surveys, and Veterinary Services, the Irrigation Adviser and the Governor of the Upper Nile Province as members, was set up to watch the interests of the Sudan. An investigation team was appointed, with H. A. W. Morrice, an engineer of the Sudan Irrigation Service, as chairman and Winder as political adviser and secretary, with support from a surveyor and veterinary member in the second year. This team achieved a remarkably comprehensive survey of the area and was able to identify the serious nature of the problems, which in general terms confirmed Winder's anxieties expressed seven years before.

Investigations in the first year of their work related to Butcher's *Jonglei*

Canal Scheme of 1936, with subsequent minor revisions, and it was only in their second year that they first had sight of the comprehensive and ambitious control proposals published in 1947 in Volume VII of the *Nile Basin*. This included plans for a storage and flood protection reservoir at the Fourth Cataract, similar works at Wadi Rayan in Egypt; a reservoir for over-year storage in Lake Tana, and a dam on the Blue Nile at Roseires. For the White Nile basin there would be the *Equatorial Nile Project* which would include substantial storage in Lake Victoria, a dam to regulate Lake Kioga, a smaller reservoir in Lake Albert, a balancing reservoir in the Nimule-Bedden reach and a diversion channel or canal through the *Sudd* (see Figures 1.2 and 1.3).

The Equatorial Nile Project

The Equatorial Nile Project contained a number of new principles of fundamental importance to local interests. First, seasonal control by annual storage would have enabled releases to be made in the 'timely season' so that water would reach Egypt when it was needed to balance the fall in discharge from the Blue Nile. This period – between mid-December and mid-June – coincides approximately with the dry season in the *Sudd* region when river levels drop and the riverain marshes are exposed and available for grazing. The 'untimely season', that is when the flow from the Blue Nile and the Atbara yield more than adequate supplies for irrigation downstream, would coincide with the time of year when river levels in the *Sudd* region are high and the floodplains under water, the seasonal inundation and exposure providing a natural, though wasteful, system of irrigated pasture of critical importance to the local economy. This natural asset would be greatly reduced in the central part of the *Sudd*, while above it there would be a virtual reversal of the natural seasons and, below, constantly high river levels.

The Equatorial Nile Project also envisaged over-year storage in the Great Lakes, allowing for the accumulation of water in years of good rainfall for release in time of shortage, the objective being the steady yearly discharge of the long-term mean. Since H. E. Hurst, one of the principal architects of this project, in applying his calculations to fifty years or more of discharge records from Lake Albert, assumed one hundred years as the period over which the desired amount of water was to be made available, this was also known as 'Century Storage'.[8]

As will be seen later, timely and untimely releases and their effects in the southern Sudan are no longer relevant, since full control and storage for Egypt is now available in the reservoir behind the Aswan High Dam. But a short description of the investigations which took place between 1946 and 1954 and the conclusions reached are important in the context of the present

Jonglei Canal Project for three main reasons. First, ecological data so provided, now amplified by further studies and survey, are in many aspects still relevant. Secondly, the conclusions of the Jonglei Investigation Team concerning the effects of this earlier scheme are often misrepresented as applying to the present canal project and a cause for greater anxiety than is necessary. Thirdly, as will be clear later, the high discharges and enormously increased flooding since 1961 has meant a massive expansion in the area of swamp, so that physical conditions in the *Sudd* region have been very different from those prevailing during the period of the Jonglei Investigation. A return to conditions similar to those before 1961 is a distinct possibility, and records of those conditions are relevant to the examination of the effects of the operation of the canal in those circumstances.

Among other things the team recommended the abandonment of the line of the canal (Line VII) which ran close to the natural channels of the river with an exit near the mouth of the Zeraf, in favour of the 'Direct Line' which would run much farther to the east, with tail works close to the mouth of the River Sobat. In doing so they demonstrated hydrological advantages which would also help to reduce losses in dry season grazing and other natural resources in the area, as well as substantially lowering the costs of cross-drainage works under the canal by constructing it higher up the catchment of the numerous watercourses running laterally into the main river channels. The argument for carrying the line of the canal even farther east was echoed some 30 years later when the alignment of the present canal came under review in 1978, as explained in Chapter 2.

The Jonglei Investigation: Phase Two, 1948–54

It was with the background of this knowledge, the preliminary identification of the main problems, and tentative suggestions for their solution recorded in the Morrice–Winder Interim Reports, that a new and expanded multi-disciplinary team assembled in the autumn of 1948. For the new team the conclusions of the Third Interim Report (JIT, 1948) revealed the enormous range and complexity of the problems. The Egyptian Irrigation Department had accumulated a mass of hydrological data over many years, but these were largely confined to the main channels of the Nile and its major tributaries. Very little was known of the intricate network of inland drainage channels fed largely by local precipitation but sometimes augmented by spill from the rivers of the White Nile system in times of high Nile or from similar conditions in the Pibor–Baro system to the east. Not much was known either, of the causes of the sheet flooding phenomenon which came to be known as 'creeping flow' (see Appendix 3; and JIT, 1947), accumulations of rainfall 'often augmented by river spill' that spread as sheet flooding mainly across the plains east of the Nile in a north-westerly direction according to the degree of slope, which in the *Sudd* area is rarely more than 10 centimetres per kilometre.

There was also the vital question of providing additional flood protection capacity upstream. In the 50 odd years of records available there had been very considerable variations in discharge, the highest between 1916 and 1918 known to the inhabitants as *pilual* – the year of the 'red water' – when high levels in the Bahr el Jebel, probably some spillage from it into the Eastern Plain, and very heavy local precipitation, caused unprecedented flooding and great damage to animal stock and cultivations. The equally high and sustained discharges of the sixties and seventies, which will be referred to in detail later, were not then envisaged or the predictions and stipulations of the Jonglei Investigation might have been different.

The Team's engineers were also called upon to analyse possible hydro-logical changes in the context of frequently changing Egyptian proposals for upstream storage and different operational regimes. Such changes were inevitable as plans for Nile control evolved, but they had to be measured against 50 years of past records in order to forecast the effects in the future. This alone was a daunting task, for comparison of any proposed operational regime against each year's records would take an engineer several days of hard work on a hand-operated calculator; it was not until near the end of the investigation in 1953 that the earliest form of computer was first available and that in London.

Precise knowledge of other aspects of the environment was equally sparse. Relatively speaking very little was known of the interaction of soils and water in depth and duration on the natural vegetation which was of such great seasonal importance to the pastoral economy. The nature of the ecology of the rain-flooded grassland plains as well as the river floodplains was a vital factor in the assessment of effects, besides the determination of remedies where these might be needed. Little also was known of fisheries resources and potential in the rivers, swamp and floodplain.[9]

The investigation was one of the first co-ordinated attempts to embark on a major environmental study in Africa, but it had to be carried out with what nowadays would be regarded as minimal technological facilities. Few aerial photographs were available. Where they were, the team's surveyors were called upon to locate them against astro-fixes. They had also to provide precise levels across floodplain and swamp under exceptionally harsh physical and climatic conditions. J. V. Sutcliffe's survey of the Aliab Valley, referred to in Chapters 5 and 8, is a case in point. Yet the organisation had one advantage which was uncommon in those days; it was multi-disciplinary, with close co-ordination of approach, and located at one centre, just north of the *Sudd* at Malakal. During four field seasons the team's members were able to work as a closely co-ordinated body; its members were in constant touch, thus facilitating debate and the exchange of different disciplinary views on the same problems. It was not large – professional staff never exceeded 11 in any one year – but there were few changes so that there was the added benefit of continuity. It was ably supported by Sudanese staff, professional and administrative; and at all times had the co-operation of local departmental and administrative officials. During all this time, too, friendly liaison and consultation was maintained with the Southern Inspectorate of the Egyptian Irrigation Department, also located in Malakal.

The first team had completed their preliminary work with the recommendation that the task should 'change from opening up the principal problems to their methodical solution'. Whether this was an over-optimistic statement or not, it was evidently taken rather too literally by the Committee in Khartoum. At all events instructions were received in the middle of that first working season that the new Team was to limit the scope of its work and 'should be asked to provide a reasonable forecast of the necessary remedial measures and of compensation expressed in terms of money'. On the insistence of the Team, the words 'and water' were added. After only four months in the field the Team had thus been asked to accept new terms of reference before the full measure of the task had been appraised (JIT, 1949). It might be possible to identify remedial measures in the form, for example, of irrigated pasture or irrigated agricultural alternatives, but much investigation would be required to cost them. Above all it would be necessary to test their feasibility by trial and experiment, an essential requirement and one which would need many years of patient research. It would also be impossible to cost the remedies until the extent of the losses they were to meet had been more accurately estimated. To do this a thorough investigation of the river channels between Nimule near the Uganda border and Kosti, 1,600 km farther north, would be needed, involving further very extensive surveys of the topography, soils, vegetation, other resources and their utilisation.

Surveys of the hinterland away from the areas likely to be directly affected were also necessary for two reasons. First because under-utilised land, with or without the provision of water points in the form of excavated tanks (*hafir*), small earth dams and reservoirs or ground water from open wells or bore holes, might be used to compensate for losses elsewhere, though this would be to use resources of potential value for natural expansion to meet, for example, the needs of increasing population. The value of this potential had also to be taken into account. Secondly, because there was little doubt that people suffering losses in grazing or other natural resources would turn to those belonging to others more fortunate, with in consequence over-concentration, over-exploitation, and as a result possibly violence.

A distinction had therefore to be drawn between those who would be directly affected by the altered river regime and those who might expect indirectly to suffer because of unwelcome pressures by the former. The total area over which the investigation had to extend to greater or lesser degrees of intensity was about twice that of the British Isles. Strangely the team were also specifically instructed not to concern themselves with economic development (JIT 1954, Vol. 1, pp. 237–40), which overlooked the plain fact that remedial measures or alternative livelihood schemes, whatever they might be called, were likely in many cases to be identical with potential development projects. In fact, full consideration was given among other possibilities to irrigated pasture or crop production, improved animal disease services coupled with more controlled pasture management, the stimulation of markets in livestock, and the modernisation and enlargement of the fishing industry. Furthermore, a major aim was to find ways and means of improving the reliability and output of raingrown agriculture in a region where crop production more often than not suffered from extremes of flooding or drought. This is still a primary development objective today. Storage and agricultural marketing organisation were also prime considerations in a region which, overall, had for long been a net importer of grain.

Owing to the proposed near reversal of the natural river regime, high in the rainy season, low in the dry season, in order to meet the timely and untimely irrigation requirements farther north, there would be distinctly different effects in three main zones. In what was called the 'Southern Zone' in the Sudan, from Nimule to canal head, this reversal would have been almost total. Dry season grazing would be largely under water and inaccessible at the time it was most needed and when it was normally exposed. It would have been subject to much less river spill during the rains, and even with lateral run-off and back-up against the river banks, which are usually higher than the land on either side, grass species might change for the worse under altered ecological conditions, particularly in depth and duration of flooding. These problems were largely confined to the reach between Rejaf and the Atem and particularly the Aliab Valley.

In the 'Central Zone', from canal head to roughly Buffalo Cape on the Bahr el Jebel and Fangak on the Bahr el Zeraf, there would be equally adverse effects, though for quite different reasons. Here a large volume of water would be siphoned off down the canal and the annual rise and fall of the natural river channels would be much reduced. In consequence there would be dramatic changes in the ecological conditions which favoured the growth of nutritive river-flooded grasses, exposed during the dry season when most needed, and a probable change in those species to others which have only limited dry season value. The migration and spawning habits of fishes, another important food resource, might also be adversely affected. The effects in this zone would have been similar to those predicted under the regime of the present canal, but much more extensive.

From those points up to Kosti, in the 'Northern Zone', different conditions were forecast. During the dry season the river would remain relatively high, coinciding with the timely season and high discharges from the canal. Owing to the Sobat flood in the rainy or untimely season, levels would be high too. Here again the Shilluk, some Dinka, and the Baggara Arabs farther north were to varying degrees dependent on pastures along the river banks and islands and these would remain under water throughout the year.

Apart from investigations and survey work to determine hydrological and ecological changes and their effects, experimental work and trials were an important feature of the programme. The possibility of irrigated pasture was given careful consideration though its economic viability was very much open to question, particularly if the quality and productivity of local cattle could not be improved. Irrigated agricultural alternatives were also considered, but it could not be assumed that crops suitable for, say, Gezira circumstances, would flourish in quite different climatic and soil conditions. Suitable varieties needed testing, water requirements determined, and in the longer term a programme of plant breeding begun to produce varieties resistant to disease, pests and weed growth. The experimental station on intermediate land just south of Malakal, later extended to the floodplain opposite, provided a great deal of information, even though the research programmes were relatively short term.

Other experiments and trials were undertaken on rain-grown crops in three separate areas corresponding to the three zones, while natural pasture trials were carried out in various parts of the area, some to determine the effects of the duration and depth of flooding in relation to soils in producing different grass species and the effects of changes in these natural water duties. Others focused on the nutritive value and palatability of raingrown grasses in the plains above river flood level (e.g. *Hyparrhenia rufa*) at all stages of growth. These grasses proved to be of little value as pasture except after burning and regrowth which only occurred so long as moisture was retained in the soil, and were a significant pasture resource only for limited periods at the end and beginning of the rains (see Chapters 10, 11, 12). Preliminary trials aimed to extend those periods were also undertaken, and experimental banking (bunding) schemes introduced to find ways of conserving water for such purposes.

The Team's final report was completed in 1953 and printed for distribution in 1954 just before the Sudan became independent (JIT, 1954). Suggestions were made for the operation of the Equatorial Nile Project, particularly in Lake Albert to provide adequate flood protection capacity.[10] The area could not be allowed to be used as a 'wash' or escape reservoir for surplus water, and areas deprived of seasonal inundation suddenly flooded again. Remedial measures in the form of irrigation schemes, for example, would be subject to disaster if this occurred. It was estimated that the livelihood of about 600,000 people would be directly affected to a greater or lesser degree by the changed hydrological and hence ecological conditions brought about by the project. A further 400,000 might expect pressures on their own resources by those facing shortages elsewhere.

Under the operation of the canal and the new river regime proposed, 36% of dry season grazing resources would no longer have been available for some 288,000 cattle or their equivalent in smaller stock. Current fisheries production would have been reduced by some 3,000 tonnes per annum and further potential by at least 20,000 tonnes. The costs of remedial measures, 'direct remedies', were calculated but between fairly extreme ranges of figures, and with very firm reservations that their efficacy had not yet been adequately established. It was noted, however, that the project would be carried out in stages over many years and there would be sufficient time for experimental work to continue. The quota of water required for remedies in the form of irrigation schemes was estimated at approximately 0.75 km^3.

The most radical proposal, taking as the basis plans for the Equatorial Nile Project given in Volume VII of the *Nile Basin* (with revisions by Bambridge & Amin), was to modify the effects of the scheme by varying seasonal discharges in such a way as to reduce losses in pasture by nearly half and losses in existing fisheries (rather than the potential) to nil. After deducting the amount of water required to service remedial irrigation schemes, the net benefit at the tail of the *Sudd* would be 6.8 km^3; the Revised Operation, as it was called, would reduce this by 0.31 km^3, only a little over 4.5% of the total estimated benefit. The Team's report on this issue concluded:

> The Project as proposed appears to us unacceptable; the Sudan is one country and any adverse effects on local interests are of more than local importance. The Revised Operation which we advocate involves very much less expenditure in capital costs. Moreover, it is designed to cause the least possible disturbance of local interests,

whatever the costs and water requirements – and they are not inequitable when compared with the total saving, which is to provide more water, much needed for irrigation in Egypt and the northern parts of the Sudan. Moreover, the Revised Operation includes a comprehensive plan for flood protection which is lacking in the project as proposed (JIT, 1954).

Negotiations between the Sudan and Egypt over these proposals and recommendations never took place. Interest shifted from the Equatorial Nile Project to the Aswan High Dam whereby water would be stored largely within Egyptian territory: Control would be at that point and the notion of 'timely' and 'untimely' discharges would no longer apply upstream. Plans for unified Nile control including storage in the Great Lakes, diversion through the *Sudd*, and reservoirs downstream were dropped. There are those who have argued that co-ordinated control at different points along the courses of both Niles, the 'Nile Valley Plan' (Morrice & Allen, 1959; see Figure 1.3), would have been a more efficient method than concentration in the High Dam. However, the technical merits of the Aswan High Dam compared with a unified overall scheme are not a matter for debate in this book.

It is now scarcely relevant, but the Jonglei Investigation Team and its recommendations are often condemned as 'negative', meaning that they aimed only at maintaining unchanged as far as was possible the physical conditions upon which the subsistence economy of the time could continue, and took no account of economic development potential in a progressive sense. The Team was not called upon to do so; indeed they were specifically discouraged from consideration of development possibilities. In fact, the team ignored these instructions and took the deepest interest in development potential and also recognised how rapidly social and economic change in the area had begun and, given stable political conditions and better communications, was likely to accelerate (see e.g. JIT, 1954, p. 575 and Garang, 1981, p. 181). Recommendations were necessarily cautious for the reasons already described: the lack of adequate experiment and trial, a much longer-term process than was feasible in the circumstances. The policy recommended by the Jonglei Investigation Team was, however, subject to the specifically stated proviso 'that it should not in any way compromise future economic development in the Jonglei area.' (JIT, 1954, Introduction and Summary, p. lxvii).

Moreover, it was the Jonglei Investigation Team that took the initiative, amid much bureaucratic and departmental opposition, in getting agreement to the establishment of a similar body, the Southern Development Investigation Team, in 1953, urging both the need for a professional approach to development studies in the southern Sudan and the merits of a semi-autonomous multi-disciplinary cadre. In this it succeeded, and while political circumstances reduced its operations to less than a year, its preliminary report was submitted in 1954 and published in 1955 (SDIT, 1954), the focus, though widened to include the southern provinces of the Sudan as a whole, being on economic development with very much the same objectives as the National Council for the Development of the Jonglei Area and its Executive Organ, established in 1974 (see Chapters 2 and 19).

As will be described in greater detail later (Chapters 4 and 8), there have also been massive and generally harmful changes in natural conditions since then in the *Sudd* region. The heavy rainfall in the East African catchment and the sustained discharges of the 1960s and 1970s into the *Sudd* have resulted in much heavier flooding and over larger areas, with consequent hardship and loss of stock and crops, comparable with the conditions of *pilual* (1916–18), of which people used to speak with such horror, but far longer sustained. A glance at Figures 8.1, 8.2 and 8.3 and Table 1.2 below will illustrate the magnitude of these changes. Given these circumstances any measures to pass surplus water through the *Sudd* by way of the canal would have the advantage of allowing some people to reoccupy lands now flooded for over 20 years, and there could be potential benefits in reducing, at least marginally, the uncertainties of an unpredictable environment. There are, of course, other losses and other gains, and the losses could be much more extensive if the recent cycle of high Nile discharges were followed by a period of low years such as occurred prior to 1961. But the adverse effects forecast for the earlier scheme under the Equatorial Nile Project, in scale and for the most part in form, should not be taken as a yardstick of what might happen under the present project. The future situation would, however, be quite different and more massive losses could be expected if that project were extended to Stage II. All these points are discussed in subsequent chapters.

Table 1.2 *The Bahr el Jebel floodplain: Areas in km² (Bor to Malakal). See also Table 5.2.*

	1952	1980	Increase
Permanent swamp	2,700	16,200	13,500
Seasonally river-flooded grassland	10,400	13,600	3,200
Total	13,100	29,800	16,700

Notes

1. Typescript report on the Hydrology of Lake Kioga by H. A. W. Morrice submitted to the Uganda Government in 1958. Morrice, Chairman of the first Jonglei Investigation Team was later successively Director of Irrigation in the Sudan, and then Irrigation Adviser. He was Hydrological Adviser to the East African Nile Waters Co-ordinating Committee from 1957 to 1963.
2. Described in Sutcliffe, J.V., *et al.*, report on 'The Upstream Effects of the Proposed Mutir Dam', submitted to the Ministry of Agriculture and Natural Resources, Entebbe, 1957.
3. *Eichhornia crassipes*; water hyacinth, the 'Million Dollar Weed'. This plant was never observed in the southern Sudan during surveys carried out by the Jonglei Investigation Team and its successor, the Southern Development Investigation Team which was working in the area until the end of 1954. Its proximity was first reported to the Uganda Government in 1956 after rapid and massive spread up the Congo had occurred. Legislation was immediately enacted to prohibit its import, though its absence in significant quantities in Uganda may be due to altitude rather than the law. It was first observed in the Sudan in 1957.
4. The main clauses of the Nile Waters Agreement of 1929 are significant:

   Save with the previous agreement of the Egyptian Government no irrigation or power works or measures are to be constructed or taken on the River Nile or its branches, or on the lakes from which it flows, so far as these are in the Sudan or in countries under British Administration, which would in such a manner as to entail any prejudice to the interests of Egypt, either reduce the quantity of water arriving in Egypt, or modify the date of its arrival, or lower its level.
   In case the Egyptian Government decide to construct in the Sudan any measures with a view to increasing the supply for the benefit of Egypt, they will agree beforehand with the local authorities on the measures to be taken for safeguarding local interests. The construction, maintenance and administration of the above mentioned works shall be under the control of the Egyptian Government. (Agreement between the British and Egyptian Governments, 7 May, 1929)

5. *The Nile Basin*, a series of reference works and planning proposals begun in 1931 and published by the Egyptian Ministry of Works. Vol. VII was by H. E. Hurst, R. P. Black and Y. M. Simaika. Subsequent revisions to the Equatorial Nile Project, part of the plans for overall control, were proposed by H. G. Bambridge and Dr Mohammed Amin.
6. Winder, J., later Governor of the Upper Nile Province in the Sudan – Typescript report (1940) available in the Sudan Archives, Durham University.
7. Personal communication with the author.
8. In other words he calculated what storage capacity would be theoretically large enough to maintain a given quota for 99 years out of 100 on the average.
9. A professional fisheries scientist was not attached to the Team, but much assistance was received from Dr E. B. Worthington, CMG, then Secretary-General of CCTA, Dr C. F. Hickling, CMG, Fisheries Adviser to the Colonial Office, Drs H. Sandon and Julian Rzóska of the Faculty of Sciences, University of Khartoum, and Dr Colin Bertram, Fellow of St John's College, Cambridge.
10. The Jonglei Investigation had long advocated that serious attention be given to the problem of flood protection:

    'When in the future, under the scheme for control of the Southern and the Equatorial Lakes, exceptionally large floods occur, provision will have to be made to dispose of them so that the Sudan is not subjected to intolerable hardship. This applies particularly to the *Sudd* Region. The method suggested is to store surplus water in the Lakes until it can be released under control with safety'. (JIT 1949)

Under the project as then designed it was clearly the intention to dispose of flood-water in the main part of the *Sudd* area at very high discharges. The need for this might arise after much of the swamp had dried out, reflooding areas by then in economic use by the local population. The Team therefore also stipulated that 'all additional flood-escape discharges must be carried in safety past the Central Zone through the *Sudd* Diversion Canal, which should be constructed with adequate capacity for the purpose' (JIT, 1954, p. lxix).

2

The Jonglei Canal Project: Stage I

Introduction

While the Jonglei Investigation Team was completing its reports, planning of the Aswan High Dam was already under way, a conception that would defer if not obviate the need for storage in the Great Lakes. Of special significance for the southern Sudan, the artificially engineered variations in flow – the reversal of the White Nile seasonal hydrological regime – would no longer be needed. The High Dam, with control and facilities for annual and over-year storage largely within Egyptian territory, would provide the solution to her main problems: 'meagre summer flows, floods, and the occurrence of low years' (Abdel Magid, 1982). To the Sudan, the project offered the opportunity to resolve the long-standing dispute with Egypt over water quotas under the very restrictive terms of the Nile Waters Agreement of 1929. Interest quickly turned away from the Equatorial Nile Project and its local effects to this overriding priority. A new undertaking was reached formally (the Nile Waters Agreement of 1959) whereby the Sudan was allocated 18.5 milliards (km^3) per annum instead of only 4, and Egypt 55.5 out of the total mean discharge calculated at the time on the basis of previous records at 84. The difference of 10 km^3 was accounted for by estimated annual evaporation and seepage losses likely to take place behind the High Dam. Future Ethiopian requirements from the Blue Nile were not specifically taken into account; nor were the modest claims of the East African countries who – under the earlier 1929 Agreement – had been prevented from taking any water at all 'save with the prior agreement of Egypt'. Possible future claims by other countries were, however, tacitly acknowledged within the new Agreement between the two Republics; they would consider and 'reach one united view regarding' claims by other riparian states, and the deduction in equal parts between them if any allotment of water claimed was conceded (Nile Waters Agreement 1959: General Provisions). The significance of future claims is discussed in Chapter 3 below.

More significant so far as the *Sudd* region was concerned, the Agreement made provision for the sharing in equal portions between the Sudan and Egypt of the benefits of any new water made available by further works in the Nile Basin, the costs of which would also be shared. By the end of the 1960s it was evident that the requirements of further irrigation development in both countries would demand more water than was available in terms of the mean of the period on which agreement had been reached.

The Permanent Joint Technical Commission for Nile Waters (Egypt and Sudan), set up under the Agreement of 1959[1] hereafter referred to as the PJTC, reaffirmed in the early 1970s the possibilities of reducing losses occurring from spill and evaporation in the Bahr el Jebel system, as well as similar opportunities in the Bahr el Ghazal catchment and the Baro-Sobat system, notably in the Machar Marshes. The potential for savings in these two latter catchment areas is not the concern of this chapter, but the first brings us back to the concept of a *Sudd* channel or canal project originally envisaged in the early years of the century, as outlined in Chapter 1.

The Jonglei Canal, whose construction is only partially complete, is referred to by the PJTC as Jonglei Stage 1, and unlike the Canal originally proposed under the Equatorial Nile Project as well as the longer-term plans for Jonglei Stage 2 proposed by the PJTC, this includes no provision for over-year storage upstream. As before, but without upstream control works, the objective of the Jonglei Canal is to reduce losses in the *Sudd* and thereby to increase the mean annual discharge of water in the White Nile at Malakal. The decision to construct the Canal was made in 1974 apparently without reference to the people through whose homelands it would pass, which thus sparked off anxiety and unfavourable reaction among southern Sudanese sensitive to the right to manage their own affairs after 17 years of civil war which had ended only two years before.

As already mentioned, to help overcome these fears, the Sudanese Government created a statutory body, the National Council for the Development Projects for the Jonglei Canal Area, with its Executive Organ (JEO),[2] charged with the duty of ensuring that the best use would be made of the Canal's coincidental but beneficial objectives. These are to improve river transport by shortening the navigational distance between Kosti and Juba by some 300 km; to improve road transport by the provision of an all-weather road along one of the Canal's embankments; to provide year-round water supplies along the line of the Canal for the benefit of the local population; and to reduce the damaging effect of flooding from the Bahr el Jebel in years of high discharge.

The Canal (Stage I) was initially planned to take an average discharge of 20 million m^3 per day, and most calculations of benefits and effects have so far been based on this figure. Its total length will be 360 km, with a ground slope varying between 7.0 and 12.5 cm per km. Cross-sectional profiles of the Canal and its related embankments at three points of its course are shown in Figure 2.1. There are three distinct reaches of the canal. The lower reach extends for 40 km upstream of the outlet at the Sobat mouth, and here water levels will depend on the natural levels of the Sobat. Except at times of high river discharge the canal water levels will be below ground level and the depth of water will decrease from 8.0 m near the outfall to 5.0 m at km 40. The bed-width will be 38 m and the side slopes 2 : 1 to 3 : 1. Between km 40 and km 309 the water level (at a discharge of 20 million m^3 per day) will be marginally within ground level, except where the canal intersects watercourses or depressions where the water level will be between 0.5 m and 1.5 m higher. The canal depth here will vary from 5.17 m to 4.6 m, depending on the longitudinal bed-slopes, ranging from 7 cm/km to 11 cm/km. The slopes of the canal banks in this reach will be gentle – 8 : 1. Above km 309 for the final 50 kilometres the water level will be well below the level of the surrounding land, a depth of 4 m, wider bed-width of 50 m, and the bed-slope 7.4 cm/km (Abdel Magid, 1985). In addition to the Canal earthworks, there will be a regulator, with ten sluice gates designed to pass 30 million m^3 per day, at the head of the Canal at Bor, and a navigation lock in a separate parallel channel alongside it to permit access to river traffic. There are also plans for a number of bridges, crossing points, berthing and passing places.

Apart from hydrological effects caused by the operation of the scheme and ecological changes that will follow along the natural channels of the Bahr el Jebel and Bahr el Zeraf, the most important physical features of the canal are therefore the alignment as it affects local resources, including villages and cultivations alongside it, access to its waters, crossing facilities, cross drainage, navigation and road communications. Each feature will be examined in some detail in Chapter 18, but needs preliminary description here.

Potential water benefits of the Canal

The nature of *Sudd* hydrology is described in Chapter 5 in terms of the balance of river inflows and direct rainfall against outflows and evaporation. Not only is the complex combination of natural channels, basins and other topographical features a major influence on the hydrology, but difficulties of access at least to the central part of the swamps have made detailed

Plate 3. The southern end of the Jonglei Canal viewed from the Space Shuttle in June 1985. The bend visible in the canal lies north of Kongor. Permanent swamp with lakes and river channels lies at top left. Open *Hyparrhenia* grassland, still showing the dark marks of dry season burning, occupies most of the lower left of the scene. Note the ill-defined drainage channels lying within this grassland. The white patches are clouds. Photo: NASA.

survey and consistent monitoring no easy task. Despite recordings of hydro-logical data by the Egyptian Ministry of Irrigation since the beginning of the century, uncertainties in the quantitative data required for accurate modelling and prediction remain. An added uncertainty is the longer-term variability of the East African climate, leading to changes in lake levels, particularly that of Lake Victoria, and consequently to the sudden and unexpected massive increases in annual discharge of the Bahr el Jebel like those that began in 1961. These recurring phenomena and their effects are described in Chapter 4.

It follows that it is impossible to predict precisely the amount of additional water that will be made available for downstream irrigation when the canal is in operation, as well as the extent of the reduction of spill and losses by evaporation in the swamps. The benefit was originally estimated as 4.7 km^3 per annum measured at Malakal, just downstream of the canal outlet or, after allowing for losses in transit, 3.8 km^3 as measured at Aswan and based on an 'average year (1923/4) representing 85% of the annual discharges of the present century' (JEO, 1975). Estimates, however, depend on the average flows taken from different periods and therefore vary (see Table 16.1). The estimate given above does not take into account the possibility that, though

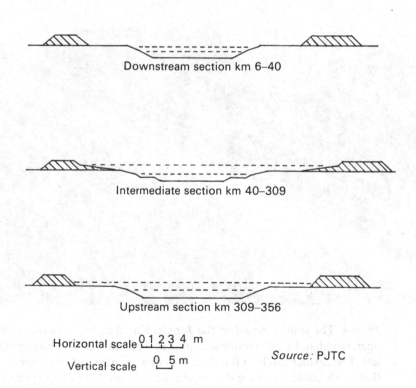

Downstream section km 6–40

Intermediate section km 40–309

Upstream section km 309–356

Horizontal scale 0 1 2 3 4 m

Vertical scale 0 5 m

Source: PJTC

Figure 2.1. Cross-sections of Canal in three main reaches.

natural river discharges might stabilise around the recent mean for the years 1961–80, which would give a much higher benefit, this is unlikely and the river may ultimately return to and stay for a long period of time at the lower discharges typical of the first six decades of the century. Other calculations have been made since, showing considerable variation; those provided by El Amin and Ezeat (1978), based on different periods of records, show figures (in km³) at the 'Tail of the Swamps' of 5.15 (1905–62), 5.42 (1905–75) and 6.66 (1962–75). These figures are the mean for those periods. Figures based on the earlier methods of calculation (85% year) would be 4.32 (1924), 4.61 (1954) and 6.51 (1975) respectively. The estimates given above were based on a proposed average canal flow of 20 million m³ per day. By 1981, however, PJTC policy had evidently changed and it was then the intention to divert 25 million m³ per day through the Canal (Mohammedein, 1982), with consequent increases in the water benefit of around 17%. Under the terms of the Nile Waters Agreement of 1959, this water would be divided equally between Egypt and the Sudan.

Plate 4. The southern end of the Jonglei Canal in 1982, running through open woodland of *Balanites aegyptiaca*. The main embankment is to the left. The smaller ditch to the right was dug to intercept drainage water so that work could continue during the wet season. The digging machine is just visible at top right. Photo: Alison Cobb.

Irrigation development plans have been set out in the Water Master Plan for Egypt (see Chapter 3) published by the Ministry of Irrigation in 1981. A similar plan for the Sudan was drawn up in 1979. Both plans indicate a massive and much needed potential for increased irrigated agriculture, and suggest a demand by the end of the century of 59.7 to 74.3 km^3 for Egypt (depending on different estimates of agricultural growth) and – a very conservative estimate – not less than 23.3 km^3 for the Sudan (*ARE, Water Master Plan*, 1981, and DRS, Ministry of Irrigation *Sudan Nile Waters Study*, 1979).

Allocations made under the Nile Waters Agreement of 1959, though calculated before the post-1961 increases in annual discharges from the White Nile, now seem insufficient to meet these targets. The relatively small increased flow as a percentage of the total mean discharge of the Nile to be made available by the present stage of the Jonglei Canal must be seen in this perspective and its inadequacy to meet ultimate downstream needs as discussed in Chapter 3. We must, however, bear in mind that there are further plans for Jonglei Stage II, involving upstream control works and over-year storage, as well as similar drainage schemes in the Bahr el Ghazal and Machar Marshes. These are discussed briefly in Chapter 20. Egypt's growing requirements can also in part be met from the reuse of drainage water which at present runs to waste in the sea. No such drainage losses occur in the Sudan, so similar savings cannot be expected there. Drainage water that could be reused in Egypt as indicated in their Water Master Plan may be as much as 4.4 km^3 by 1990 and 5.4 km^3 by 2000. On the other hand, the Plan suggests that, if the benefits of increased hydro-electric output were included, the advantages of the Jonglei Canal Project would be greater in economic terms than the reuse of drainage water downstream at Aswan.

It appears to be the case, however, that the Canal project as a whole has never been submitted to a conventional cost : benefit analysis, in which the respective projected national irrigation benefits are weighed in the balance against both the capital investment in and operating costs of the canal as well as the local gains and losses, which would be virtually impossible to calculate. Estimates of existing local benefits in terms of grazing, fisheries and other resources in the *Sudd* region under natural uncontrolled 'run of the river' conditions would probably not be possible in quantitative economic terms in any event. It must be conceded, however, that strictly on those terms and without taking into consideration other criteria, the comparison in returns between local and downstream usage would not favour the former, unless irrigation in the Jonglei area (or elsewhere in the south) proves technically feasible and viable. The issue at present is how much water can be drained from the *Sudd* without serious and irreparable damage to the local economy and its potential expansion.

As a prelude to a financing decision, the European Development Fund commissioned a brief appraisal of the project as a whole and the structures in particular, but this was not called upon to attempt any such precise comparative analysis. Estimated costs (at 1978 prices) had been given as £S92 million, made up of £S14.7 million for structures and training works, £S43.5 million for canal excavation, £S18 million for local development works, and £S15.8 million for contingencies (Mohammedein, 1982). These were revised to £S73.2 and £S90.0 for structures and excavation respectively at 1982 prices but for obvious reasons, since hostilities have halted all work on the canal since 1983, these figures are no longer relevant. The costs of repair to existing works and completion of the remaining excavation and structures are bound to be substantially higher in the future.

Nevertheless, it is evident that the PJTC were satisfied at the time (1983) that additional water from the canal would be sufficient to irrigate some 600,000 feddans (252,000 hectares) in the northern Sudan, or alternatively about half that amount coupled with the intensification of agricultural production in the existing Gezira Scheme. Taking into account costs as then estimated, the return was calculated at something in excess of 300 US$ per 1,000 m^3 based on a projected cropping pattern to include medium and long staple cotton, groundnuts and wheat. For the Sudan, benefits from improved navigational facilities between Khartoum and Juba need also to be taken into account, particularly in respect of speedier services and reduced freight charges. Passenger traffic, however, had declined dramatically from the mid-1970s onwards,[2] most people preferring the greater speed and relatively greater reliability of road transport during the dry months of the year. The possibility of river–road competition in the future is referred to later in Chapter 18. For Egypt, though different methods of analysis were used, the estimated returns were much lower and varied very widely according to the soils in the areas chosen for development. Egypt would, however, have the additional benefit of increased hydro-electric power generated by extra water from the canal through the turbines at Aswan.

Construction

Construction work on the Jonglei Canal began in 1978, and was undertaken by a French consortium CCI (*Compagnie de Constructions Internationales*), whose interest in embarking on the work arose from the fact that they possessed the largest mechanical excavator in the world, which they had previously had constructed in Germany to dig the Jhelum Link Canal in Pakistan.

Work progressed extremely slowly for the first two years, as the contractors tried to find solutions to the many technical problems that initially beset them, mainly concerned with adapting the machine to cope with local conditions. The soil was too hard in the dry season for the teeth of the bucket, and too soft for anything in the wet. Wet season operation finally became possible when the buckets had been suitably modified and when an immensely complicated system of advance drainage works had been devised to divert rain water and creeping flow away from the anticipated 80 km or so to be excavated during the rains. After only 80 km of progress in the first three-and-a-half years of operation, the rate of excavation accelerated and in the next year-and-a-half the digger reached 240 km, but has made no progress since November 1983, when the SPLA[3] brought the works to a halt.

The machine is shown in action in Plates 5 and 6. Able to shift up to 3,500 tons of earth per hour, the machine deposits the spoil from the excavated

Plate 5. The digging machine in 1982. The excavating unit, with the bucket wheel, is on the left. The spoil is transferred along the gantry to the unit on the right, which spreads it to form the embankment. Photo: Alison Cobb.

canal on embankments, to which it is guided by an integral mobile conveyor belt. Embankments will also act as dykes, retaining the water within the canal in places where it will be above the surrounding land in the central portion of its length. The final profile of the canal (see Figure 2.1) with gently sloping banks along most of its length, which will make animal access into the water relatively easy, is to be completed by draglines and bulldozers.

Perhaps the most striking feature of the canal's construction was that CCI succeeded in resolving so many daunting mechanical and logistical problems, in particular in bringing the equipment to the site and maintaining an adequate supply of fuel at a time when the country was beset with transport difficulties and shortages. A host of other practical and technical difficulties were overcome in exceptionally intractable terrain and harsh climatic conditions. Local labour was effectively and sympathetically mobilised. The halting of all construction work in 1983 was a sad reward for a remarkable engineering achievement.

Alignment

Under the Equatorial Nile Project the canal was to start at Jonglei, a small village site on the river Atem, one of the branches of the Bahr el Jebel below Bor. An earlier line – Line VII – was to run much closer to the natural channels, but, as we have seen in Chapter 1, on the recommendation of the Jonglei Investigation Team at the time, a 'Direct Line' to run straight from Jonglei to the Sobat mouth, was agreed. With the wisdom of hindsight an even more easterly alignment might with advantage have been recommended as more appropriate, but it must be remembered that the edge of the seasonally inundated marshes did not at that time extend anywhere near as far eastwards as it did from the early 1960s onwards.

When plans for a canal to drain the *Sudd* were revived in the early 1970s, three possible offtake sites were considered, two in the vicinity of Jonglei; one near Bor. These considerations included an examination of the merits of an offtake with and without a barrage; and, in the case of the Jonglei sites, with or without banking to prevent spill along the Atem distributary. For a number of reasons the Bor offtake site was chosen though not before excavation had started at the northern end. For one thing the bulk of the dramatically increased river-flows from 1961 onwards have since flowed through the Aliab Valley to the west; for another, to prevent the blocking of the Atem by floating vegetation, notably papyrus and *Eichhornia crassipes* (water hyacinth), would require extensive and very costly embankment to a point some kilometres above Bor, and there would be problems of sedimentation downstream of the Jonglei site. While the selection of Bor as the site for the head works of the canal required an additional 100 km of excavation, the above reasons and hydraulic considerations indicated that this was the most economic and advantageous alternative location. This assumption was confirmed only after further study in 1979. The length of the canal was thereby increased from 260 to 360 km. A coincidental advantage would be that the canal route would be even more direct, with consequent navigational savings which in the long term might partly offset the cost of extra excavation. Construction had begun following the 'Direct Line' planned earlier while the new extension and alignment were being surveyed; hence the noticeable bend that occurs near Mogogh some 100 km from the tail.

In the meantime in 1977 the Jonglei Executive Organ had wisely commissioned a small team of social scientists to examine the alignment and associated features, such as the need for, type, and location of crossing points, in the context of local needs and interests. Their comprehensive survey revealed the manifest local advantages of running the canal well to the east of any earlier proposals. Their recommendations (Hoek, B.v.d. *et al.*, 1978) were that the eastern line from Bor to Kongor, which by then had been agreed,

Plate 6. Another view of the digging machine, with the excavating unit and bucket wheel to the right. Photo: John Goldsworthy.

should be projected northwards along the east side of the Duk ridge to join the planned alignment south-east of the administrative centre of Atar. The result would be to reduce the number of crossings required, thus allowing the majority of essential cattle migration routes to continue to be used unobstructed. The important centres of Duk Padiet and Duk Payuel would be close to perennial water supplies and, moreover, fewer roads would be cut by the canal. Inundation of existing permanent villages and cultivations from creeping flow from the south-east would be reduced, for the canal would act as a barrier. Moreover, the distance would be shorter than the western alignment, local advantages seemed indisputable, and the costs were thought to be no greater. This proposed alignment was to some extent a compromise on the direct line from Bor to the Sobat mouth, which would run even farther east, as suggested by Euroconsult earlier. Consideration of this earlier proposal was also recommended by the consultants appointed by the JEO, but that alternative was, it was alleged, by then already too late. The advantages of a much more easterly alignment had also been stressed by El Sammani and El Amin (1978) and by Payne, Senior Consultant, UNDP (undated report, probably in the same year). Yet, although the choice of options in 1979, particularly along the southern stretch of the canal, was still open, in the event a more westerly alternative was chosen – better than the original line, but with fewer local advantages than the alignment favoured by these various consultants. This decision has never been fully explained and was the cause of considerable resentment among the inhabitants of the area, which will be referred to later.

Canal crossings: bridges, ferries and swimming points

Apart from the bridge to be constructed across the regulator and lock at Bor, it is proposed that three bridges will be provided at km 25, km 125, and km 250. For navigational purposes these will require clearance to a minimum height of 15 m above maximum water level in the canal. Bridge designs have not been completed but specifications indicate a carriageway requirement of 7 m for motor transport, with footpaths on either side to allow for the crossing of livestock and pedestrians. In addition, 12 motorised ferries are planned at approximately 20–30 km intervals along the canal.

The JEO social scientific team mentioned above, after an exhaustive survey of livestock migration routes, pointed out that the easterly route they had recommended would reduce the need for crossing points. The alignment now adopted will mean that the canal will have to be traversed by people and livestock at more points than provided for by the bridges and ferries included in the plan, requiring more of these than the costs might warrant and therefore simpler alternatives in the form of 'swimming points'. For crossing facilities of this kind a design has been suggested in some detail by the same consultants and independently by Euroconsult (Hoek, B.v.d. *et al*; Euroconsult, 1978). This would involve low gradient of the banks at the points of entry and exit, and greater canal width to allow the safe take-off and landing of animals on either side, as well as adequate distance to allow for drift downstream, and space to enable 25–30 animals to enter at the same time. These would be minimum requirements. Rough stone bases would be needed to prevent the animals slipping and to protect the bottom of the entry pools, and measures would need to be taken to prevent slippage and erosion of the banks by the use of stone or gravel on the slopes or alternatively, planting suitable vegetation to bind the soil (Euroconsult, 1978).

From the local point of view and the unimpeded operation of a migratory pastoral economy in which freedom of movement is all-important, the provision of sufficient crossing points is essential, and local response to the enquiries conducted in great detail by the consultants was unanimous in demanding such facilities. The need to take into consideration migration routes to and from dry season grazing grounds, which embody well-established and closely guarded rights, is a vital one and this and logistical problems, including disease control and the problem of adequate grazing for livestock waiting to cross, will be referred to again in Chapter 18.

Cross drainage

The principal reason for the abandonment of the original Line VII (see Chapter 1) under the Equatorial Nile Project in favour of the 'Direct Line' was the problem of cross drainage of the many channels and water courses joining the Zeraf River from the east. This was accepted by the Egyptian Government, for the Direct Line would reduce constructional difficulties and costs. The new alignment extends from the Jonglei latitude as far south as Bor, and, except for the lower reach, runs east even of the original Direct Line and hence cuts these lateral watercourses much higher up in their catchments where their courses are less well defined. The amount of water that flows through them is therefore less evident, except where the canal now crosses the Khor Atar at a point north of Mogogh. Here a siphon underpass was at one time considered but not, it seems, pursued. Instead, since the water in the Canal will, at normal flows, be above the surrounding land at this point, it has been proposed to pipe water under the embankment, into the western portion of this watercourse, thus maintaining its flow (Associated Consultants, 1981).

The volume and discharges involved in creeping flow or sheet flooding on the east side of the Canal (but see JIT, 1947) have not been calculated and for the time being no provision to meet this hazard has been announced. It is held by the PJTC that observation of flows after the canal is excavated but before it is put into commission will give sufficient time to introduce adequate cross drainage measures should they be required. This conclusion seems doubtful and is at variance with all earlier recommendations, and indeed more recent consultant advice (RES, 1983).

Navigation and transport

Berthing places are to be constructed in bays by widening the canal where required. One of the obvious reasons for the rapid development of Bor as a town is its port facilities, which have long served the hinterland on the east bank. The siting of berthing places at other points along the canal is therefore likely to have economic implications, with a potential for development, first of small trading centres, and subsequently perhaps small towns. The reach of the canal below km 40 is too narrow, expecially when the canal is not operating to full discharge capacity, to allow river craft to pass each other. Widening of the canal at 3 km intervals in this section is therefore provided for.

The shortening of the distance between Kosti, Malakal and Bor, and improved speed of river communications could do much to stimulate traffic on the ailing river steamer services[4] in the southern Sudan by reducing river freight costs. Against this must be set increased trade and passenger competition from road transport services along the line of the canal. Distances in this regard will not be substantially less than the present road route, but improvement in road standards may be expected. The proposal to construct an all-weather road along the Canal embankment will also offer links with surrounding areas in the dry season, when road transport there is feasible. A substantial investment will therefore be necessary to improve the quality of the existing road system, or to create new ones. For a communications development programme of this kind to be effective, the Canal road needs to be on the same side of the Canal as the majority of the people. It was stated by the PJTC at the JEO Conference at Bor in November 1983 that the road would be on the east embankment along its entire length, which would be much to the disadvantage of many people in Kongor District, who almost all live to the west of the Canal and who had previously been led to believe that the road would cross, by a bridge at the latitude of Kongor town, on to their side, the west side, of the Canal and continue from that point to Canal head at Bor (see Chapter 18).

Management

As we have seen, the Permanent Joint Technical Commission, with Egyptian and Sudanese representation, is in charge of overall management of the project and responsible for all engineering plans. The National Council for the Development Projects in the Jonglei Canal Area is responsible for the formulation and execution of economic and social development plans for the inhabitants of the Jonglei area, as well as any measures required to mitigate adverse effects. The Jonglei Executive Organ (JEO) is charged with implementation of these objectives with the emphasis on local development.

It has been suggested that a Jonglei Canal Authority be established to manage the operation of the water intake, maintain the structures, ensure the operation of the canal as a navigation channel, including the management of bridges, ferries, crossing points, and the road along the banks. Such a body would also collect, collate, and analyse hydrological and morphological data relevant to the operation of the canal and the monitoring of its local effects. The great importance of the role of such a body, taking local interests into account, not least in determining the throughput of water allowed down the Canal, is discussed again in Chapter 20.

Notes

1. The terms of reference of the Permanent Joint Technical Commission are set forth in the fourth section of the Nile Waters Agreement of 1959 as follows:

Technical co-operation between the two republics

1. In order to ensure the technical co-operation between the Governments of the two Republics, to continue the research and study necessary for the Nile control projects and the increase of its yield and to continue the hydrological survey of its upper reaches, the two Republics agree that immediately after the signing of this Agreement a Permanent Joint Technical Commission shall be formed of an equal number of members from both parties; and its functions shall be:

   (a) The drawing of the basic outlines of projects for the increase of the Nile yield, and for the supervision of the studies necessary for the finalising of projects, before presentation of the same to the Governments of the two Republics for approval.

   (b) The supervision of the execution of the projects approved by the two Governments.

   (c) The drawing up of the working arrangements for any works to be constructed on the Nile, within the boundaries of the Sudan, and also for those to be constructed outside the boundaries of the Sudan, by agreement with the authorities concerned in the countries in which such works are constructed.

   (d) The supervision of the application of all the working arrangements mentioned in (c) above in connection with works constructed within the boundaries of Sudan and also in connection with the Sudd el Aali Reservoir and Aswan Dam, through official engineers delegated for the purpose by the two Republics; and the supervision of the working of the upper Nile projects, as provided in the agreements concluded with the countries in which such projects are constructed.

   (e) As it is probable that a series of low years may occur, and a succession of low levels in the Sudd el Aali Reservoir may result to such an extent as not to permit in any one year the drawing of the full requirements of the two Republics, the Technical Commission is charged with the task of devising a fair arrangement for the two Republics to follow. And the recommendations of the Commission shall be presented to the two Governments for approval.

2. In order to enable the Commission to exercise the functions enumerated in the above item, and in order to ensure the continuation of the Nile gauging and to keep observations on all its upper reaches, these duties shall be carried out under the technical supervision of the Commission by the engineers of the Sudan Republic, and the engineers of the United Arab Republic in the Sudan and in the United Arab Republic and Uganda.

3. The two Governments shall form the Joint Technical Commission, by a joint decree, and shall provide it with its necessary funds from their budgets. The Commission may, according to the requirements of work, hold its meetings in Cairo or in Khartoum. The Commission

shall, subject to the approval of the two Governments, lay down regulations for the organisation of its meetings and its technical, administrative and financial activities.

2. The National Council for the Development Projects for the Jonglei Canal Area was established in the Sudan by Presidential Decree in October, 1974. Terms of reference were summarised by the Commissioner of the Jonglei Executive Organ in an address to the Royal Geographical Society in London on 5th October, 1982 as follows:

> '(a) to promote studies relating to the effects of the Jonglei Canal Project;
>
> (b) to endeavour to derive the maximum benefit from conditions created by the project for inhabitants of the Canal area and to mitigate any adverse effects which might arise;
>
> (c) to plan and to implement integrated economic and social development in the canal area.'

As the executing agency of the National Council, the Executive Organ (JEO) is in effect charged with the responsibility to assure:

> (a) comprehensive socio-economic planning for the canal area:
>
> (b) the promotion, supervision, or conduct of studies as appropriate;
>
> (c) the procurement of funds and the design and administration of development works in agriculture, social services, and even industry (Awuol, 1982).

3. For SPLA see note 1 to Chapter 11.

4. For example, passenger traffic had dropped from 216,000 in 1976 to only 25,000 in 1981.

3

Water use and demand in the Nile Basin

Introduction

In the mid-1970s it appeared that Egypt and the Sudan, the two heaviest users of Nile water, had placed their agricultural development plans on a collision course. The Sudan had set its sights on the rapid expansion of irrigated agriculture and especially the development of sugar cane production under irrigated conditions. Egypt, desperate for new agricultural land, planned to increase its cultivated surface by nearly 50% through desert reclamation. If the efficiency with which irrigation water was then used in the two countries were to remain unchanged, then a time would come before the turn of the century, when the supply of Nile water would not be sufficient to meet Sudanese and Egyptian demand.

In these more optimistic times, it was pointed out that two factors could postpone or obviate the impending crisis: 1. some combination of greatly increased efficiency in water use through plant genetics and improved cultivator practices, and 2. economic stagnation in the Sudan or Egypt (Waterbury, 1979 p. 241). The first factor has not come into play but the second most certainly has. Since the late 1970s, the Sudanese economy has gone into severe recession. The large flows of Arab investment and aid that had been expected failed to materialise, especially after President Nimeiri endorsed the Camp David peace accords signed by Egypt, Israel, and the United States in March, 1979. The second oil shock after 1979 thoroughly upset the country's balance of payments. New dissidence manifested itself in the southern provinces after 1980 and, in the north, severe drought afflicted the rainfed areas beginning in 1983. Under these sad circumstances the expansion of irrigated agriculture in the Sudan has been postponed indefinitely.

Egypt's economy, by contrast, has been fairly buoyant, but its agricultural sector has stagnated. The demand for water has grown with increased cropping intensities, but that has probably been offset by the loss of agricultural land to urban and village encroachment. There is still an official commitment to reclaim something like 800,000 hectares of desert land within 15 years, but so far the technical assistance and external aid necessary for the projects have not been forthcoming in the desired amounts.

In short, the crisis foretold some time ago, has not yet materialised. We should hope that it is not, in Martin Adams' words, 'a crisis postponed' (Adams, 1983), for its absence will testify only to the continued economic stagnation of the Sudan and Egypt. It is the intention in what follows to examine the major facets of water use throughout the Nile basin and how they may evolve over the medium term. The discussion will be largely about Egypt and the Sudan which between them account for over 90% of all water drawn from the Nile. But, as we have seen, there are seven other riparian states in the basin, and their demand for water will inevitably grow.

Egypt: supply, demand and development

Each of the national consumers of Nile water presents unique characteristics in the shaping of its real or potential demands. In Egypt, demand is and will be conditioned primarily by three basic factors: the rate and type of land reclamation, the re-utilization of drainage water, and the intensity of cropping on cultivated land. Before examining these factors, however, it should be stressed that to treat Egypt and the states further upstream as separate entities, is merely to conform with the effective units in which policy is made and implemented. Although there are nine co-riparians in the basin that *could* be regarded as parts of a large basin-wide unit for water resource planning, few steps have been taken in that direction. The state of our knowledge is such that any basin-wide planning exercise may be premature. But it has been in the interests of none of the major participants, including external donors, to try to see the system as a whole. Rather, a fragmented view has been more convenient for it is sustained by an underlying assumption of abundant water supply. No one wants to call that assumption into question until each national participant knows fairly exactly where its interests lie and what cards it has to play.

As far as Egypt is concerned, the sanguine outlook of the middle 1970s, when international bodies of aid and technical assistance such as the FAO or USAID foresaw no possible water supply constraints before the year 2000 (Waterbury, 1979 pp. 210–11), has given way to more cautious and more detailed assessments that look to the 1990s as a period of possibly conflicting demands (IBRD, 1981 *Irrigation Subsector Report*; ARE, MOI, 1981 *Egyptian Water Master Plan – Main Report*).

This caution has not yet seeped into official pronouncements. Hasballah Kafrawi, Minister of Housing and the Development of New Lands, stated in September 1981, that:

> the Ministry had prepared its major plan for the reclamation of 2.8 million feddans* and shown it to various specialists and consulting firms after it had been confirmed that there was sufficient water for such projects until the year 2000 (*al-Ahram*, 30 Sept. 1981).

The Egyptian Water Master Plan (EWMP) undertaken by the Ministry of Irrigation in collaboration with the IBRD and UNDP, consists of 17 volumes containing detailed analyses of each of the variables affecting water supply and demand in Egypt (ARE, 1981). While the reports are sound in their analysis and conclusions, there are some problems in the definition of certain parameters that are either unexplained or stem from assumptions that are not fully explained. It is worth exploring some of these problems with the *caveat* that the margin of error involved is considerable. A case in point concerns estimates of the expected average annual discharge of the Nile.

In 1959, as a precondition to the construction of the Aswan High Dam, Egypt and the Sudan negotiated an Agreement on the Full Utilisation of the Nile Waters. A mean annual discharge of 84 km^3, based on the period of 1900–59, was used as the basis for allocating water between Egypt and the Sudan. That mean, as it turns out, looks too modest when compared with the data for an entire century. The period 1900–59 comprised a series of fairly low floods. Since 1960 there is evidence of a new rise. A century of readings does not give one much to go on when dealing with the Nile, but the mean for the period 1880–1980 is 89.7 km^3, or, on average, 6 km^3 more per annum than was provided for in the 1959 Agreement. Thus the crisis in conflicting demands may be delayed because there may be more water available than was commonly believed. Still, after 20 years of increase, the three years 1982–4 saw the annual average discharge at Aswan at 70 km^3, as the drought in the Ethiopian catchment of the Blue Nile and Atbara intensified.

Under the terms of the 1959 Agreement, Egypt is entitled to release 55.5 km^3 of water each year through the sluice gates and turbines of the High Aswan Dam. For some years Egypt has been using even more than this amount, but no one disputes that an enormous amount of waste is involved. How to measure that waste and to ascertain what are real needs as opposed to actual consumption, and therefore what could become available for

---

*2.8 million feddans = 1.18 million hectares

productive use if waste were reduced, are challenges of crucial importance. How one measures the waste and makes assumptions about recovering it, directly determines any estimate of overall national water balance.

One may break down Egypt's average use of its 55.5 km³ in the late 1970s as follows:

		km³
	Total available	55.5
1.	crop needs	30.0
2.	crop use	45.0
3.	drainage (2–1)	15.0
4.	conveyance losses*	6.7
5.	navigation and power release	2.5
6.	industrial and domestic	1.3
	Total 2+4+5+6	55.5

Note: * These consist of surface evaporation in the canal delivery systems, as well as seepage from unlined canals. Some seepage finds its way into the aquifer and is potentially recuperable.

This is an artificial accounting exercise to give orders of magnitude. In fact Egypt's water use has been higher than its allotted share because of three factors:

(a) Egypt was able to borrow 1.5 km³ annually from the Sudan because the latter was unable to utilise its full share. It is not clear if these 'loans' persist, but they had the effect of raising Egypt's total share to 57 km³.

(b) The return flow of drainage water into the Nile between Aswan and Cairo may have averaged 3–4 km³ annually.

(c) About 2.5 km³ of drainage water in the Delta was being reused in agriculture annually.

These sources raised water available by 7–8 km³ annually or to a total of 62–3 km³. The additional water found its way into industrial and domestic consumption at levels much higher than most estimates, as well as into land reclamation and more intensive cropping.

The essential point here, however, is agricultural waste. Over the period

1964–72, the first eight years of the so-called High Dam era, the effective crop use of irrigation water grew from 27 km^3 to 33 km^3. All of this increase could have been accounted for by expanded rice cultivation made possible by the water stored at Aswan. At the same time the amount of drainage water grew from 14 km^3 to 16 km^3. The crop use of irrigation water thus became somewhat more efficient rising from 48% to around 50 or 51% (El Gendi & El Ghomri, 1977, p. 5). Despite this improvement, it is still the case that far more water is poured on fields than the crops under cultivation can use.

In Table 3.1 are reproduced the basic parameters for the medium-growth scenario from the EWMP, assuming land reclamation at an average rate of 42,000 hectares per year, for the period 1980–2000. What assumptions underlie these projections? First, the one source of demand that is changeless is that of the old lands at 29.4 km^3 for the entire 20-year period. It is not clear from the EWMP reports how this figure was obtained. It is assumed that in 1978 the cultivated area in Egypt totalled 2.44 million hectares, and, with a multiple-cropping intensity (MCI) of 1.9 the cropped surface was equivalent to 4.63 million hectares. (This simply means that for every field cultivated an average of 1.9 harvests were obtained per year.) Average water consumption then works out to 12,067 m^3 per cultivated hectare and 6,305 m^3 per cropped hectare. These figures must be based on certain notions of crop consumption that are not made explicit. They do not represent much greater efficiency in water use because, as Table 3.1 shows, over the period, drainage is lowered only from 16 km^3 to 13 km^3. They do not square at all with estimates by I. Z. Kinawy (February 1976 as presented in J. Waterbury, 1979 p. 221). These, on

Table 3.1. *EWMP Water Demand Medium growth scenario land reclamation at 42,000 hectares per year, 1980–2000 (km^3)*

Source of demand	1980	1985	1990	2000
Agriculture				
old lands	29.4	29.4	29.4	29.4
reclaimed lands	--	4.8	8.5	11.5
Municipal, net consumption	1.8	1.9	2.2	3.5
Industrial, net consumption	0.3	0.5	0.8	1.4
Navigation and spills	3.8	1.6	1.6	1.6
Evaporation	2.0	2.1	2.1	2.2
Drainage	16.0	15.7	14.2	13.4
Unaccounted	0.7	0.3	0.1	0.1
Total demand	54.0	56.3	58.9	63.1

the basis of an MCI of 1.6 in the early 1970s, yield water consumption of 14,495 m^3 per cultivated hectare and 9,126 m^3 per cropped hectare (cf. M. T. Eid *et al.*, 1966; Abou al-Atta, 1976). Moreover, the EWMP presents tabular material for only eight crops and does not include the area devoted to each (see Main Report and Technical Reports 2 and 17). The consumptive use estimates thus appear too low by perhaps as much as 3 to 4 km^3 per year. Crop consumption of water is probably higher than estimated by the EWMP. For example, the IBRD Subsector Report (IBRD, 1981) estimates annual consumption at 31 km^3 (for ex. Table 12) explaining the difference in a cryptic note to the effect 'Subsector Review includes water consumption unallocated by EWMP'. Perhaps the authors have in mind the likelihood that cropping intensity is going to increase substantially. Fitch and Abdel Aziz (1980) put that case convincingly. They chart the increase to the mid-1970s and make an estimate for 1985–90:

Year	MCI
1952	1.60
1962	1.76
1974	1.86
1985–90	2.1–2.45

The authors note that if the area under sugar cane is deducted (because sugar cane maturation takes about eleven months) the MCI of Upper Egypt is already 2.12. Moreover, the MCI is highest on small farms below 4.2 hectares, using family labour. Much of the Egyptian countryside has moved toward that kind of farm unit (Harik, 1979). As it is likely that the area under cotton will be reduced (a crop that ties up the land for about nine months and which the EWMP regards along with sugar cane and citrus as a 'permanent crop') and the area of vegetables increased, and that tile drains replace open field drains, cropping intensity will increase as indicated above. If, for example, 2.4 million hectares are cultivated in 1990 with an MCI of 2.3, the cropped area would be 5.5 million hectares, and crop water consumption might be as high as 35.8 km^3 with 14 km^3 allotted to drainage.

Egypt's demand for water will increase dramatically as a result of land reclamation. Herein lies another anomaly in the EWMP's data presentation. Note that in Table 3.1 for the year 1980 no water is assumed to be used on new lands. Yet it has now been firmly established that a gross area of 383,000 hectares had been reclaimed between 1960 and 1972, of which about 250,000 were actually under cultivation (Waterbury, 1979 pp. 136–43). This area is in addition to the 2.44 million hectares of old land, and water use upon it has been very high, around 24,000 m³ per hectare per year depending upon the stage of reclamation that it is in. If the 250,000 hectares use on average 16,667 m³ each year, 4.2 km³ would be consumed. That figure for what one might call the 'old new lands' should remain fairly constant for the entire 20-year period.

The EWMP assumes water use in all future reclamation projects at a rate of 12,850 m³ per hectare per year, a figure that in its modesty bears no resemblance to Egypt's performance on the 'old new lands'. Even this figure is not used consistently. On p. 89 of the Main Report the figure of 12,850 m³ per hectare per year (5,400 m³ per feddan per year) is given while on p. 32 Technical Report 2 a figure of 10,000 km³ per hectare per year (4,200 m³ per feddan per year) is offered as the norm, based on the cropping pattern of 1978. The IBRD Subsector Report is far more realistic in estimating 'a command area diversion requirement of 21,900 m³ per net hectare (9,200 m³ per net feddan').

Both the EWMP and IBRD Subsector Report calculate water demand according to varying rates of land reclamation: a range of 21,000–75,600 hectares per year to the year 2000. The figures in Table 3.1 assume an average rate of reclamation over 20 years of about 4,200 hectares per year at 12,850 m³ per hectare. If this Subsector Report estimated average of c. 21,400 m³ per hectare were used, the gross water use for the new lands would look as shown.

	EWMP (km³)	Subsector Report km³
1985	4.8	8.3
1990	8.5	14.4
2000	11.5	19.2

These differences are enormous. That for 1985 alone, nearly 4 km³, is about twice what Egypt expects to gain from the Jonglei Canal project. The difference for the year 2000 is the equivalent of all that Egypt stands to gain from all the proposed Upper Nile projects. The point here is that the margin for error in the assumptions we make cannot be very wide. Mis-estimating such factors as crop water duties, return flow of drainage water, and water demand on reclaimed land has serious consequences for the measurement of water constraints that become tighter with each passing year. It used to be that Egypt and the Sudan made allocation decisions 'give or take 5 km³'. Now it is a case of give or take half or a quarter km³.

In the Nile Basin, while agriculture accounts for at least 80% of all water consumption, it is still important to establish accurate estimates of domestic and industrial use. Here again the EWMP errs on the low side. It makes some very optimistic assumptions about the reuse of industrial and municipal waste water and fails to estimate rural, non-agricultural consumption. If waste water is not reused on the scale projected, and if about $1 km^3$ in rural consumption is added in, the estimates in Table 3.1 for industrial, municipal, and unaccounted uses be too low by as much as 3 to $4 km^3$.

There is no single estimated variable in the EWMP that differs more from the author's own estimates of the mid 1970s than that of evaporation and conveyance losses. The EWMP puts these at $2 km^3$ in 1980 rising minimally to $2.2 km^3$ by the year 2000. The slight increase will be brought about by the extension of the irrigation grid to the new lands, causing an increase in evaporation and seepage from the canals. There is some consensus now that surface evaporation rates at the High Dam reservoir are of the order of 2.7 to 2.9 metres per year (M. H. Omar & M. M. El Bakry, 1970, EWMP, Report 14 p. 108) possibly averaging as much as $13 km^3$ per year. It is reasonable to expect similar rates on the downstream networks, above all because of the multiplier effect of aquatic weeds, such as water hyacinth, for which average increases in evaporation rate range between 1.3 and 6-fold (Pieterse, 1978), at shallow canal depth, and so on. So evapotranspiration alone should claim at least $3 km^3$ annually. Yet the EWMP allows for only $2 km^3$ for *all conveyance* losses. Conveyance losses in systems similar to Egypt's can be estimated at roughly 15% of total supply delivered – that is a minimum of $30 km^3$ for agricultural consumption plus $15 km^3$ per year for drainage. Conveyance losses would, thus, total about $6.7 km^3$ per year instead of the $2–2.2 km^3$ put forth by the EWMP. Even this revised estimate may be too conservative. In 1976, the Ministry of Irrigation put conveyance losses at $8 km^3$ (Abou El Atta, 1976) and USAID at $11.2 km^3$ (USAID, 1976 p. 89).

The remaining variables in Table 3.1 require little comment. The small reduction in drainage water may be somewhat conservative, but if we take drainage water as a proportion of total water delivered to old and new lands, it declines from 46% in 1985 to 32% in the year 2000.

The question of water supply requires analysis of fewer variables, but the magnitude of the unknowns is if anything greater. Supply depends on:

1. The 1959 Agreement allocation: $55.5 km^3 + 1.5 km^3$ on loan
2. Actual discharge at the High Dam Reservoir
3. Return flow of drainage water in Upper Egypt

4. Re-utilisation of drainage water in lower Egypt
5. Implementation of Upper Nile projects

The author's own estimates (JW) for supply and demand for 1980 and 1990 are compared with those of the EWMP in Table 3.2.

On the demand side, higher crop water duties and a higher MCI are assumed than those used in EWMP. The estimates for new lands assume 250,000 hectares under reclamation in 1980 and 672,000 in 1990. The rate of water use is put at $19,000 \, \text{m}^3$ per hectare in 1980 and $17,000 \, \text{m}^3$ in 1990 – probably an underestimate. The need to revise upwards municipal, industrial and conveyance losses has already been stressed above, while a somewhat greater efficiency in water use is assumed than by the EWMP.

On the supply side for 1980 it is assumed that the Sudanese loan of $1.5 \, \text{km}^3$ is in effect, in addition to which another $3 \, \text{km}^3$ is assumed to have found its

Table 3.2. *EWMP and JW estimates of water demand and supply in Egypt;*
1980 and 1990 (km^3)*

	1980		1990	
	EWMP	JW	EWMP	JW
Demand				
Old lands	29.4	32.4	29.4	33.0
New lands	—	4.8	8.5	11.2
Municipal net loss	1.8	3.0	2.2	4.0
Industrial net loss	0.3	1.0	0.8	2.0
Navigation	3.8	2.5	1.6	1.6
Unaccounted for and evaporation	2.7	6.7	2.2	7.0
Drainage	16.0	15.0	14.2	14.2
Demand total	54.0	65.4	58.9	73.0
Supply				
At Aswan	57.5	60.0	61.7	58.9
Drainage reuse	—	2.5	5.4	6.0
Drainage return flow	—	4.0	—	4.0
Supply total	57.5	66.5	67.1	68.9
Balance	+3.5	+1.1	+8.2	−4.1

Note: * More detailed calculations are to be found in Waterbury, 1982.

way downstream from the High Dam. This is due to real discharge above the assumed mean. For 1990 that Sudanese loan is assumed to have disappeared, but it is also assumed that the first phase of the Jonglei Project will bring a net increase of 1.9 km^3 for Egypt, in addition to which discharges will continue above the average for the first half of the century. This assumption is open to question in view of the recent series of low floods in the Blue Nile. While the EWMP suggests that other Upper Nile projects will be implemented by 1990, this scarcely seems feasible within that time scale or perhaps ever. That explains the difference between the two estimates for supply at Aswan: 61.7 km^3 vs. 58.9 km^3.

The EWMP has undoubtedly advanced our knowledge of the basic variables affecting water supply and demand in Egypt. But the Plan appears to have some important shortcomings. These are, on the demand side:

(a) an under-estimation of the likely MCI and crop water duties on the old lands
(b) an under-estimation of water requirements for the new lands
(c) an optimistic assumption as to the amounts of municipal and industrial waste water that will be re-utilised
(d) an under-estimation of conveyance losses

On the supply side, the shortcomings are:

(a) an over-estimation of the likely yield of the Upper Nile projects
(b) an under-estimation of return flows of drainage water in Upper Egypt

Only the last is an error that reduces estimated supply. All the others either reduce demand or increase supply. If these turn out to be real errors, planning that takes place in the light of them will be seriously flawed.

Sudan: supply and demand

The IBRD Agricultural Sector Survey (1979, Vol. II p. 114) gave out the same bad news for the Sudan as for Egypt: 'Crucial elements of irrigation policy such as transmission losses, supply–demand balance, on-farm water management, irrigation efficiency remain after 50 years only theoretical estimates'. Given that assessment, it is small wonder that estimates of those same parameters were at considerable variance. The same report suggested, for example, that water supply on the old established cotton schemes of Gezira-Managil was and is about 12% below crop requirements at crucial points in the growth cycle. At the same time it was posited that as much as 30% of the water delivered was not used by crops (IBRD, 1979:

Annex 3 p. 14). The MOI Nile Water Study (1979, Main Report: 47) by contrast had this to say:

> The present efficiency of irrigation water use in the Sudan is generally very high ... . (F)ield irrigation efficiencies are high and canal transit losses low. Waste of water due to spillage and escapage is generally small.

How much irrigation water was being used in, say, 1980? Estimates range from 13 km^3 per year to 18 km^3, the difference being far from trivial. Part of the problem lies in estimating the portion of the command area actually under cultivation. Here the range is from 1.13 million hectares for 1978/9 (DRS, MOA, 1979) to about 1.34 million hectares for 1977 (MOI, Blue Nile: Vol. 3, Report IV, Table 2.8). It is not impossible that the combination of the deteriorating irrigation grid at Gezira–Managil and the Sudan's deepening economic crisis actually led to a decline in the cropped area.

Apart from the area under cultivation, estimates of water use depend upon crop water requirements. As was the case with Egypt, the values assigned by different sources to different crops can be quite confusing. These are sometimes calculated according to the Penman E_o crop factor method which yields estimates of plant requirements that are sometimes much lower than actual application (Waterbury, 1979). However, in large irrigation schemes it is impracticable to deliver water in precisely the quantities needed by crops, when they need it and where they need it. And of course it is rather much to expect peasant tenants to respect theoretical norms in water use, although the water charges being introduced in the Gezira scheme may push the cultivators in that direction. In fact at Gezira–Managil, Khashm al-Girba and other large, state-run irrigation projects average water deliveries to the command areas are between 9,700 m^3 and 12,600 m^3 per cultivated hectare per year. With the cropped area probably fluctuating between 1.13 and 1.34 million hectares, water *use* ranges from 14.3 km^3 to 16.9 km^3 per year.

Demand at that level seemingly gives the Sudan a good deal of leeway in developing new irrigated agricultural schemes, while remaining within the 20.5 km^3 per year alloted to it under the 1959 Agreement. A decade ago that might not have been the case. Minister of Irrigation, Saghayroun al-Zein, put Sudan's real crop use at 18.2 km^3 in 1975 (Zein, 1975 as cited in J. Waterbury, 1979 p. 233), on the basis of 9,900 m^3 per hectare of the command area. Similar figures were put out by the Ministry of Irrigation (*Control and Use of the Nile Water*, 1975) as well as by the Gezira Research Station (Mehdi El Bashir, 1981). If one adds in industrial and domestic water consumption, it can be seen that the Sudan was approaching the full utilisation of its share.

Moreover several large sugar-cane schemes were on the drawing boards. Sugar-cane consumes great amounts of water, 28,000–40,000 m^3 per hectare per year. New sources of water supply, either through the Upper Nile projects and/or more efficient on-field use were seen as urgent requirements. In fact only the 33,600-hectare Kenana sugar scheme has become operational.

It can be safely posited that by 1982 agricultural water use in the Sudan was about 17 km^3 per year. That figure includes conveyance losses but not evaporation at reservoirs. What can we expect if and when the Sudan emerges from its spell of agricultural stagnation? The MOI Nile Waters Study (Vol. 1: 4–13) and the IBRD Agric. Sector Report (Vol. II: 136) contend that the Sudan's existing command area of 1.7 million hectares can be raised to over 2.1 million with a total water requirement of 20.5–21.8 km^3 per year. According to the Nile Waters Study a command area of 3.15 million hectares could be cultivated with 30 km^3 per year.

What these estimates mean is that the rate of water use per command area hectare remains 9,500 m^3 and that cropping intensity remains around 0.75. This assumption is flawed. In other sections of these same reports it is clearly stated that at Rahad the MCI is close to 1 and that rehabilitation of the Gezira–Managil complex should bring the MCI up to 1 there. Private pumping schemes should achieve an MCI of 1.5. Let us then make a simple adjustment. Taking the 1977 command area we arrive at:

1,699,740 hectares × MCI(1) × 12,600 m^3 per cropped hectare = 21.4 km^3

In addition, a little over 1.26 million hectares in new irrigated projects have been targeted for development. Of these the highest priority is attached to 0.67 million hectares in the following projects:

Rahad II	126,000 hectares
Gezira intensification	113,500 hectares
Kenana II	126,000 hectares
Kenana III	126,000 hectares
Jebelein-Renk	76,500 hectares
Renk-Gelhak	97,000 hectares
Total	665,000 hectares

A somewhat more distant hope is a 252,000-hectare project on the Upper Atbara. Let us assume an MCI of 1 for this area as well:

665,000 hectares × MCI(1) × 12,600 m^3 per cropped hectare = 8.4 km^3

The grand total would then be

Existing schemes:	1,699,740 hectares	-21.4 km^3
Priority projects:	665,000 hectares	$- 8.4 \text{ km}^3$
Total	2,364,740 hectares	-29.8 km^3

Leaving aside for the moment the question whether or not the Sudan will find financing for its priority projects, the water demand estimates are not excessive. They do not, for instance, make any allowance for the heavy demands of new sugar-cane schemes. There are currently 86,900 hectares under sugar-cane using at a minimum 2.1 km^3 per year. Nor do these figures include about 2.5 km^3 in likely evaporation losses at reservoir sites. If the Upper Atbara project were to be implemented, these figures would have to be adjusted upwards by about 3.5 km^3 per year. Sudan's agricultural water use would then be equivalent or superior to that of Egypt's today.

If all the Upper Nile projects to increase usable discharge are brought into being by the turn of the century, the Sudan would have a little less than 30 km^3 per year to meet a demand of upwards of 32 km^3 without taking into account potential extractions in the Upper Atbara. The latter figure does not include industrial and domestic use. The possibility that those projects will be implemented is very remote. What could, and most likely will, keep supply and demand in equilibrium is the Sudan's financial inability to bring off new agricultural schemes as well as its managerial difficulties in increasing the existing area under cultivation.

The Sudan faces, or may face once its creditworthiness is restored, an important strategic choice that has serious implications for Egypt's water supply. All the Upper Nile projects currently under way (Jonglei I) or under study (Jonglei II, Machar swamp drainage, Bahr el Ghazal channelling) would serve to increase discharge on the White Nile. Yet it is probably in the best interests of the Sudan to place priority on development of the Blue Nile (including the Dinder and Rahad) and Atbara (including the Seteit-Amgharib) systems.

Storage sites on these systems are adjacent to existing agricultural schemes and major urban concentrations. But, most important, irrigation water from Blue Nile storage sites can be delivered in part by gravity flow whereas any scheme upstream of Khartoum depending on White Nile water would require pumping arrangements that are costly in fuel and thus very expensive, a problem already encountered at the Kenana sugar scheme. Moreover, existing dams and reservoirs on the Blue Nile-Atbara (Roseires, Sennar,

Khashm al-Girba) will require protection against further siltation. Finally, until the Sudan's oil reserves are brought into production, and perhaps even then, the expansion of hydropower generation will be important to the country's balance of payments. There are no good hydropower sites on the White Nile system within the Sudan below the Bedden Rapids upstream of Juba.

In this sense any increases in discharge on the White Nile are not, from the Sudan's point of view, interchangeable. Even though Jonglei I has been started, the Sudan may have little reason to pursue the other upper Nile projects upon whose water benefits Egypt is already counting. The recrudescence of Southern dissidence not only brought Jonglei I to a halt but rendered impolitic other schemes that will deliver few tangible benefits to the south. It is for these reasons that it can be argued that Egypt should assume no additional water beyond what Jonglei I can provide and perhaps not even that. As for the Sudan, once its demand for water begins to grow again, it may seek to meet it from further storage in the Blue Nile and Atbara, perhaps in co-operation with Ethiopia.

Ethiopia: the great unknown

Eleven rivers arising in the Ethiopian highlands flow across its borders to Somalia and the Sudan. Each year these rivers discharge about 100 km^3 to Ethiopia's neighbours. By far the largest river is the Blue Nile (known as the Abbay in Ethiopia) which on average delivers about 50 km^3 to the Sudan each year, or about 60% of the total discharge of the main Nile. In addition, in the south-west the Baro and Pibor rivers that form the Sobat river, and in the north-west the tributaries of the Atbara, supply respectively 14% and 13% of the Main Nile's discharge.

In the Blue Nile catchment area, variously estimated at 174,000 km^2 (Hurst *et al.*, 1947 Vol. VIII) and 204,000 km^2 (Bureau of Reclamation, 1964) total annual rainfall on the highlands may be in the order of 208 to 244 km^3. On average only a fifth to a quarter of that rainfall (i.e. *c.* 50 km^3) is captured annually by the river itself and delivered downstream to Roseires in the Sudan. It is one of the features of the Blue Nile that its source, Lake Tana, provides less than 10% of its total discharge while the major tributaries downstream of Tana – especially the Didessa, the Dabus, the Fincha, and the Balas – provide most of the rest.

Ethiopia's potential for hydropower development in this catchment is enormous, but precisely because of the steep slopes involved rapid siltation of the reservoirs is a very real problem. Run-of-the-river hydroelectric schemes, involving very little or no storage, would be poorly suited to most of

Ethiopia's rivers that are highly seasonal in their discharge. Further, the best dam sites are at considerable distance from major energy-consuming areas such as Addis Ababa and Harar.

It is merely a question of time before Ethiopia turns its attention to harnessing the resources of its western watershed. How and when it will proceed to do so are questions of vital importance to its downstream neighbours. Since the first propositions on Century Water storage were put forth after the turn of the century, various schemes have been drawn up to use Lake Tana for overyear storage. But the interest of the Ethiopian government was never more than lukewarm. Its regard was always resolutely eastwards, toward Somalia and Djibouti, toward the Red Sea and the Indian Ocean. However, in anticipation of the day when the Ethiopians would tackle the western watershed, the US Bureau of Reclamation began in 1959 an extensive survey of its water resources. 1959, be it noted, was the year in which Egypt, with Soviet financing, began construction of the Aswan High Dam.

The Bureau of Reclamation studied the location of 26 dams and reservoirs that could provide water for both irrigation and hydroelectric power. Between 371,000 and 433,000 hectares were identified as suitable for irrigation. Power projects theoretically capable of generating 38 billion kWh were examined as well (USDI, 1964, Vol. 1 p. 97). In all, the survey estimated that if all 26 projects were implemented the annual water requirements for irrigation and storage losses would reduce the discharge of the Blue Nile at the Sudanese border by 5.4 km^3 (USDI, 1964, Vol. 1 p. 41). Even in the early 1960s that would have meant a major reduction in the supply available to Egypt and the Sudan. Today, such a reduction would be near catastrophic.

Ethiopia, both before and since the overthrow of Haile Selassie in 1974, has done little to implement any of these projects. A run-of-the-river power station, known as Tis Issat, was built on the Blue Nile 25 km downstream of its outlet from Lake Tana. The Fincha River project, involving a large irrigated perimeter and a 100 MW power station was completed in the mid-1970s. Only the latter project has involved any draw-down in the Blue Nile discharge.

None the less, the revolutionary Ethiopian regime, led by Mengistu Meriam, indicated as early as 1977 that it had big plans for the Blue Nile and the Sobat. At the UN Water Conference at Mar del Plata, Argentina, in March 1977, the Ethiopian country paper (Ethiopia, 1977) announced that over the short term as many as 91,000 hectares in the Blue Nile basin, and 28,000 in the Baro, would be brought under irrigation. Over the medium term, the total water abstraction might reach 4 km^3 per year.

In 1981 at the UN Conference on the Least Developed Countries, Ethiopia presented a summary of its ten-year investment programme. It listed some 50 irrigation projects for the entire country, comprising up to 704,000 hectares, of which 381,000 are in the Blue Nile basin and 15,000 in the Baro-Akobo basin. Among these projects top priority is to be given to twenty covering 337,000 hectares. Few of these, however, appear to involve the Blue Nile or the Sobat. There will be two hydroelectric projects at Fincha and Lake Tana, and two irrigation projects, one on the Dabus covering 15,000 hectares and the other at Ribb-Gomera covering 12,000 hectares. These were scheduled for implementation between 1981 and 1985. Between 1986 and 1990 only two relevant projects are planned. The East and West Megech irrigation perimeters, covering 11,600 hectares and utilising water pumped from Lake Tana, and the Upper Balas hydro-electric project. Thus despite ambitious long-term goals, the next ten years will see only a handful of modest water control projects on the Blue Nile.

Ethiopia's population is expected to grow from 31,000,000 in 1979 to *c*. 41,000,000 in 1989. A move out of the over-populated, over-farmed highlands towards the alluvial plain around Homera along the Sudanese border is already under way. Geographically and hydrologically it is part of the heavy cotton clay zone that extends all the way to Khartoum and it is watered by the same rivers: the Blue Nile, the Rahad, the Dinder and the Seteit-Atbara. The south-west Baro region which drains into the White Nile system through the Sobat, and which is blessed with high rainfall and good soils, will also be developed. In short, Ethiopia's eastern fixation is likely to give way after a decade to much greater concern for the west and southwest. How that concern will become manifest is of great interest. As the Ethiopians declared at Mar del Plata, they would welcome an accord on the utilization of the Nile with their downstream neighbours, but in the absence of such an accord they reserved the sovereign right to carry out their projects unilaterally.

The Lacustrine states

Three Equatorial lakes – Victoria, Albert/Mobutu, and Kioga – could all figure in long-term, over-year storage schemes to deliver an additional 4–5 km^3 downstream to the Bahr el Jebel. In 1974 over 200 possible storage combinations were being made ready for computer analysis (Hydromet, 1974, Vol. I, 903). Any of them will require the accord and co-operation of six countries in the catchment of the Equatorial Lakes hereafter referred to as the Lacustrine states: Kenya, Uganda, Tanzania, Zaïre, Rwanda and Burundi. Of these, Uganda, with control over portions of Lakes Victoria and Albert and total sovereignty over Lake Kioga, would be the most directly affected. There is a certain irony in the fact that all this modelling, potential engineering and diplomatic manoeuvring would at best yield an amount of water that is less than one-fifth of 1% of Lake Victoria's average volume (2,700 km^3). It is the simple fact, however, that the *Sudd* swamps and the existence of fairly flat stretches along the Albert and White Niles preclude the delivery of larger volumes of water than those presently contemplated owing to the risk of spillage. Likewise there are no alternative storage sites to the lakes farther downstream.

There is some question whether or not in the future the inevitable resort on the part of the Lacustrine states to waters that normally run off into Victoria will gradually reduce the amount of water stored there. In the October 1961 meetings between representatives of the three Lake Victoria states and Egyptian and Sudanese members of the Permanent Joint Technical Committee on the Nile which was set up following the Sudan–Egyptian Nile Waters Commission of 1959, the East Africans put their future needs at 5 km^3 (existing needs were put at 1.7 km^3), a figure the PJTC rejected for lack of supporting data. It is quite conceivable, however, that Tanzania, Uganda, and Kenya will bring some half million hectares under irrigation by the year 2000 (Hydromet, 1981), and that the on-field use and losses in storage of the irrigation water will reduce flows into Lake Victoria by 6–7 km^3 per year. That represents only about 5% of precipitation over the lake and less than one-fifth of 1% of lake volume. None the less, even this small amount could gradually drop the lake's level by a few centimetres each year, all other factors being equal. Moreover, as agriculture develops around the lake shore, there will be an incentive to keep the lake level low in order to promote proper drainage. An off-setting factor, however, will be that as swampy areas in the lake catchment are drained, evaporation will be reduced and run-off into the lake increased. All in all, Egypt and the Sudan will have to consider financial compensation to the Lacustrine states for keeping the lakes' (including Albert's) levels higher than would locally be desirable. Given the value of additional water to the agriculture and hydropower of the downstream states, such compensation should be manageable.

This eventually will arise only if Egypt and the Sudan decide to proceed with Jonglei II and seek increased storage in the Equatorial lakes. That eventuality is, however, remote because of the disturbances in the southern Sudan, the difficulty of obtaining international finance, and the political divisions among the Upper Nile states themselves. But were it to come to pass the best site for increased storage would be Lake Albert/Mobutu. It is a moot point whether or not Albert would have to be operated in tandem with

Victoria. Although the possible combinations are numerous, it is estimated that the gross amounts of water that would have to be stored to assure annual discharges of 26 to 44 km³ at Mongalla would fall in the ranges represented in Table 3.3.

It is possible and probably desirable that all storage be confined to Lake Albert. There are two basic reasons for this: first, the lake shores are sparsely inhabited, and, second, the surrounding slopes are so steep, raising the lake's level would increase volume far more rapidly than surface area. With the heavy rains of the early 1960s the lake level in fact rose from 620 to 624 metres a.s.l. That increased its volume from 155 to 181 km³ and its surface area from 5,700 to 6,100 km². It might be necessary to raise the lake level to 640 metres a.s.l., or 30 metres on the Butiaba guage, in order to assure discharges at Mongalla of 38 km³. At that level the volume of the lake would be 280 km³ and its surface area 6,880 km². This enormous gain in storage could thus be brought about with minimal disruption, relative to other storage sites, to existing populations and land-use and with little increase in surface evaporation. While storage at Lake Albert may be technically feasible, the benefits of increased storage for Uganda and Zaïre, which share the lake, are not at all obvious. Egypt and the Sudan could offer them a new source of hydropower, but the area in the vicinity of Lake Albert has very little demand for such energy. Perhaps annual rents could be paid to both countries. But neither country will be in a hurry to accommodate the needs of their downstream neighbours, especially when they feel handicapped by their relative lack of expertise in assessing the costs and benefits of hydraulic projects. The caution they will exhibit is yet one more obstacle to implementation of any Upper Nile projects beyond Jonglei I.

Table 3.3 *Ranges of Storage Capacity at Victoria and Albert needed to assure indicated discharges at Mongalla.*

Combined storage at Victoria and Albert (km³)	Regulated discharge at Mongalla		Jonglei II drawoff
	km³/year	millions m³ per day	millions m³ per day
100	26.3	72	50
100–200	27.7	76	50
200–240	38.4	105	50
240 +	43.8	120	50

Source: adapted from Hydromet, 1974, Vol. 1, pt. II, 903.

Conclusions

If the projections advanced earlier in this chapter are correct, both about the unlikelihood of further Upper Nile projects and about anticipated Egyptian and Sudanese demands, a supply and demand balance sheet for the year 2000 might look something like this:

Supply (as calculated at Aswan)	
Mean discharge assumed in Nile Waters Agreement	84.0 km^3
Actual increase in century mean	4.1 km^3
Benefit of Jonglei Phase 1	3.8 km^3
Total	91.9 km^3
Demand (as calculated as at Aswan)	
Egyptian	68.9 km^3
Sudanese	32.0 km^3
Ethiopian	2.0 km^3
Total	102.9 km^3

The deficit implied in these figures cannot of course be turned into reality: some or all of the consumers will have to reduce their demands. These figures also make the assumptions, realistic or not, that, on the one hand, abstractions from the Equatorial lakes will have a negligible effect on the downstream water balance and, on the other, that Sudanese agricultural development plans made in the 1970s will eventually proceed and that Ethiopia, hard pressed to find solutions to its crippling agricultural problems, will turn its attention to its western watershed.

In short, what these figures indicate is that it will be very difficult to accommodate the reasonable national agenda of the principal riparians for water use. Continued economic stagnation or decline may postpone the crisis, but one can scarcely take comfort in such an eventuality.

Seen in this light, two facts become evident about the present Jonglei Canal project. Firstly, that whatever they may do to improve the efficiency of water use and agricultural productivity, the governments of Egypt and the Sudan will be looking with gathering urgency as the century draws to a close at all conceivable ways, however socially, politically or economically unrealistic, of increasing the utilisable discharge of the Nile. Secondly, the sense of urgency is all the more acute in view of the severe droughts of the early 1980s in much of the Nile watershed, particularly Ethiopia. Both Egypt and the Sudan have had to reduce demand for Nile water by restricting the area under cultivation. Under these circumstances, the Jonglei I project, while contributing only 4% of the net benefit of the river at Aswan, acquires great international importance.

Note

This chapter was written before the severe reduction in the natural yield of the Nile registered in 1986 and 1987 that will likely force Egypt to reduce its overall consumption of Nile water.

Part II

The natural environment of the *Sudd* region

4

The influence of Lake Victoria: climatic change and variation in river flows

Introduction

The outflow from Lake Victoria at the Ripon Falls provides the main source and the steady component of the flows of the Bahr el Jebel. These outflows are modified by storage in Lakes Kioga and Albert, and they receive a seasonal component above Mongalla, but the very large storage available in Lake Victoria dominates the inflows to the *Sudd*.

The main features of the lake hydrology are the importance of the lake area and the precipitation over it and the sensitivity of the system to changes in rainfall because evaporation and average rainfall are almost equal. Tributary inflows are small compared with lake rainfall, but are sensitive to periods of high rainfall; lake evaporation is relatively constant from year to year. Thus lake levels and outflows, highly damped by storage, are liable to long-term fluctuations and these have been observed both in the historic record since the late nineteenth century and in the prehistoric record preserved in lake sands and sediments (Piper *et al.*, 1986).

This chapter therefore tries to give a very general picture of the lake as an indicator of climatic change, to help in the interpretation of recent changes in Nile flows, and thus in the area of the *Sudd*, and in consequent assessment of possible future trends.

The origin and early history of the lake

Lake Victoria originated in Middle Pleistocene times as a result of uplift of the earth's crust along a line running NNE–SSW, parallel to, and about 50 km east of, the margin of the western Rift Valley, which prevented the outflow of west-flowing rivers. First the Kagera and then the Katonga rivers were reversed to form a proto Lake Victoria. The lake thus formed spilled over the lowest col on its northern watershed at Jinja to enter a tributary of the Kafu-Kioga and flow into Lake Albert. Subsequent downcutting of the Jinja spillway is indicated by lake terraces of uncertain age believed to have formed in relation to water levels successively about 18 m, 13 m, and 3 m above the 1956 level of Lake Victoria (Bishop, 1969). The present lake is therefore substantially shallower and smaller than it was previously.

Towards the end of the Pleistocene, between about 18,000 and 12,000 years ago, when climate on the East African plateau was much drier than it is now (Street & Grove, 1976), Lake Victoria seems to have been no more than 26 m deep (Livingstone, 1976). Pollen from the basal material in cores taken from Pilkington Bay, about 20 km south of Jinja, indicates that grasses dominated the vegetation of the surrounding area for some millennia before 12,500 years ago (Kendall, 1969). The lower part of a core 9.9 m long, taken from the floor of the Damba Channel (about 50 km east of Entebbe) at a depth of 32 m, contained abundant gypsum crystals (implying dry lake bed conditions) in material deposited between 16,000 and 12,000 years ago (Stager, 1984). Throughout this time there was no outlet from the lake; it occupied the lowest parts of the flat floor of a closed basin. Lake Albert too was without an outlet prior to 12,500 BP (Harvey, 1976).

About 12,000 years ago the chemical characteristics of lake-floor sediments, their diatom content and ^{18}O values show that Lake Victoria filled up until it began to overflow. The lake may have ceased to spill for some centuries about 10,000 years ago (Kendall, 1969) and then came a period of maximum discharge until about 7,000 years ago. Less humid and more seasonal conditions are indicated by the diatom content of sediments deposited after 7,000 BP and both Victoria and Albert responded to an increase in aridity after 3,000 BP (Stager, 1984). However, there is no indication in the sediments that flow out of the lake has been interrupted over the last 10,000 years.

A cave at Hippo Bay near Entebbe contains beach sand about 2.5 to 3 m above the 1959 level of the lake. Water-rolled fragments of charcoal in the sand have yielded an age of 3,720 $\pm$ 120 years BP (Y-688). Bishop (1969, p.107) states that 'the level of Lake Victoria between 1961 and 1964 is of interest as, at their maximum, the lake waters were within two feet of the dated beach gravel in Hippo Bay cave. This cave fill is so unconsolidated that even a few weeks would suffice to remove the whole deposit. It is certain that at no time since 3,720 years ago has Lake Victoria risen to re-occupy the 10 to 12-foot cliff notch'.

The extent of the basin

Lake Victoria lies within the depression between the two arms of the Rift Valley and the basin is broadly rectangular with drainage from the sides. Though the north-eastern sector of the lake catchment (50,000 km²) is relatively steep and forested, the south-eastern tributaries drain a flatter area (50,000 km²). In the south-western sector the Kagera (60,000 km²) drains the mountains of Rwanda and Burundi, but passes through a complex system of swamps and lakes. To the north-west the Katonga drainage has been reversed and choked by swamps and with the Ruizi contributes little from an area of 30,000 km². The tributary basins total 190,000 km², compared with the lake area of 67,000 km². Thus the lake is a relatively high proportion of the total catchment.

Rainfall

The seasonal distribution of rainfall over Lake Victoria and its catchment depends on the migration of the Inter-Tropical Convergence Zone, but is also influenced by lake breezes. The two rainy seasons, in March/May and November/December, are general, with additional rainfall in July and August to the north-east of the lake.

The pattern of annual average rainfall over the lake and its catchment is clear. Rainfall is high to the north-east of the lake, and lower to the south; rainfall also tends to be higher at rainfall stations near the lake shores, like Bukoba. The long-term mean rainfall over the lake itself is estimated from gauges around the lake and on islands at 1650 mm (de Baulny & Baker, 1970), so lake rainfall is higher than rainfall over the greater part of the tributary area. However, because average lake rainfall and lake evaporation are almost equal, tributary runoff into the lake is a significant contribution to the balance and its variability.

Lake inflow

Although the total area of its catchment is three times the area of the lake itself, the runoff from the tributaries is on average a relatively small contribution of its total water supply. However, as runoff is the residual between catchment rainfall and evaporation, it is sensitive to rainfall fluctuations and has greater relative variability than rainfall.

Measurements of runoff into the lake cover a shorter period than the other components. Regular measurements on the largest tributary, the Kagera, started in 1940. Records for four Kenya tributaries are available from 1956, and most of the remaining tributaries were brought into a WMO programme of measurement and investigation from 1969 (Kite, 1981). Thus the measurements have only recently been sufficiently complete to make reasonable estimates of the total runoff from measured flows. Although the drier catchments to the south of the lake provide more variable runoff than the wetter catchments, total runoff has been estimated from 1956 on the basis that the four tributaries provided the same proportion of the ungauged flow as in the period of nearly complete measurements. The seasonal distribution of tributary flows follows the rainfall, with the Kagera flow damped by lakes and swamps.

Comparison of annual tributary flows with annual lake rainfall totals (Table 4.1) confirms that tributary inflows are relatively more variable than

Table 4.1. *Annual water balance of Lake Victoria, 1956–78*

Year	Lake rainfall (mm)	Tributary inflow (mm)	Lake outflow (mm)	Lake (Jinja) level (m)
1956	1,667	288	291	10.91
1957	1,659	270	300	11.02
1958	1,513	218	294	10.94
1959	1,652	199	275	10.84
1960	1,723	262	305	10.87
1961	2,486	326	307	11.94
1962	1,755	539	577	12.39
1963	2,146	517	669	12.91
1964	1,917	483	753	12.88
1965	1,631	260	699	12.48
1966	1,768	320	641	12.32
1967	1,803	320	564	12.31
1968	2,206	487	646	12.58
1969	1,787	315	677	12.36
1970	1,872	412	661	12.45
1971	1,601	301	588	12.17
1972	1,961	298	562	12.35
1973	1,651	298	573	12.05
1974	1,516	313	523	11.97
1975	1,761	283	499	12.04
1976	1,742	215	521	11.82
1977	1,836	435	551	12.13
1978	1,966	531	587	12.56
Mean	1,810	343	524	
SD	229	106	152	

lake rainfall. The 1956–78 tributary inflow averages 343 mm over the lake (67,000 km^2) with a standard deviation of 106 mm, whereas lake rainfall averages 1,810 mm with a standard deviation of 229 mm; thus the variability of annual inflow is 31% compared with 13% for rainfall. The annual series demonstrates the marked increase in tributary flows after the high rainfall of 1961–4, and the fact that tributary flows, particularly from the Kagera, have remained somewhat higher than the 1956–60 period.

Evaporation

Variations from year to year in the evaporation from a large equatorial lake are unlikely to be great compared with variations in rainfall and lake inflow. Monthly mean evaporation from Lake Victoria has been estimated by heat budget methods (Kite, 1982); the annual total is 1,593 mm. These are compared in Table 4.2 with 1956–78 monthly average rainfall, inflows and outflows, expressed as mm over the lake.

Lake level and lake outflow

The lake outflow was controlled naturally by the Ripon Falls before the construction of the Owen Falls dam which was completed in 1954. Thereafter the dam was operated to reproduce the natural flows as a function of lake level. Thus the lake outflow can be estimated directly from lake levels using a rating curve based on measured flows and model testing.

Lake levels have been read at Entebbe gauge since 1896, and at Jinja since 1913; the equivalent series at Jinja is given in Figure 4.1. Whereas the range before 1961 was 10.0–12.0 m on the Jinja gauge, with the upper limit approached on only four occasions, the lake has remained in the range 11.5–13.5 m since 1962, with peak levels in 1964 and 1979. There is an annual cycle with a peak reached usually around April or May, but the pattern is not uniform due to varied seasonal rainfall distribution and the damping effect in

Table 4.2. *Monthly estimates of lake water balance, 1956–78 (mm over 67,000 km^2)*

	J	F	M	A	M	J	J	A	S	O	N	D	Year
Rainfall	138	161	220	288	210	61	36	67	92	144	210	181	1,810
Inflow	20	16	23	40	48	31	30	31	29	23	25	26	343
Evaporation	119	112	139	154	151	166	175	137	109	114	107	110	1,593
Outflow	43	39	43	43	48	47	46	44	43	43	41	44	524

runoff. The annual range is about 400 mm, which is small compared with the terms in the balance.

There is evidence that the lake reached high levels in the late nineteenth century. Garstin (1901, p.49) states that 'there appears to be no doubt that, for many years previous to 1901, the mean levels of this lake have shown a steady fall, independent of the annual fluctuations due to the rainfall. This general lowering of the water levels was noticed both by Sir H. Stanley and by Mason Bey at the time of their respective visits'. In a later report Garstin adds 'Lastly, there is the statement of Mr McAllister, Sub-Commissioner of the Nile Province ..., that the French Fathers at Muanza, in the south of the lake, had a register showing from marks on the rocks that there has been a fall of some 2–4 metres during the last 25 years' (Garstin, 1904, p.21). Sir Henry Lyons, who accompanied Garstin to the White Nile in 1901 and 1903, and who 'made a special study of the levels of the Victoria lake', contributed an appendix to Garstin's later report (1904, App. III). He concludes that the lake was very high in August and September 1878, fell from 1878–92 and rose in 1892–5. This analysis was repeated in Lyons (1906) and is the primary source

Plate 7. The Owen Falls Dam at Jinja, Uganda. Here the Nile leaves Lake Victoria. The Ripon Falls, over which the river formerly left the lake, are now submerged above the dam, which was completed in 1954. Photo: Lesley McGowan.

of information. Lamb (1966) deduces from the evidence that the 1878 level was higher than 1964.

Evidence from the Bahr el Jebel supports information about high lake levels in the late nineteenth century and in particular of the peak in August 1878. Hurst & Phillips (*Nile Basin*, Vol. V, 1938) point out that the average discharge of the main Nile at Aswan (of which the bulk is derived from the Blue Nile) from 1871 to 1898 is much greater than the average from 1899 to 1936. They discuss Lyons' (1906) evidence of lake levels from which one can perhaps 'infer that from 1870 to 1900 high floods seem to have been more common than in the period following 1900'. Lyons himself points out (p.103) that Lado on the Bahr el Jebel was flooded in 1878 and the river was still very high in December 1878, suggesting an unusually high level in Lake Albert.

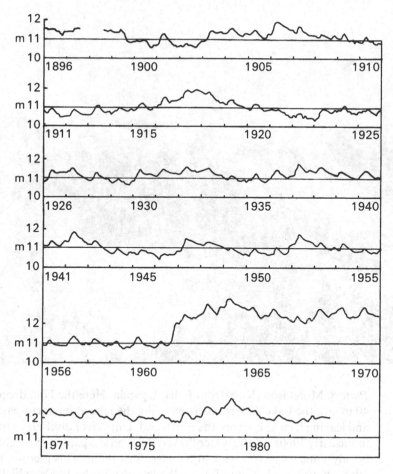

Figure 4.1. Levels of Lake Victoria, measured at Jinja, Uganda, 1896–1983. Gauge readings in metres.

The peak level at Gondokoro was recorded on 22 August 1878, and there is ample evidence that this flood was exceptional. For example, Hope (1902) quotes an account of blockages: 'in 1878 the White Nile rose to an unusual height, and enormous quantities of vegetable debris were carried off by the current. A formation of bars (blocks) on an unprecedented scale was the result, and communication between the Upper and Lower Nile was not restored until 1880'. Nuer oral history (Chapter 9) also suggests that very high floods occurred at this time.

Increased flows in the period before 1895 would also reconcile the discrepancy between early accounts (1870–85) of immense herds of Bari cattle above Mongalla and the shortage of grazing reported there in the 1950s (Jonglei Investigation Team, 1954). It has been suggested (Sutcliffe, 1957, 1974) that the high flows of the last century could have produced dry season

Plate 8. Murchison (Kabalega) Falls, Uganda. Here the Nile drops about 40 m into the Lake Albert basin, entering the lake at its north-east corner and leaving it close by from the north end. Until river discharges increased in the early 1960s, all the water flowed through the gap on the right, which is only some six metres wide at its narrowest point. This picture, taken in 1968, shows an additional fall on the left, dry in the 1950s, which carries some of the extra discharge. Photo: Lesley McGowan.

grazing in this area, and the recent high flows have in fact provided grazing for large herds.

A number of authors have considered the topic of Nile flows as an expression of the wider question of longer-term African climatic change; one such general review being that of Nicholson (1980). Herring (1979a) has used the 13 centuries of Nile records made at Rodah, near Cairo, which permit an approximate differentiation between Blue Nile (autumn) floods, and White Nile (summer) floods, and has related these to oral chronologies of the movements of people in the southern Sudan and Uganda (see also Chapter 9). Harvey (1982) has reviewed evidence of the distribution of shells in raised beach deposits, which permit dating of the periods when Lake Turkana flowed outwards into the Pibor River, at the eastern edge of the Jonglei area. Together all these lines of evidence indicate that substantial climatic change has been a recurring, if irregular feature, of the White Nile Basin, and that events such as those of 1875–95 and again since the early 1960s are in no sense unprecedented, and are no more abnormal than the low flows experienced between 1900 and 1960.

Recent lake balance

A comparison of lake rainfall and tributary inflow with lake outflow and changes in storage (Table 4.1) suggests that evaporation, the only

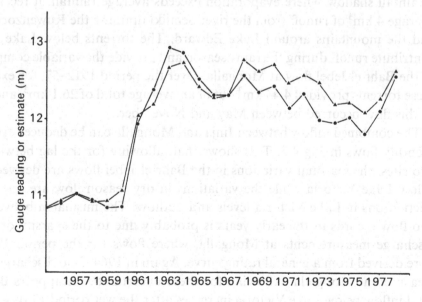

Figure 4.2. Comparison of observed (●) and deduced (▲) levels of Lake Victoria over the period 1956–78.

missing term in the balance, is reasonably constant from year to year at about 1,550 mm. This is about 40 mm smaller than the evaporation estimated from heat budget considerations; an alternative is that lake rainfall is under-estimated. A comparison of deduced and observed lake levels over the period 1956–78 (Figure 4.2) using a constant evaporation of 1,550 mm, suggests that measured variations in lake rainfall and inflow have been sufficient to account for the marked rise in lake level and its subsequent behaviour. Although the lake rainfall declined from an average of 2,080 mm in 1961–4 to an average of 1,780 mm in 1965–74, there was sufficient storage within the Kagera basin and the lake itself to ensure that the decline in lake level after 1964 has been gradual.

Inflows below Lake Victoria

Between Lake Victoria and Mongalla there are contributions from the catchments of Lake Kioga and Lake Albert and from the torrents between Lake Albert and Mongalla. Because the outflows from Lakes Kioga and Albert have not been measured as precisely as the outflows from Lake Victoria or the inflows to the *Sudd* at Mongalla, it is easier to deduce the gains over the whole region than over individual catchments.

However, evaporation from Lake Kioga exceeds local runoff in dry years, while in wet periods, like 1917–18, inflow from local rainfall results in a considerable gain between Jinja and Masindi Port. Although Lake Albert is in a rainfall shadow where evaporation exceeds average rainfall, it receives on average 4 km^3 of runoff from the river Semliki draining the Ruwenzori range and the mountains around Lake Edward. The torrents below Lake Albert contribute runoff during the rainy season and provide the variable component of the Bahr el Jebel flow at Mongalla. Over the period 1912–57, for example, these torrents provided 4.44 km^3 out of an average total of 26.1 km^3, and most of this flow occurred between May and November.

The combined inflow between Jinja and Mongalla can be deduced from the monthly flows in Fig 4.3. This shows that, allowing for the lag between the two sites, the seasonal variations in the Bahr el Jebel flows are derived from below Lake Victoria while the variations in dry season flow are due to the fluctuations in Lake Victoria levels and outflow. The mismatch between the two flow records in the early years is probably due to the sparsity of actual discharge measurements at Mongalla, where flows for the period 1905–20 were derived from a general rating curve. Again in 1964–7 no discharges were measured at Mongalla. Allowing for these uncertainties it appears that the total inflow below Lake Victoria increases after the wet period 1961–4. There is a contrast between the floods of 1916–18, when the high inflows to the *Sudd* were due to the torrents, and those of 1961–4, when the Lake Victoria outflows contributed more to the rise.

This is confirmed by comparing decade mean flows at the two sites (see Table 4.3). Although annual flows show no relationship between the lake contribution and runoff from the lower catchment before 1961, there is a striking increase in Lake Victoria outflow and in the net gain above Mongalla

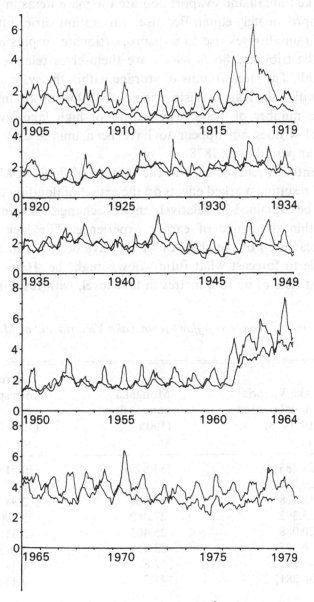

Figure 4.3. Monthly flows (in km³) of the Nile at Jinja, Uganda, and Mongalla, Sudan, 1905–80.

after 1961. The apparent high gains before 1920 are probably due to measurement problems at Mongalla.

Future flows

It has been shown that Lake Victoria outflows are extremely sensitive to moderate changes in rainfall over the lake and its tributaries, because average lake rainfall and evaporation are the main items in the lake balance and are approximately equal. Because evaporation varies little from year to year, high rainfall gives rise to a disproportionate surplus and also greatly increases the tributary flows which are themselves relatively more variable than rainfall. The large volume of storage within the system, within the lake itself and within the Kagera basin, ensures that this surplus influences lake outflow for a number of years. The period of high lake levels and outflows after 1961–4 does not appear to have been unique and there is evidence of a similar period after 1878.

Thus the apparently extreme change in the pattern of flows at Mongalla since 1964, and the resulting marked effects on the extent of flooding from the Bahr el Jebel, can be explained by relatively modest changes in rainfall and appear to fall within the range of earlier experience. The lack of any meteorological explanation of the changes which have occurred in the past makes it impossible to forecast what future flows might be. However, the evidence shows that rises of up to 2 metres in lake level, with resulting large

Table 4.3. *Comparison of decade mean flows from Lake Victoria and at Mongalla*

	Lake Victoria annual outflow (1901–78) km³	Mongalla annual flow (1905–80) km³	Difference in corresponding years (1905–78) km³
1901–10	22.283	(33.875)	(10.513)
1911–20	20.830	32.285	11.455
1921–30	18.228	21.209	2.981
1931–40	23.392	27.082	3.690
1941–50	20.088	25.403	5.315
1951–60	20.335	23.950	3.615
1961–70	41.502	52.889	11.387
1971–80	(36.881)	47.761	(10.456)
Total	25.149	33.014	7.178

increases in lake outflows, are not exceptional and have probably occurred twice in 100 years. On the other hand, falls to lower levels would take up to ten years to occur under long-term average rainfall conditions because of the attenuating effect of the large lake storage volume.

5

Hydrology of the Bahr el Jebel swamps

Introduction
The purpose of the Jonglei Canal is to save water which is currently evaporated in the swamps of the *Sudd*. At present only half the inflow of the Bahr el Jebel at Mongalla emerges from the tail of the swamps, and the remainder spills from the river into permanent and seasonal swamps and evaporates. Because the total volume of water evaporated is proportional to the areas of flooding, the area of swamps must decrease if water is to be saved. However, there is an important distinction between the permanent swamp and the areas seasonally flooded and uncovered which provide dry season grazing. The places most hydrologically important in the *Sudd* are shown in Figure 5.1.

The aim of this chapter is to give a general description of hydrological processes in the *Sudd*. These include river inflows, transfer of water to the swamps by spillage, drainage beside and back to the river, rainfall gains and evaporation losses over the flooded area. In order to quantify these processes it has been necessary to develop a model described here and in Chapter 16 to estimate the areas of permanent and seasonal swamp and to assess the effect of the canal on those areas. That analysis was based on a simple reservoir model, using measured values of hydrological inputs and outputs wherever possible.

The study has covered the historical period 1905–80, during which natural fluctuations in river flow from the East African lakes have caused considerable variations in the areas of both permanent and seasonal swamp.

Previous studies
Before outlining present knowledge of the hydrology of the Bahr el Jebel, it may be useful to describe how this knowledge has grown. The process

may be divided into three phases, those of exploration, measurement and hydrological study.

The first phase has been described often. Egyptian expeditions opened up the White Nile from 1840, establishing trading posts above the swamps; early

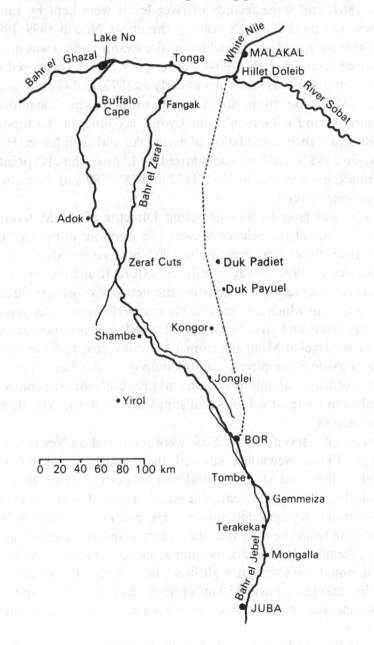

Figure 5.1. Map of the study area, showing localities mentioned in the text.

accounts include those of Werne (1849). Speke and Grant reached the Ripon Falls from Zanzibar in 1862, and in Gondokoro they met Baker, who reached Lake Albert in 1864. In 1874 Gordon prepared a map of the river up to Lake Albert. Although Lombardini (1864) estimated the loss by evaporation in the swamps in 1864, and some records of river levels were kept by Emin and others, it was not until Garstin's visits to the upper Nile in 1899–1903 that river flows were measured above and below the swamps and it was noted that the flow halved (Garstin, 1901, 1904). Although Willcocks proposed the use of the East African lakes as reservoirs as early as 1893, and Garstin suggested in 1904 a canal from Bor to the Sobat mouth, the main scientific information from the early period is Garstin's and Lyons' accounts of the topography (Lyons, 1906), and their compilation of early lake and river levels. High and low years from 1848 to 1883 are summarised by Lyons, who also pointed out that *Sudd* blockages occurred in 1863, 1872 and 1878–80 and were associated with rises in water level.

Lyons, who was later to become acting Director of the Meteorological Office and Director of the Science Museum in London, introduced current meters for river flow measurement in 1902 and recruited Hurst to the Egyptian service in 1906 during a visit to Oxford (Sutcliffe, 1979). It was Hurst and his colleagues who established the network of gauges throughout the Nile basin, from which the remarkably comprehensive measurements of rainfall, water levels and river flows were derived. For instance, continuous readings of river level at Mongalla from 1905, converted to flows by regular calibration, provide a complete record of inflows to the *Sudd*. The mass of hydrological records, with interpretation and proposals for Nile control, have been published in some 40 volumes and supplements of *The Nile Basin* from 1931 to the present.

Two studies more specific to the *Sudd* were published by Newhouse (1929) and Butcher (1938). Newhouse stressed the importance of the torrents between Lake Albert and Mongalla, and pointed out that it was more urgent to start hydrological measurements, for which a run of years was essential, than to undertake topographic surveys. He made a distinction between channels flowing from the river into the swamps, and those returning to the river, calling them heads and tails; by comparing flows measured at Mongalla and Bor, he noted the losses at high flows but offered no explanation. He estimated the total loss as over 13 km^3 or about half the inflow, and thought that the swamps were 'not a reservoir where water is stored, but a sink where it is wasted'.

Butcher (1938), while stating that a study must be largely statistical, discussed the losses in terms of area and evaporation. He made use of air photography to estimate the swamp area as 7,200 km², 60% of it on the right bank. He estimated evaporation rates from a tank filled with papyrus but these estimates could only explain less than half of the loss. He noted that series of discharges down the river show rapid variations in flows as spilling leaves the river and returns. The loss between Mongalla and Bor, for instance, was explained by spill over and through the river banks into a complex network of channels known as the Aliab which bypasses Bor and rejoins the river at Lake Papiu. This led to the concept of latitude flow, where flow is measured in a number of parallel channels. Flows at Mongalla were compared with those at Jonglei latitude and Peake's latitude, but the low estimate of evaporation (1,533 mm gross) led to the conclusion that there had to be considerable spill to the east below Bor and towards the Bahr el Ghazal below Shambe. It is worth mentioning here that when the estimated evaporation is increased to realistic levels, these spills are no longer required to balance losses between latitude sites. Butcher concluded that the Bahr el Jebel and its subsidiary channels form a gigantic module discharging a constant flow, and that the river disposes of surplus flow by a series of spills to the surrounding swamps whence subsidiary channels return some flow to the river.

Hurst & Phillips (1938) summarised the topography and hydrological measurements, but also discussed the water balance in terms of the equation of continuity. This states that the inflow and local rainfall must equal the outflow and evaporation, after allowing for the changes of storage estimated from the swamp area and rise or fall of river level. They deduced that the evaporation estimate would have to be increased by 30% or more to provide a seasonal balance. They also derived a relation between Mongalla inflow and swamp outflow with varying lag.

While these measurements and analyses led to the proposals for the Jonglei Canal as part of the Equatorial Nile Project, further hydrological analysis has since been made possible by more realistic and higher measurements of evaporation from swamp vegetation (Migahid, 1948), and by theoretical advances by Penman (1948) which enabled evaporation to be estimated from records of temperature, humidity, wind speed and insolation. Indeed Penman (1963) pointed out that swamp evaporation measurements corresponded with open water evaporation.

The hydrological background

A brief description of our present knowledge of the hydrology follows. The inflow to the swamps combines the fairly steady outflow from the East African lakes, which varies from year to year according to lake levels, and the seasonal and variable discharges of the rain-fed torrents above Mongalla normally leading to floods between May and October. The effects of the variations in the proportion of the flow contributed by the various sources on the chemistry of the river water is dealt with in Chapter 6.

The first source is heavily damped by lake storage but periods of high lake levels and high lake outflows have alternated with periods of lower levels and outflows; a dramatic rise of flows which began in 1961 was discussed in the previous chapter. The second source, the torrents above Mongalla, fluctuates rapidly during the rainy season. Thus for six months of the year the river flow at Mongalla is fairly steady, and for the rest of the year it is larger, more variable and carries more sediment. The average annual inflow for 1905–80 is 33.0 km^3.

The longer-term fluctuations in the levels of the East African lakes, and the resulting outflows, discussed in Chapter 4, are an important feature of the hydrology of the *Sudd* (Figure 5.2). Because the average rainfall over Lake Victoria is almost equal to the evaporation, the lake system is sensitive to changes in rainfall and tributary inflow. Thus an increase in rainfall of about 20% over the average in 1961–4 led to a 2.5 m rise in the level of Lake Victoria and a doubling of the outflow. This has not yet returned to earlier levels. The flows at Mongalla show a marked contrast between the low flows recorded between 1905 and 1960, when the mean annual discharge was 26.8 km^3, and the higher flows of 1961–80, when it was 50.3 km^3.

As we have seen in Chapter 4 there is evidence from rainfall records, lake levels and Nile records (Lyons 1906; Nicholson 1980) that the historical period was preceded by high flows from about 1875 to 1895, and earlier by alternating periods of high and low flows. Without physical explanations for changes of regime, it is not possible to forecast when the present high flows might revert to lower values.

The Nile system within the swamps

Below Mongalla the channel carrying capacities are less than the high flows, and the alluvial channels themselves are higher than the surrounding floodplain. Thus excess flows leave the river through spill channels, and by flow over the banks, and inundate large areas on either side of the river. This flooding is limited by higher ground only in the south of the swamps. High river flows coincide with the rainy season within the swamps, when evaporation is comparatively low. Much water is lost from the swamps by evaporation and transpiration, and as a result the outflow from the swamps is relatively constant, with the seasonal cycle heavily damped, and totals only about half the inflow.

The combined effect of these processes is that varying areas are inundated permanently or seasonally, with the uncovering of the seasonal swamp coinciding with the dry season. The areas of permanent swamp reflect the longer-term variations in flow from the East African lakes, while the seasonal swamps depend not only on seasonal inflows, but also on the fluctuations of

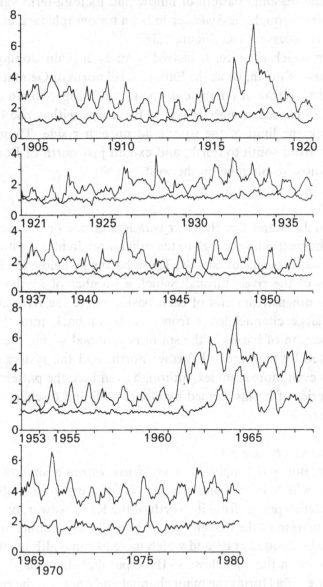

Figure 5.2. Inflows to and outflows from the *Sudd*, 1905–80, in km³ per month. Inflows measured at Mongalla; outflows derived by subtracting River Sobat flows at Hillet Doleib from White Nile flows at Malakal (from Sutcliffe & Parks, 1987).

the torrent inflow and the annual cycle of balance between rainfall and evaporation within the swamps.

The topography of the *Sudd*
Introduction
Given the seasonal pattern of inflow and its long-term variations, the passage of water through the *Sudd* depends on topographical factors, and some description is necessary (see Figure 5.3).

The area within which the river is incised is an even plain sloping gently north or slightly east of north, while the Bahr el Jebel north of Gemmeiza runs west of north and therefore at an angle to the ground slope. North of Juba, the river runs in an incised trough, bounded by scarps with a rise of a few metres which mark the limit of the woodland on either side. These scarps decrease in height from south to north, and extend just north of Bor on the east bank, and almost to Shambe on the west.

Juba–Bor
Between Juba and Bor the river wanders in one or more channels from one side of the restraining trough to the other, and divides the floodplain into a series of isolated basins or islands. These basins lie below the levels of the alluvial banks of the river, through which a number of small channels carry spill. At the downstream end of each basin, where the river and high ground meet, a large channel leads from the basin back into the river. Whereas the succession of basins in the south is confined within the trough, they are not limited by high ground further north, and the system of river channels becomes even more complex, although even here the pattern can be interpreted as a series of basins formed by lateral spill and discharging back into the main river.

Bor–Jonglei (Figure 5.4)
Between Bor and Jonglei the river flows within a shallow trough about 15 km wide, which, in the southern part, is sufficiently well defined for there to be little lateral spillage from it. North of the Khor Adwar, on the east bank, the eastern margin of the trough becomes locally less distinct and there is a belt of seasonally flooded grassland which increases in width northwards. Throughout this section the river flows in three parallel channels.

At Bor the Bahr el Jebel forms the main channel and abuts on the east bank of the trough. To the west lies the more complex channel of the Aliab. This is joined, north of Lake Fajarial, by an outflow from the Bahr el Jebel to form what was, before the 1960s, the main navigation channel. This flows northwards, close to the west bank of the trough, and does not rejoin the new navigation channel until Lake Shambe. The new navigation channel, sometimes called the Bahr el Jedid ('New River'), follows a fairly central course

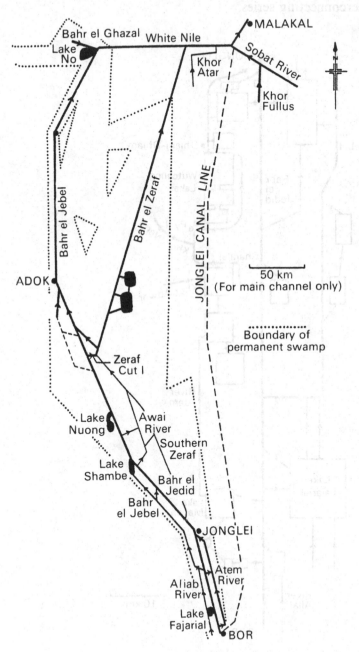

Figure 5.3. Schematic representation of the main flow channels and lakes in the *Sudd* Region (after SES, 1983).

108   *The Jonglei Canal*

through the swamps. Four major eastward outflows from it form the Atem system, which flows through and past a series of lakes and rejoins the Bahr el Jedid at Jonglei. Between the three channels there are many lakes, some forming interconnecting series.

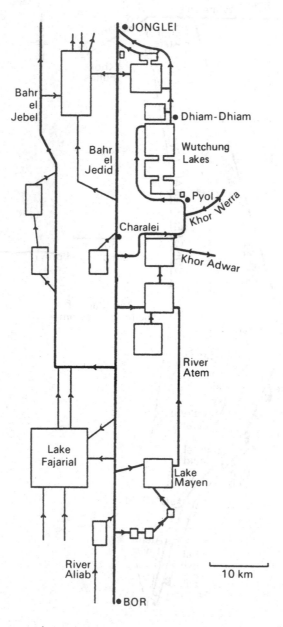

Figure 5.4. Diagram of the river channels and lakes between Bor and Jonglei (after SES, 1983).

Jonglei–Shambe (Figure 5.5)

As in the previous section, the main channel and lake complex continues to remain within a band about 15 km wide, but is no longer

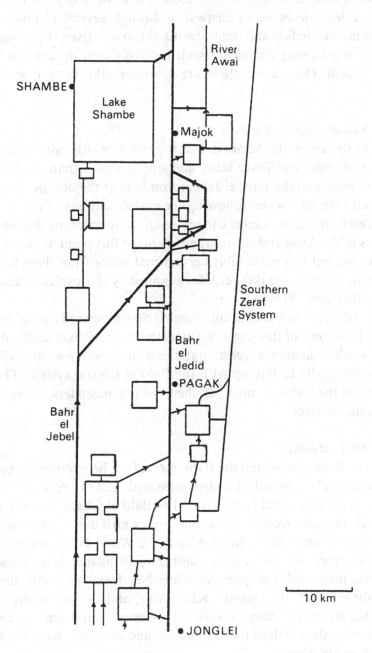

Figure 5.5. Diagram of the river channels and lakes between Jonglei and Shambe (after SES, 1983).

confined within a trough on the eastern side. Because of this there is extensive eastward spillage, and large areas of seasonally flooded grassland and permanent swamp result.

The Bahr el Jebel continues to flow along the western edge of the trough. The Bahr el Jedid loses water eastwards through several channels. Some of this rejoins the Jedid, and some forms the Awai river, but some joins ill-defined north-flowing channels which probably link up eventually with the Zeraf system. Once again there are extensive lake systems within this section.

Shambe–Adok (Figure 5.6)

In this reach the *Sudd* attains its greatest width, with vast, largely inaccessible swamps, and fewer lakes and side channels than farther south. The Bahr el Jedid and the Bahr el Jebel recombine at the outflow from Lake Shambe, and with little water following the eastern courses of the Awai and southern Zeraf, the Bahr el Jebel carries a high proportion of the total flow. The courses of the Awai and the southern Zeraf at this point are not clear on satellite images, but it is reasonably certain that water from these forms the Bahr el Zeraf, to which the Bahr el Jebel is joined by the artificial Zeraf Cuts, dredged in 1910 and 1913.

South of Adok, a number of small channels flow westwards out of the Bahr el Jebel. At least some of this water re-enters the main channel south of Adok, but some satellite images suggest that there may be some loss of water westwards, eventually to link up with the Bahr el Ghazal system. There are fewer lakes in this section, most of them within meanders, close to, but isolated from, the river.

Adok–Malakal

The Bahr el Jebel and the Bahr el Zeraf are here separated by 50–60 km of swamp and seasonally flooded grassland. Lakes are very few, and virtually none are connected to the river. The Bahr el Ghazal enters from the west through the very shallow Lake No, but contributes little to the flow. Here the river becomes the Bahr el Abiad (the White Nile), and turns and starts to flow from west to east; a number of elongated lakes parallel its course. Three major affluents join the White Nile from the south; the swift-flowing Bahr el Zeraf, the sluggish Khor Atar, and the seasonally swiftly-flowing Sobat River. The Sobat shows huge seasonal variations which affect the main river to the north of the confluence, and also hold back the flow of the White Nile upstream.

The flooding regime

Between Juba and Bor the basins act as reservoirs in series, which each store water when the river rises and return it to the river downstream when it falls. Farther north the river banks and the carrying capacity of the

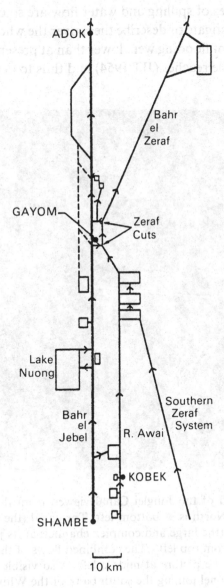

Figure 5.6. Diagram of the river channels and lakes between Shambe and Adok. Note different scale to Figures 5.4 and 5.5 (after SES, 1983).

river channel are both lower, so that spilling is more continuous. Because there is no topographic limit and because the gradient on the east bank is away from the river, the flooding extends further in periods of high flow and much of the spill does not return to the river. The net result of these successions of spills is that the seasonal succession of rises and falls of inflow is damped down to a fairly steady outflow. Only prolonged periods of very high inflow, as in 1961–6, are reflected in large fluctuations in outflow.

Within the swamps the processes of spilling and water flow are so complex that nobody has been able to investigate or describe them over the whole area. At a time when inflows and levels of flooding were lower than at present it was possible in 1951–2 to survey sample reaches (JIT 1954) and thus to determine

Plate 9. The northern end of the Jonglei Canal viewed from the Space Shuttle in January 1986. North is at bottom left. The canal (the straight line) joins the White Nile (the large and complex channel) at its junction with the Sobat (entering from top left). The combined flows of the White Nile and the Sobat leave the picture at middle left. Also visible are the Khor Atar (the broad channel joining the south bank of the White Nile), and the Khor Lolle (the northernmost of the several parallel channels of the White Nile west of its confluence with the river Sobat). Photo: NASA.

in some detail how spill leaves the river, how it flows down the swamps and how some returns to the river and the rest is lost by evaporation. An account of the sample reach is given in Chapter 8, but the findings may be generalised here.

At moderate flows and levels, river spilling flows through the alluvial banks of the main river in a succession of spill channels; some of these are deep and become part of the main river system while others are simply formed from cattle watering points or hippopotamus access to land which carry spill at higher river levels. When river flows and levels are high, widespread spilling occurs over the banks of the main river.

There is some evidence that spill channels develop from small breaks in the bank, grow into major subsidiary channels during periods of higher floods, and finally silt up at their exit from the main river and become obsolete channels with their own high alluvial banks. These obsolete channels, like the river Aliab, then form barriers to the lateral movement of spill.

The lateral flow of water can be illustrated by cross-sections surveyed in 1952 (Figure 5.7), where the water gradient is clearly demonstrated. At the same time water which has spilled either through channels at medium flows or over the bank at high flows drains north, parallel to the main river through drainage channels or downstream through the swamps. Where an individual basin, like the Mongalla basin near Gemmeiza or the Aliab valley at Lake Papiu, is pinched out by the main river swinging to the limit of the floodplain, a channel pierces the bank of the main river carrying the excess drainage from the basin back into the river channel.

Thus the basins of the floodplain, especially above Bor, act as a series of reservoirs which receive flood waters from the river and return them lower

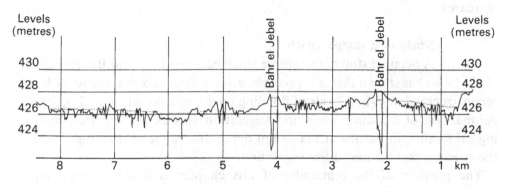

Figure 5.7. Cross-section of the Aliab and Bahr el Jebel valley, surveyed in 1951–2. Dotted lines show high and low water levels (after JIT, 1954).

down. The amount in passage increases as the river rises and decreases as it falls. In fact the return channel is a sensitive indicator of flooding conditions within each basin. The large channel returning to the river at Gemmeiza carried no flow in March 1952 when the Mongalla basin was receiving little spill, but was flowing strongly in March 1982 when higher river flows were causing large-scale spilling upstream.

Whereas the river has eroded below the woodland upstream of Bor and flooding is confined by high banks, these limiting banks disappear down-stream of Bor on the east bank and Shambe on the west and there is no lateral constraint to the spread of flooding. Because the antecedent land slope between Mongalla and Shambe is generally east of north, while the Bahr el Jebel flows west of north, there is a tendency towards tributary inflow to enter the river valley from the south-west while drainage flows away from the river to the north-east. Therefore, although the spill above Bor flows to the west of the main river, much of the flow past Bor returns to the river and the direction of immediate spill is deceptive. The overall topography suggests that most of the spill is to the east, and this is confirmed by the relative distribution of the flooded areas. Although the boundary between Bahr el Jebel spill and flooding from the Bahr el Ghazal tributaries is indeterminate, most of the area that flooded after 1964 as a result of increased flows lies to the east of the river.

It is impossible to give a simple summary of the whole maze of channels in the main swamp, but the pattern of flooding appears similar to that already described for the upper reaches, with spill spreading from the river to form a large evaporating surface and some drainage channels like the Atem flowing parallel to the river and rejoining it downstream. The pattern is complex, but the overall situation may be described by a simple hydrological model, where the swamp is represented by a reservoir which rises and spreads as inflow increases.

Study of a sample reach

A detailed study of a sample reach between Juba and Bor (Sutcliffe, 1957, 1974) has shown that it is possible, given inflow and outflow records, to reconstruct volumes and levels of flooding over a number of years. Further, a comparison of flooding levels and vegetation species showed that hydro-logical factors, in particular maximum depth and range of flooding, control the species composition of the vegetation.

The purpose of the remainder of this chapter is to develop a simple hydrological model to monitor the behaviour of the *Sudd* over the historical period. In essence, the previous study of a sample reach was extended to the whole area, and inflow and outflow records for the whole swamp were analysed together with rainfall records and evaporation estimates to deduce the flooding regime. This should reflect and illustrate the effect of the increased Nile flows after 1961, in terms both of permanent and seasonal swamp. (As described later in Chapter 16, by incorporating the effect of the Jonglei Canal on historical inflows to the swamps and on projected outflows, the model can be adapted to estimate the effect of the project on areas of flooding.) The model uses actual records as far as possible, in the belief that the results should provide a reasonable estimate of areas of flooding.

Available records
River flow
The river flows required for this model are the inflow at Mongalla where the Bahr el Jebel flows in a single channel, and the outflow from the tail of the swamps, deduced from the difference between the White Nile at Malakal and the Sobat at Hillet Doleib. These flows are available from 1905 to 1980. The flow of the Bahr el Ghazal at Lake No has not been taken into account, as most of this is spill from the Bahr el Jebel.

It is interesting to compare rating curves relating level to flow at Mongalla over the period of record. For a given river flow there was a continuous rise in water level from 1905 to 1960, but during the high flows of 1963–4, the rating changed abruptly and the water level and the bed level fell by a metre through erosion. There were a number of changes in river channel and spill channels during this period, in response to the high flows, and the areas of flooding and swamp vegetation changed markedly.

Rainfall
Monthly rainfall records at eight stations near the swamps were used to derive a swamp rainfall series for the period 1905–80. The record at Fangak was rejected after comparison with other stations, as there was a marked discrepancy between the early and late records. Long-term rainfall variations over the area were studied by calculating 7-year moving averages; these are compared with river inflows and outflows in Figure 5.8. Whereas the early period of high flows around 1917 corresponds with a period of high rainfall, the prolonged period of high flows since 1961 is not reflected in the rainfall over the *Sudd*.

Evaporation
The water losses from the *Sudd* are directly related to the rate of evaporation and the modelling of the *Sudd* depends on a reasonable estimate

of this factor. Experiments have been carried out to measure the evaporation, from papyrus grown in tanks in the swamp, but it was difficult to maintain vigorous growth within the tank. Penman (1963), discussing the experiments by Migahid with tanks filled with papyrus and open water, suggests that, if the diurnal weather cycle observed is maintained, transpiration from the papyrus and evaporation from the open lagoon will be nearly equal.

The most satisfactory method of estimating evaporation is through the

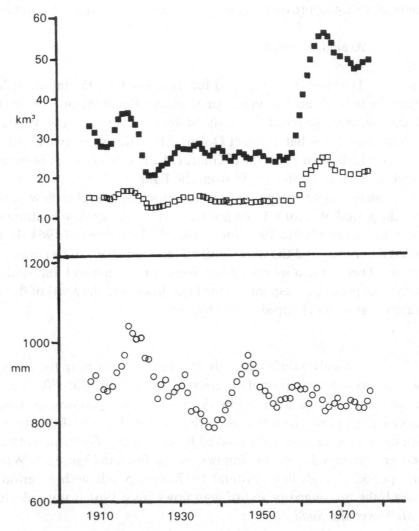

Figure 5.8. *Sudd* inflows (■) and outflows (□) compared with rainfall over the *Sudd* (○). All figures are 7-year moving means plotted at centre year. Inflows and outflows derived as in Figure 5.2 (from Sutcliffe & Parks, 1982).

Penman method, which combines the energy balance approach and the aerodynamic estimate to use standard meteorological records of temperature, humidity, sunshine and wind speed. Estimates for Bor of 2,150 mm/year should be reasonably typical of the swamp, and as the evaporation rate should not vary much from year to year the monthly averages have been used to estimate the evaporation from flooded areas. Although this estimate could be adjusted if necessary, the modelling provides a test of the evaporation estimate and in fact confirms that the estimated open water evaporation rate is reasonable from the whole swamp area.

Table 5.1. *Average rainfall, 1941–70, and open water evaporation (mm)*

	J	F	M	A	M	J	J	A	S	O	N	D	Year
Rainfall	2	3	22	59	101	116	159	160	136	93	17	3	871
Evaporation	217	190	202	186	183	159	140	140	150	177	189	217	2,150

The monthly average rainfall is compared with open water evaporation in Table 5.1 which illustrates the seasonal cycle.

Areas of flooding

A final important item of information is the area flooded on specific dates, which can be used to test the areas estimated by the hydrological model. The area of flooding cannot be measured directly but can be estimated from air photography or from satellite imagery, or deduced indirectly from vegetation maps compiled at different times.

Measurements of flooded areas were found from four separate dates. The first was based on air photography taken in 1930/1, from which maps were produced and planimetered to give a swamp area of 8,300 km² from Mongalla to Lake No, of which 845 km² was between Mongalla and Bor. A study of the contemporary accounts of this measurement (Butcher, 1938; Hurst & Phillips, 1938) and a comparison of the maps with detailed surveys, suggests that these areas correspond to the mean flooded area of that year. Satellite imagery provides a potentially powerful method of estimating areas of flooding, but without some field survey it is not easy to determine the precise limits of flooding when these are obscured by vegetation. A map of flooded areas was compiled by Mefit (1977) using LANDSAT imagery of February 1973, supplemented by observation from the air. The area of flooding was measured from this map as 21,300 km² from Bor to Malakal, and this should correspond with the flooded area in February 1973. The contrast with the

1930/1 figure is evident; the doubling of the flooded areas reflects the increase in outflows from the East African lakes and the inflows at Mongalla after 1961.

In some ways the areas of permanent and seasonal swamp may be deduced more easily from the vegetation distribution which in turn reflects the contemporary pattern of flooding. The use of satellite imagery to map vegetation also requires some field observations, and a map has been compiled from systematic aerial reconnaissance supported by satellite imagery of 1979/80 and observations in the field (RES, 1983). The areas of different species were measured from this map, making reasonable estimates of areas relevant to the Bahr el Jebel. These provide an estimate of the current areas of permanent and seasonal swamp, which must correspond to the flooding over a small number of years before the survey. The total area of permanent swamp below Bor is 16,200 km^2, with additional seasonal swamp of 13,600 km^2 to give a total of 29,800 km^2.

An earlier map of vegetation distribution is included in JIT (1954; Vol IV, Map 7; see also Figure 8.2). The comparison between this map and the present situation (see Figure 8.3) gives a striking indication of the increase in flooding over the past 30 years, but the map was based on general observation and not on air photography, and was not claimed to be more than approximate and indicative. The area of permanent swamp shown below Bor is only 2,700 km^2, whereas the area of seasonal swamp is 10,400 km^2 to give a total of 13,100 km^2. Because the perimeter of the seasonal swamp could be approached by land, while the boundary of the permanent swamp could only be surmised from a river steamer, the total area is likely to have been more reliably estimated than the area of permanent swamp, which appears low by comparison with other estimates. These estimates are summarised below (Table 5.2) and may be compared with areas predicted by the hydrological model.

A hydrological model of the system
Development
Hydrological modelling of an area of swamp can be based on measurements or estimates of river flow into the area and outflow from the area, rainfall on the swamp and evaporation from the flooded area. The model simply uses the equation of continuity which treats the swamp as a reservoir whose storage volume is cumulative inflow less outflow. In order to estimate the direct rainfall on this reservoir and the evaporation from it, the area flooded with a given volume of storage is required. This corresponds to the area–capacity curve of the reservoir.

The model requires reliable and continuous river flow records for the selected area. Because no river flow measurements below Mongalla are completely representative of the total flow down the river and floodplain, the whole of the *Sudd* from Mongalla to the Sobat mouth is the only area which can be studied over a long period using direct flow measurements. These records, together with a relationship between area and volume of flooding, and estimates of rainfall and evaporation over the flooded area, have been used in a sequential reservoir model to provide estimates of areas flooded during the period of historic hydrological records.

The equation of continuity[1] states that, over a given time interval, the increase in volume of flooding equals the inflow less outflow together with rainfall less evaporation over the area flooded, while allowing for soil moisture recharge over newly flooded areas.

With the time interval taken as one month, nearly all the terms in the equation are known for the period 1905–80. There are measurements of inflow and outflow, the monthly rainfall series has been estimated from records, and the evaporation may be taken as the mean open water evaporation for the calendar month. The soil moisture recharge may be estimated on

Table 5.2. *Measured areas of flooding* (km^2)

Source	Type of information	Date	Measured areas of permanent (P), seasonal (S) and total (T) swamp[b]	
			Below Mongalla	Below Bor
Air survey maps	Planimetered maps	1930–1	(T) 8,300	7,500
JIT report Map 7	Vegetation maps from reconnaissance	1950–2	(P) 2,800 (S)11,200 (T)14,000	2,700 10,400 13,100
Mefit SPA Regional dev. study Hydrographic map	Flooding map from LANDSAT	Feb 1973	(T)22,100	21,300
Range Ecology Survey (1980) Interim rept	Vegetation map from LANDSAT and from low level aerial reconnaissance[a]	1979–80	(P)16,600 (S)14,000 (T)30,600	16,200 13,600 29,800

Notes: [a] After development of the hydrological model, this map was revised and new measurements derived; these are given in Chapters 7 and 8.
[b] Explanations for inconsistencies between the areas estimated here and later, in Chapters 8 and 16, are given in Chapter 8.

physical grounds as 200 mm at the beginning of the wet season, and decreased by cumulative net rainfall for areas flooded during the wet season. The flooded area required to estimate rainfall and evaporation for the swamp is strictly the mean of the initial and final values for the month, but this may be dealt with by an iterative procedure. Thus there exist for each month a set of records and an equation linking these records and two variables, the change of storage volume and the change of flooded area. There should also exist a functional relationship between storage volume and flooded area.

Direct determination of this relationship would require a detailed topographic survey of the whole area, combined with surveys of different flooding surfaces. Such a survey does not exist for the main swamps, but topographic and hydrological surveys in three sample areas provide quantitative evidence on volumes and areas of flooding. For the reaches Malakal to Melut and Melut to Renk on the White Nile (JIT 1954; Vol. III) and for the reach Mongalla to Gemmeiza on the Bahr el Jebel (Sutcliffe, 1957) there is a linear relation between volume (V) and area (A) within the range of information. The sum of a number of linear relations will also be linear and may be written as $A = kV$ since the volume is zero when the area is zero; it seems reasonable to use and test this relationship for the whole *Sudd*. As it implies an exponential relationship between area of flooding and level, with a constant average depth of flooding over the swamp of $1/k$ metres, the value of the constant k may be estimated physically as 1.0 though other values could be tested.

Once a relationship between area and volume of storage is available, the water balance equation contains only one unknown, and the volume and area of flooding at the end of the month can be deduced from the values at the beginning of the month, given records of inflow, outflow, rainfall and evaporation. As demonstrated in the end note, a second iteration is used to ensure that the rainfall and evaporation terms are adjusted to the mean flooded area over the month.

Application

Starting from an initial storage value of 8 km³ on 1st January 1905, the storage and flooded area were estimated at monthly intervals to the end of 1980. The predictions of flooded area are plotted in Figure 5.9, and are compared with monthly gauge levels at Shambe lagoon near the centre of the swamps up to 1966; after about 1964 the relation of the channel to the lagoon changed and the levels are not a consistent index. The comparison shows that the seasonal variations of flooding are reasonably well reproduced.

The six measurements of flooding on specific dates are compared with the corresponding estimates in Figure 5.10. The areas measured from swamp vegetation maps were assumed to correspond to a three-year period of flows.

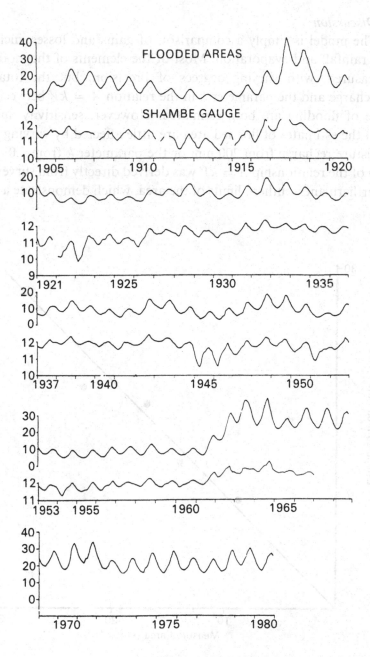

Figure 5.9. Areas of flooding, estimated using the model (see text) below Mongalla (km² × 10³), and monthly mean Shambe gauge level (metres). Shambe gauge records ceased in 1967 (from Sutcliffe & Parks, 1982).

This comparison demonstrates a reasonable fit and suggests that the simplified model represents the physical situation within acceptable limits.

Discussion

The model is simply a comparison of gains and losses, including river flows, rainfall and evaporation. Most of the elements of the model are directly measured, with varying degrees of precision, but the total soil moisture recharge and the parameter k in the relation $A = kV$ between area and volume of flooding are both estimated. However, sensitivity analyses showed that the estimates of flooded area are little affected by varying either the soil moisture recharge from 200 mm or the parameter k from 1.0.

The form of the relationship $A = kV$ was derived directly from survey data from the northern and southern limits of the area, which demonstrate a linear

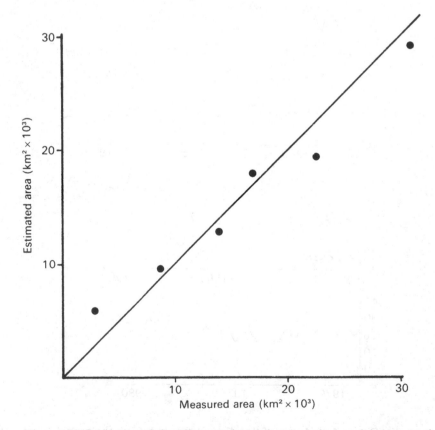

Figure 5.10. Areas of flooding, estimated using the model (see text), plotted against measured areas for six occasions (see text and Table 5.2), (from Sutcliffe & Parks, 1987).

relation between A and V; the mean depth $1/k$ is then constant and can be estimated as 1.0 m. However, it could be argued that the predictions would be sensitive to the form of the $A : V$ relationship. This was tested by substituting $A - kV^x$ while maintaining realistic values by varying k and x together to fit the mean values of A and V from the previous trial.

A set of curves was tested with values of x ranging from 0.5 to 1.1. The effect of reducing x from 1.1 to 0.5 is to reduce the mean area of seasonal swamp steadily from 8,200 to 4,100 km^2 while the total area of flooding is little changed. However, the fit between observed and predicted areas of flooding deteriorates as the value of x is reduced from 1.0 to 0.5. These limits can be interpreted in terms of geometry. A value of 0.5 is equivalent to a plane sloping evenly towards the river, while a value of 1.0 corresponds to an exponential relationship between level and area, with the area of flooding increasing with level in such a way that the overall mean depth remains

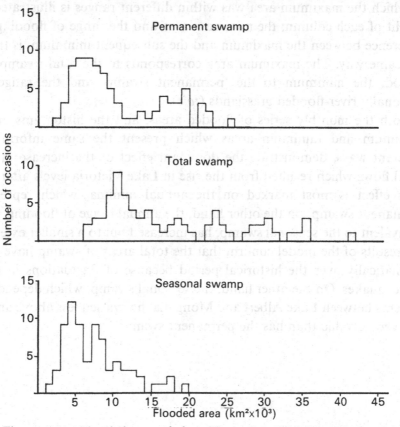

Figure 5.11. Flooded areas below Mongalla, 1905–80, estimated from measured inflows and outflows using the model (see test). Bars represent the number of occasions on which flooded areas fell within the limits shown (from Sutcliffe & Parks, 1987).

constant. These values therefore represent the range of likely geometry. Thus a value of 1.0 provides the best fit between observed and predicted areas of flooding as well as corresponding to the available survey data; it is interesting that a similar relation and k value were derived for the Okavango swamp in Botswana.

Thus, although the model is shown to be sensitive to the form of the equation linking area and volume, which should be borne in mind as more data become available for calibration, the form which is derived from physical evidence gives a reasonable fit to the measured areas of flooding. One can therefore deduce that the model gives a reasonable representation of the flooding regime, within the limits of historical experience. The results of the model may be expressed either as a monthly series of flooded areas (Figure 5.9) or may be summarised in some statistical form. For example, Figure 5.11 presents the 75 years of flooded areas in histogram form. The number of years in which the maximum area was within different ranges is illustrated by the height of each column; the minimum area and the range of flooding, or the difference between the maximum and the subsequent minimum, is treated in the same way. The maximum area corresponds to the total swamp in each period, the minimum to the 'permanent swamp' and the range to the seasonally river-flooded grasslands (*toic*).

Both the monthly series of flooded areas, and the histograms of annual maximum and minimum areas which present the same information in different ways, demonstrate the dominant effect of the increase in Bahr el Jebel flows which resulted from the rise in Lake Victoria levels after 1961–4. This effect is most marked on the annual minima, which represent the permanent swamp; on the other hand, the annual range of flooding, which is equivalent to the seasonal swamp, has increased but to a smaller extent. Thus the results of the model confirm that the total areas of swamp have changed dramatically over the historical period because of fluctuations in the East African lakes. On the other hand, the seasonal swamp, which depends on the torrents between Lake Albert and Mongalla, has varied less above and below its average value than has the permanent swamp.

Notes

1. The water balance for an area of swamp may be expressed by the equation of continuity for a time interval δt:

$$\delta V = \{Q - q + A(R - E)\} \, \delta t - r\delta A$$

where δV is increase of volume of flooding

Q is inflow
q is outflow
R is rainfall
E is evaporation
A is flooded area
r is soil moisture recharge, which is positive when δA is positive and zero when δA is negative.

With the time interval taken as one month, most of the terms in the expression are known for the period 1905–80. The inflows and outflows are measured, the monthly values of the rainfall R have been estimated from surrounding records by the percentile method, and the evaporation E is taken as the mean open water evaporation for the calendar month. The soil moisture recharge r was estimated as 200 mm at the beginning of the wet season, and decreased by $\Sigma (R - E)$ for preceding months when rainfall exceeded evaporation. Thus for each month there exist a set of records and an equation linking these records and two variables, the change of storage δV and of flooded area δA. There also exists a relationship $A = f(V)$ between flooded area and storage volume, for which evidence suggests the form $A = kV$ with k estimated as 1.0.

Starting the analysis at the beginning of month i, with an initial storage V_i and area $A_i = kV_i$, the net evaporation can be taken as $(E - R)A_i$ over the initial area; the monthly inflow Q_i and outflow q_i are known and thus

$$V_{i+1} = V_i + Q_i - q_i - A_i(E_i - R_i) - r(A_{i+1} - A_i)$$

$$= V_i + Q_i - q_i - kV_i(E_i - R_i) - rk(V_{i+1} - V_i).$$

Hence $V_{i+1}(1 + rk) = V_i(1 + rk) + (Q_i - q_i) - kV_i(E_i - R_i)$ where Q_i, q_i, E_i and R_i are tabulated for each month while r varies with net rainfall from an initial value of 0.2 m.

This equation provides an estimate of V_{i+1} and thus A_{i+1}. Because the evaporating surface is strictly the mean of the initial and final values for the month, these estimates were used to adjust the evaporation estimate to the mean flooded area to give:

$$\frac{A_{i+1} + A_i}{2}(E_i - R_i)$$

in a second iteration of the equation.

6

The biological effects of variation in water quality

Introduction

The waters of a river can be expected to change both physically and chemically during their journey to the sea. The 600-kilometre long traverse of the *Sudd* by the White Nile results in substantial deoxygenation of its waters, and in quite marked alterations in the chemical composition of its major constituents. The decrease in available oxygen affects many aquatic organisms, and in particular influences directly the distribution of fishes within the system. Other chemical and physical changes result in less obvious but still important shifts in acidity, nutrient availability, and water clarity, which in turn affect other organisms living in the series of ecological zones which comprise the *Sudd*.

The origins of the Upper Nile waters

The characteristics of the swamp system, and especially of the interplay between the flowing waters of the main and subsidiary channels with the stagnant or slow-moving waters of open lakes and vegetation-covered zones, reflect to a large part the long-term and seasonal variations in the quantity and quality of water entering the head of the *Sudd* between Mongalla and Bor. As we shall see, the three main components of the input have very different physical and chemical properties, so that short-term variations in their relative importance in the flow can give rise to 'pulses' of water, each with distinctive characteristics, passing successively down the river.

Lake Victoria

Lake Victoria is the main source, though its contribution varies seasonally with peaks of discharge which correspond to the long and short rainy seasons in the catchment. These peaks are, however, damped by lake storage. The water of Lake Victoria is dilute and clear, with relatively little dissolved or suspended material, because much of it is derived from rain that falls directly on to the lake surface. The turbidity of water leaving the Victoria system is further reduced by sedimentation during its sluggish passage through the swampy Lake Kioga, where there may also be a net loss by evapotranspiration.

Lake Albert

Water from the Western Rift Valley joins the Victoria Nile at the mouth of Lake Albert (L. Mobutu). It carries the drainage from the volcanic belt south-east of the Ruwenzori range, and from the connected series of lakes, Albert, Edward and George. The basic soils that cover much of this extensive catchment provide runoff rich in dissolved minerals; this drainage thus contrasts sharply with the dilute waters of the Victoria Nile. Typically, the concentration of the solutes in the waters leaving the western catchment is approximately seven times that in water from Lake Victoria (Beauchamp, 1956). The western input also displays a greater seasonal variation in the volume of discharge and a generally higher initial turbidity and water colour. The combination of these factors can lead to pulses of chemically enriched water moving into the *Sudd*.

The torrents

Highly seasonal, fast-flowing tributaries, originating partly in northern Uganda to the east of the Nile and partly in the southernmost hills of the Sudan, join the Nile between Lake Albert and Juba, and provide a third element in the flow of the Bahr el Jebel. These 'torrents' (JIT, 1954) display very variable and short-term discharges, often taking the form of flash floods resulting from the steeply inclined catchments. The water of these torrents is characterised by moderately low concentrations of dissolved salts combined with very high turbidities due to suspended silt and clay. These rain-fed turbid pulses are primarily derived from intense but short-lived soil erosion, and the washing out of previously deposited sediment from the beds of tributaries of the torrent system.

Variations in water quality at Bor

Evidence for changes in the water quality of the main river channel at the entrance to the swamps is available from water analyses carried out at Bor (see Figure 6.1). The onset of the rains in the southern Sudan results in an early peak of river flow and in a very marked rise in the turbidity of the river.

Much of this dramatic increase in the load of suspended material reflects the input from the seasonal torrents; electrical conductivity[1] displays a correlated decrease. A further dilution of the river water, shown by a second drop in conductivity, marks the peak of discharge in late August, as the effect of the rains which have fallen on the Victoria catchment reaches Bor. During the survey, a third peak of river flow at Bor occurred during the last week of November, produced by unusually heavy late season rains in East Africa. Each peak in turbidity was lower than its predecessor as the amount of easily available and transportable material declined through the season.

The final surge of turbid material occurring in late December was accompanied by peaks in the concentrations of dissolved bicarbonate, calcium, nitrate and phosphate as well as by a marked change in the planktonic algal flora carried by the river. The nature of this chemical and biological change was such as to suggest the passage of a long pulse of water originating in the Lake Albert catchment. There is some evidence for a similar pulse passing Bor in early September of the study year, though its effects are partially masked in many of the chemical determinations by the generally high Victoria discharge at the time. Many of the pulses are likely to be of very brief duration and would only be detected by a programme of very intensive monitoring. Thus, an abrupt discontinuity in conductivity noted on 4 September, 1982, during a longitudinal transect of the main channel probably represents such a movement of a discrete water mass downstream.

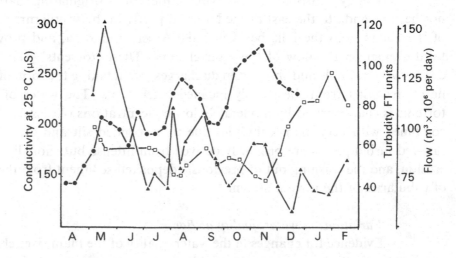

Figure 6.1. Water flow (●), conductivity (□) and water turbidity (▲) at Bor between April 1982 and February 1983 (after SES, 1983).

Water quality in the *Sudd* – the changes along the river

Introduction

It is apparent that the waters of the White Nile which flow out of the swamp region differ in many respects from those which enter it. The slow flow[2] dictated by the extreme flatness of the region results in a deposition of particles previously held in suspension during the descent from the uplands. This sedimentation will in turn lead to an increased water transparency providing an environment more conducive to photosynthesis and hence to a productive food chain within the open waters. However, the vigorous growth of papyrus and other components of the emergent vegetation of the swamps may be expected to draw heavily upon available nutrients. A situation may result in which much of the growth potential is sealed into a rapidly recycling 'grass' system, leaving the open waters warm, clear but deficient in usable nitrogen, phosphate, and sulphur, and perhaps in several vital trace elements.

Two approaches to an investigation of water quality within the *Sudd* are practicable. Traverses of the entire length of the navigable portions of the principal channels and their associated lakes can provide a series of 'spot' analyses such that an overall assessment of water quality at the time of sampling may be revealed. Alternatively, events within the system may be examined in finer focus by a comparison of analyses performed over a period of time at two or more fixed stations sufficiently adjacent within the overall system to be sampled more or less simultaneously, yet sufficiently apart that the water mass travels far enough for effects of interactions between river and swamp to be discerned. Both approaches were attempted during the 1982–3 study period and between them yielded valuable insights on the interplay between flowing and retained water.

The river Atem as a model system

The river Atem forms a parallel system to the main channel in the southern area of the *Sudd*. The twin northward flows are connected in at least three places in each of which the mass movement of water is from the main channel to the Atem. The two flows differ in one important respect: whereas the main channel flows between well-defined banks for 88 km from the original divergence to the Jonglei confluence, the Atem flows for some 100 km through extensive lakes and areas of papyrus swamp and can, over much of its length, exchange water and hence aquatic organisms with them.

Seasonal patterns of change

Apart from a period in December and January, the conductivity of the water reaching Dhiam-Dhiam, a village near the final confluence of the Atem and the main channel, was higher than at Bor, but, in general, chemical conditions in the Atem closely paralleled those in the main river. This suggests that the modifying effects of the swamp in this relatively short traverse are of limited magnitude. However, whilst evaporation contributes to the concentration of the dissolved substances present, at least part of the observed changes can be attributed to an interplay between river water and the swamp. The timing of events will vary from year to year, but a seasonal pattern may be discerned.

The behaviour of the Atem during the study period closely resembles the seasonal changes found by Gaudet (1979) within North Swamp, Lake Naivasha. Gaudet concluded that three phases are distinguishable in the relationship between the water of the Kenyan papyrus swamp and that of the Malewa River, which traverses the system before discharging into the lake. Following that author's terminology the sequence of events in the Atem is as follows:

> *Stagnation:* late dry season (end of January to mid April), charac-
> terised by low river flow and partial isolation of swamps.
> Decomposition and evapotranspiration cause a rise in ionic
> concentration within them. Any water percolating from the
> swamps into the lakes and flowing channels is highly
> concentrated.
>
> *Flow-through:* most of the wet season (mid-April until about mid-
> October in most years). Pulses of water sweep into the swamps,
> initially flushing out old, ion-rich water so that for the first half
> of the season materials move from the swamps to the river and
> lake system. Overall there is a general reduction of ionic levels in
> the *Sudd* during flow-through. The swamps eventually fill with
> dilute water from the river and from direct rainfall on to their
> surface.
>
> *Drying out:* end of wet into dry season (October to end of January).
> Initially the swamp will contain dilute river water accumulated
> during flow-through. Further dilution will occur as the swamp
> vegetation removes dissolved materials; dilute water drains back
> into the lake and river systems, such that in 1982–3 a short-lived
> reversal of the ionic concentration at Bor and Dhiam-Dhiam
> was observed. The implication of this sequence is that dissolved
> substances, including plant nutrients, can be both taken up and
> released by the swamp.

Changes in individual parameters

The passage of the water through the swamp between the two stations reduces turbidity but adds to water colour. Turbidity due to the coarser particles present in the flowing water is reduced by increased sedimentation as the current diminishes within the swamp area. Much material is also deposited on the suspended root masses of fringing vegetation, particularly of the water hyacinth. The major increase in colour occurs during flow-through when old water, rich in dissolved organic matter, is flushed from the swamps. Alkalinity is also enhanced at this time. In general, however, the acidity of the river water increases during its passage through the swamps. This appears to be the result of increased production of carbon dioxide associated with decomposition processes in the swamps.

The level of nitrate in the river water is generally low. Removal both by fringing vegetation and by planktonic algae depresses the concentration even further. However, a brief upsurge of usable nitrogen occurs during the early flow-through period when the swamps are flushed by oxygenated water, converting ammonia to nitrate and hence leading to a temporary increase in this nutrient in the river. The behaviour of total iron, inorganic phosphate and sulphate also relates to levels of oxygenation. Iron increases in the swamp, probably due to the formation of soluble ferrous iron under conditions of oxygen poverty. Phosphate is reduced both by the growth demands of the swamp vegetation and by adsorption[3] on mud surfaces during flow-through. The latter process is reversible however, so that, when levels of dissolved oxygen fall during stagnation, phosphate is released. Sulphur is depleted during swamp passage by its conversion to insoluble sulphides under reducing conditions.

The river transect

The extent to which the local changes in water quality seen in the River Atem are mirrored in the main channel during the entire passage through the *Sudd* can be gauged from the results of a sample series taken in August and September 1982 during two longitudinal transects between Bor and Malakal, a river distance of approximately 730 km.

Conductivity and ionic balance

Conductivity rose at varying rates between Bor and the Sobat confluence but the overall increase was less than expected when water loss, and hence the resultant enhancement of the concentration of dissolved material, is taken into account. This suggests that during the period of observation the swamp acted as a net accumulator of ions. The individual components largely followed their performance in the shorter Atem transect. Thus, when an attempt is made to produce a balance sheet of ionic concentrations relative to the quantities of water entering and leaving the *Sudd*, it appears that at the time of sampling the outflow showed a gain in silicon, substantial losses of sulphate, iron, phosphate, nitrate and ammonia and some loss of calcium. It also appeared that the water had probably retained a balance of hydrogen carbonate, chloride and magnesium.

The progress of change along the course of the river
Some of the chemical changes which characterise the passage of the river waters through the swamp are illustrated in Fig. 6.2. The fate of selected components of the water mass can be followed through the series of Maucha plots which represent water samples taken at a series of locations between Bor in the south and the confluence of the Bahr el Jebel and the Sobat downstream of the swamp system.

The river water at Bor is characterised by high levels of turbidity, phosphate and nitrate, moderate amounts of dissolved carbon dioxide and phytoplankton, and by relatively low concentrations of ammonia and dissolved silica.

Samples taken in the River Atem after passage through 80 km of swamp show that the water has already become clearer and that anaerobic decomposition within the swamp has slightly enhanced the levels of carbon dioxide. The reducing conditions operating within the swamp probably also explain the lower levels of nitrate, the principal oxidised form of nitrate in fresh water. Conversely, the concentration of silica has increased, a further indication of reducing conditions since silica incorporated into sediments is released as oxygen concentration drops at the mud/water interface.

Lake Wutchung, a satellite of the Atem (Figure 5.4), shows a decreased influence of the swamp. It acts as a settling basin, so that turbidity is further reduced and the open still water allows development of a crop of phytoplankton. Enhanced levels of carbon dioxide and silica, characteristic of the swamps, are conspicuously absent.

Water which reaches Kobek (Figure 5.6) after a substantial interplay with the swamp displays a pattern of concentrations contrasting markedly with that of the water originally arriving at Bor. The stream is of remarkable clarity, high in dissolved carbon dioxide, and also showing high levels of silica and very low nitrate levels as a result of the oxygen-deficient nature of the environment.

The Bahr el Jebel at Buffalo Cape exhibits its dual nature. Some of its waters have followed the main channel and have remained unaffected by the

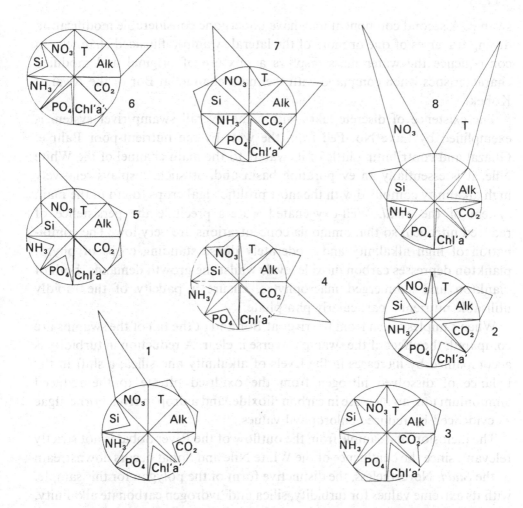

Figure 6.2. Maucha plots of nutrient levels at eight sites within the *Sudd*. (1: Bahr el Jebel at Bor; 2: Lake Wutchung; 3: River Atem at Dhiam-Dhiam; 4: Awai River at Kobek; 5: Bahr el Jebel at Buffalo Cape; 6: Lake No; 7: White Nile at tail of the swamps (nr Tonga); 8: River Sobat.) The diagrams are constructed so that the distance from the centre to the circumference is equal to 50 units, and the enclosed area is proportional to the concentration of the constituent.

Symbol		Units
T	Turbidity	Formazine Turbidity Units
Alk	Alkalinity	meq l^{-1} × 40
CO_2	Carbon dioxide	mg l^{-1} × 10
Chl 'a'	Chlorophyll 'a'	µg l^{-1} × 2
PO_4	Total dissolved phosphate	µg l^{-1} PO_4-P
NH_3	Ammonia	µg l^{-1} NH_3-N
Si	Silica	mg l^{-1} × 5
NO_3	Nitrate	µg l^{-1} × 5

swamp. A second component may have undergone considerable modification during traverses of one or more of the lateral swamp-influenced channels. In consequence the water mass displays a mixture of original and modified characteristics when compared with the main channel at Bor or the Awai at Kobek.

The existence of discrete lakes within the overall swamp/river system is exemplified by Lake No. Fed from the west by the nutrient-poor Bahr el Ghazal and contributing little of its waters to the main channel of the White Nile, it is essentially an evaporating basin and, as such, displays relatively high alkalinity combined with the most prolific algal crops found in the main system of the *Sudd*. Well-oxygenated waters preclude the persistence of reduced nitrogen, so that ammonia concentrations are very low. The combination of high alkalinity and moderately dense standing crops of phytoplankton depresses carbon dioxide levels whilst the growth demands, both of plankton and submerged macrophytes, ensure a paucity of the rapidly utilisable nutrients, particularly phosphate.

When samples taken from the river at Bor and at the tail of the swamps are compared, the effect of the swamp traverse is clear. A reduction in turbidity is accompanied by increases in the levels of alkalinity and silica; a shift in the balance of dissolved nitrogen from the oxidised nitrate to the reduced ammonium ion, an increase in carbon dioxide, and a drop in river-borne algae as evidenced by reduced chlorophyll values.

The inclusion of a sample from the outflow of the River Sobat is not strictly relevant, since the confluence of the White Nile and Sobat occurs downstream of the *Sudd*. Nevertheless, the distinctive form of the polygon for this sample, with its extreme values for turbidity, silica and hydrogen carbonate alkalinity, serve to emphasise the gradual and unspectacular nature of the changes within the swamp traverse.

Previous river transects

Five longitudinal surveys have been carried out in the past, two by Talling (1957*a*) in June and December 1954, one by Bishai (1962) in December–January 1960–1, and two by a Russian expedition in December–February and April–May 1963–4, (Kurdin, 1968; Monakov, 1969). The 1954 study was the most detailed, and was conducted along the full course of the White Nile as a background study to the biological investigations of the Hydrological Research Unit of the then University College of Khartoum. When the results of the current (1982–3) investigations are compared with Talling's findings in 1954, it appears that despite the considerable increase that has occurred in the volume of water flowing through the system in recent years, and even though the area of permanent swamp has become greatly enlarged (see Chapter 8), the main channel characteristics have remained fairly constant. However, seasonal changes in concentrations of ammonia, iron, phosphate and in the degree of sulphate depletion found during the survey were less marked than hitherto. The explanation for this apparent damping of seasonal fluctuations in the chemical composition of the out-flowing water may lie in the less extreme levels of deoxygenation encountered in the most recent studies. The observed situation may result from a complex equilibrium between the greater input of oxygen-rich river water currently entering the system and the opposing effect of the increased area of swamp through which the river now percolates. An additional circumstance which has the potential to alter both absolute ionic concentrations and the relative values for individual ions is the establishment of water hyacinth since Talling's surveys were made (see Chapter 8).

Water quality in the *Sudd* – where changes occur
The sites of water quality changes

Since a water mass following the main channel between Bor and Malakal could pass through the *Sudd* quite quickly (12 days would seem a reasonable estimate), the effects of evaporation on water following this direct route are likely to be small. It follows that most of the observed changes take place in those regions where the water is diverted away from the river into lagoons, lakes and the interstitial waters of the papyrus and *Typha* zones. Bodies of open water are numerous and somewhat varied, although all share two important features: shallowness and a greater or lesser influence on their water quality from the surrounding emergent swamp vegetation.

Classification of lakes

Further investigations of the open waters of the *Sudd* could lead to a system of lake classification which might prove of value in estimating, for example, sustainable fish yields, and the breeding areas of specific organisms. Such a classification is likely to be based upon lake area, degree of connection with a flowing river channel and the length of the lake–swamp interface coupled with the area of swamp catchment associated with the open water. Any attempt at a detailed classification is certainly premature, but the following examples may suggest the way in which such a classification might operate.

Large lakes
These, including lakes such as Shambe, Fajarial, Nuong and No, are of considerable surface area, close to and connected with the main channel. However, exchange of water between lake and river is insignificant in relation to the volume of the lake,[4] and the effect of the fringing vegetation on water quality appears to be minimal. Lakes in this category must have extended retention times and a consequently large proportion of 'old' water. Their ecology is most likely to be influenced by the stirring effects of the wind. These lakes may prove to be among the more productive water bodies of the system, coupling adequate, if patchy, growth of submerged macrophytes with moderate crops of phyto- and zooplankton in the well-oxygenated waters of relatively low turbidity. Long-term growth potential in these waters stems from periodic enrichment due to turbulent stirring of nutrients from the bottom mud during storms.

Chain lakes
These are lakes, mostly in the medium-sized range, which usually form series (chains) parallel to the principal flowing channels. They are characterised by having a substantial degree of water exchange with a river system (and often with each other). Influence exerted by the fringing vegetation is intermediate; it is clearly greater than in the previous type, but still only moderate in comparison with other open water bodies. These lakes have short retention times and are characterised by 'young' water which may become very turbid during the period of peak river flow. Lakes in this category are very common in the southern half of the 'permanent' system, and a wide range of sub-categories within the general class could be constructed from the lakes of the Wutchung, Dhiam-Dhiam, Jonglei, Pagak, and Kobek series. Lake Wutchung, a typical example of this common lake type, is considered below.

Other lake types
Small 'blackwater' lakes. These are small isolated lakes, lacking visible connections to a flowing river channel or other water body and too small to be affected by wind action. Often 'lakes' of this kind are no more than pools enclosed in the swamp vegetation, whose influence upon the water quality in them is paramount.

  Too little is known of lateral lakes, the large isolated lakes of the upper Bahr el Zeraf area and the lakes within the meanders north of Lake Shambe, for meaningful discussion.

Swamp–river interactions – evidence from Lake Wutchung
Chemical changes
A seasonal study of chemical and algal changes in Lake Wutchung is of some significance. Chemically it exhibited two contrasting aspects over the year. From early February to the end of August, there was a 'lake phase', during which the lake and the river behaved as distinct entities, and the lake developed ionic concentrations much above those of the river. Thereafter, the diluting effect of river flood culminated in a brief river phase when the waters of the lake and the adjacent River Atem became virtually indistinguishable. The lake was less turbid than the river, particularly during the lake phase. Short-lived, wind-generated turbidity could, however, disturb an otherwise favourable light climate for plankton and water plant growth. Low carbon dioxide concentrations, characteristic of the lake, were the principal indicators of a relatively low level of influence from the papyrus swamp.

Changes in the phytoplankton
During the lake phase, standing crops of phytoplankton, though modest, were appreciably higher than those in the River Atem but an equivalence obtained in the river phase when the lake lost plankton to flowing water. Regular determinations both of the algal species present in the plankton, and of the percentage abundance of groups of related species, suggested that many algae could be used as indicators of the recent history of the water in which they are found.

When the planktonic algae of the lake are compared with those of the river, lake populations generally exceed those found in the latter but there are some notable exceptions. A more or less continuous sequence of algal maxima, each characterised by the prevalence of a group of closely related species, occurs through the year in both lake and river populations. When a 'species group' exhibits a peak in both lake and river these are usually out of phase. A lake peak preceding that at the river station downstream suggests a lake origin for that group of organisms. In the converse situation it would seem likely that the lake has received a river-borne inoculum from upstream. A third element comprises algae which are more prevalent in the lake but are present at extremely low levels for most of the year. Their peaks occur between September and December during the mid to late flow-through phase of the swamps. These algal associations are characterised by species of the genera *Phacus, Trachelomonas* and *Euglena*, all members of the same class and characteristic of waters of high organic content. It seems likely that they are derived from the flushing out of the interstitial waters of the papyrus swamp. Similar behaviour has been recorded in open water for species of the desmid genus *Closterium*, normally associated with waters of moderately high acidity, and dinoflagellates of the genus *Peridinium*, which also appear to be favoured by swamp conditions. Within the open waters of the lake the 'swamp' algae respond to the twin benefits of higher light intensities and lack of competition for nutrients to produce reasonably large standing crops when the other algae are scarce.

Swamp–river interactions – evidence from the Awai River

The manner in which the influx of phytoplankton from differing ecological zones can influence the algal population of a flowing river, and hence indicate changing proportions of 'young' and 'old' water in the stream, is illustrated by data from two sites 40 km apart sampled within 48 hrs during a journey in the Awai River sub-system.

At the point where the Awai leaves the main channel a normal Nile plankton is found. Within 40 km, however, during which the braided flow of the Awai has passed between and through lakes and lagoons, deeper into the

Plate 10. The turbid waters of the channel of the Bahr el Jebel, showing spillage into the lagoon on the right. Patches of sandy deltaic deposits are visible at the inflow of the spill channel. Farther away are small isolated lakes and pools, entirely surrounded by swamp vegetation, containing clearer 'black' water. Photo: Alison Cobb.

permanent swamp, all four recognisable river species groups have fallen both in proportion and in absolute biomass. In part this loss has been made up by additions, apparently from the adjacent lakes and swamps. *Peridinium* spp. dominated the sparse phytoplankton of the 'blackwater' Lake Kobek, and, like the euglenophytes discussed, were almost certainly flushed into the lake and hence into the river flow from the enclosing swamp. Desmids of the genus *Closterium* were prevalent in seasonal pools and very probably also have a swamp origin within the *Sudd*.

The productivity of the waters of the *Sudd*

The sites of production

When attempting to understand the relationship between water quality and aquatic primary production within the *Sudd*, one is immediately struck by the fact that hitherto almost all productivity studies of inland waters in tropical Africa have so far been done on lakes. The mosaic of flowing channels, open water and swamp which comprises the *Sudd* provides a more complex environment in which it is extremely difficult to isolate nutrient availability, planktonic associations, and the constant input and export of materials, so as to reach conclusions as to performance and the reasons for it within discrete units of the system.

The general impression, based on chemical analyses and the average algal counts in diverse samples taken, is of a system comprising a flowing river of generally high turbidity and low aquatic primary productivity, often flowing

Table 6.1. *Composition of phytoplankton in the River Awai at its origin and at 40 km downstream*

Algal genus	Awai-Bahr el Jebel divergence	Km 40	Apparent source
Melosira	11.2	9.4	River
Microcystis	23.6	6.1	River
Merismopedia	13.8	1.7	River
Pediastrum	26.1	3.7	River
Nitzschia	2.0	9.5	Lake
Surirella	2.8	39.9	Lake
Peridinium	0.4	6.2	Swamp
Closterium	0	7.5	Swamp
Total chlorophyll *a* (mg l^{-1})	3.7	1.02	

into, and through, numerous lakes where nutrients are limiting if not actually deficient; that is, they are adequate to allow modest growth rates but insufficient to permit high standing crops. These lakes and flowing channels are embedded in a massive swamp whose interstitial waters are so oxygen deficient as to be uninhabitable by all but specially adapted populations. In this unpromising series of habitats it would seem superficially as though only emergent vegetation operating a closed pool of resources can flourish, by means of efficient, rapid, nutrient recycling. In contrast to this implied paucity of aquatic production there is ample evidence of the presence of large fish populations, relatively few of whose members can directly utilise the biomass of the emergent macrophytes. Our knowledge is very imperfect, but it seems entirely possible that the key to the apparent paradox lies within the shallow waters of some of the larger lakes, whose connections to the main channel are sufficient to allow for transport of material from lake to river, yet not so large as to prevent a distinct 'lake' phase to exist for all or for most of the year. These may prove to be the 'factories' whose output drives much of the open water economy of the *Sudd*.

   In this respect it is of interest to examine more closely the modest crops of floating algae found in water samples. The amount of living algal material present in a water volume can be estimated from a measurement of the concentration of the photosynthetic pigment chlorophyll *a* (see, e.g. Talling, 1957*b*). Similarly, the suitability of a water mass for the growth of algae and submerged vegetation can, in part, be indicated by calculating the transparency of the water column.[5] The minimum extinction coefficient, *Ev* (min) (the rate of attenuation of the most penetrating wavelength of light) is a useful measure of the opacity of a waterbody; the lower the value recorded the more transparent – and hence suitable for plant growth – is the water column. When a measure of chlorophyll present is plotted against the values of *Ev* (min) for a range of waters in the study area a reasonably clear pattern emerges (Figure 6.3.). The flowing channels demonstrate high extinction coefficients with low chlorophyll contents, a combination indicating severe light limitation for photosynthesis. The swamp-derived waters of the 'blackwater' lakes and swamp channels show high transparencies and low chlorophyll contents. This suggests strongly that in these waters one or more nutrients limit photosynthetic production. The residual samples, where the highest rates of algal growth are found, are all substantial lakes whose size limits the effects of the fringing swamp and whose predominant flow is *into* rather than *from* a flowing channel. Thus algae and invertebrates produced in the lake are used by lake fish, and both fish and potential food are donated to the rest of the system. Such lakes also provide the best habitat for the growth of submerged macrophytes, which offer ideal environments for the development of fish populations (see Chapter 14).

Although quantitative data are lacking, field observations suggest that these lakes are the major sources of water hyacinth. The hyacinth generated in the lake is dispersed into the river during windy periods and upon colonising

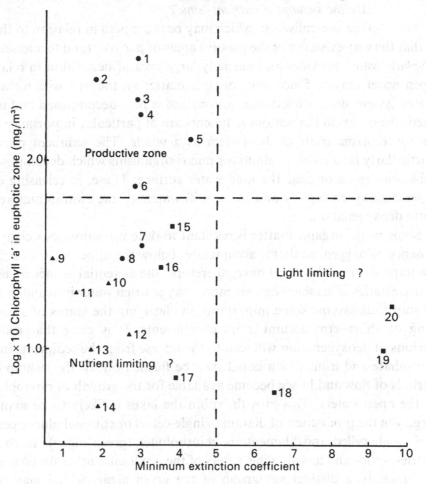

Figure 6.3. Phytoplankton concentration (as chlorophyll 'a') in relation to light penetration in waters of the *Sudd*. ● – open lake waters; ▲ – 'black' waters; ■ – river channels. Sites: 1: Lake No; 2: Lake Wutchung 1; 3: Lake 'Teardrop' (Adok-Lake No); 4: Lake of the Islands (nr. Lake Nuong); 5: Lake Mayen; 6: Lake Shambe; 7: Lake Dhiam-Dhiam 2; 8: Lake Dhiam-Dhiam 1; 9: Lake Kobek 3; 10: Northern Zeraf; 11: Lake Kobek 2; 12: Khor Lolle; 13: Southern Zeraf; 14: False Channel; 15: Khor Atar; 16: Bahr el Jebel, 385 km. N. of Bor; 17: Bahr el Jebel, 528 km N. of Bor; 18: Bahr el Jebel, 755 km N. of Bor; 19: River Sobat; 20: River Atem, turbid phase.

the river margins is capable of vigorous growth (see Chapter 7). The site and extent of hyacinth production is important, since like the beds of submerged macrophytes, this floating water plant provides major shelter and grazing for the fish population.

Are the swamps a nutrient 'sink'?

One generalisation which may be attempted in relation to the *Sudd* is that the very existence of the massive areas of papyrus (and to a lesser extent *Typha*) swamp provides an unusually large area of deposition in relation to open water masses. Since most organic matter synthesised within the fresh-water system sinks towards the bottom and is here decomposed and minera-lised, the events in the bottom sediments are of particular importance for the turnover of materials in the system as a whole. The sediment provides a particularly favourable medium for micro-organisms which decompose vege-table matter on or near the mud/water surface. These, in releasing energy, consume oxygen rapidly and in the still waters of the swamp may result in total deoxygenation.

Some of the organic matter is resistant to decomposition, especially in the absence of oxygen, so that it accumulates below the upper layer of sediment. Swamps such as the *Sudd* have, therefore, the potential to act as nutrient traps. Studies of seasonal movements of oxygen-rich water into and out of the swamp thus assume some importance in clarifying the status of swamps as long or short-term accumulators of nutrients. It is clear that prolonged periods of deoxygenation will lead to the release from the sediments of silica, phosphate and iron, which could then be flushed out of the swamp during periods of flow and hence become available for the growth of phytoplankton in the open waters. This growth within the lakes is likely to be manifested largely in the production of diatoms, single-celled or colonial blue-green algae and single-celled and filamentous chlorophytes (green algae). In the more nutrient-poor and less alkaline waters of the predominantly rainfed areas it is the desmids, a distinct sub-group of the green algae, which may become prominent. The interior of the swamps with their oxygen poor, organically rich waters, seem to be characterised by the motile single-celled algae of the Euglenophyta.

The *Sudd* compared with other African swamps

The topography of the African continent, with its prevalence of basins related to tectonic movements, provides the opportunity for massive swamp development. A considerable degree of uniformity might be expected within the swamps of Africa combining, as they do, shallow, slow-flowing, waters with closely packed aquatic vegetation exhibiting a high rate of continuous growth and rapid microbial activity, leading to swift decomposition of that portion of the ageing biomass which is not buried beneath settling sediment.

In practice, however, within this broadly uniform environment the swamps of Africa exhibit a range of varying characteristics. Apart from those associated with Lake Naivasha (see p. 130) and probably the oxygen-poor areas of papyrus fringing Lake Bangweulu (see e.g. Debenham, 1947), it would be difficult to equate the seasonal pattern of events demonstrated for the *Sudd* with those operating in other African swamp systems. For example, the extensive swamps of the middle Niger differ from those of the White Nile in two important respects. For reasons unknown, papyrus is virtually absent from west Africa. Thus the swamps of the Niger contain totally different plant associations from those of the rest of the continent. Secondly, and almost certainly coincidentally, the Niger swamps are bathed in extremely dilute waters. This lack of dissolved chemicals is probably due to the low solubility of the pre-cambrian granites of much of the catchment area (Beadle, 1974).

In Lake Chad, the extensive areas of papyrus within the southern basin have considerable water exchange with the open lake, a situation which renders unlikely all but ephemeral periods of oxygen depletion (see e.g. Hopson, 1969). The recent retreat of open water within the Chad basin may lead to the development of a more anoxic swamp and to further increases in salinity of the residual open water.

A similar high level of dissolved salts distinguishes the Okavango Swamps, east of the Kalahari desert, from those of the *Sudd*. The swampy areas of the Okavango have formed in a closed basin that has no drainage. The subsequent accumulation of salt, though modest enough to permit the development of papyrus swamp is none the less much higher than those occurring in those areas of the *Sudd* most subject to dry season evaporation.

The situation found in the swamps of central Zaïre represents, in contrast, a more sustained flow-through of dilute rain water. Although the onset of the rainy season causes an annual increase in flow of the main channel and of its tributaries, the water depth remains fairly constant. This is because the sytem drains humid tropical forests on both sides of the equator whose combined seasonal rainfall is never very low. Thus, although seasonal flooding of swamp does occur, its relative extent is less than that within the *Sudd*.

Conclusions

The extent to which the *Sudd* represents a significant resistance to the flow of the White Nile must vary between seasons and over extended periods as phases of high discharge from equatorial East Africa alternate with phases when rainfall and subsequent discharge are depressed.

It is unlikely that a simple relationship exists between high rates of flow and a diminution of the effects of the swamp. Although increased flow into the system may serve to scour out the channels and result in a faster and more direct passage of water from head to tail of the swamp-bounded area, there is also a distinct possibility (see Chapter 8) that increased flow may lead to the formation of *sudd* blockages, as sizeable portions of bordering swamp become detached and move downstream. The effect of such dams is to cause extensive spillage of river water from the channels and thus to increase the interaction between oxygenated water and the stagnant, deoxygenated sediment/water mixture lying beneath the closed canopy of the swamps.

If the main channel remains unobstructed, then increased flow probably results in a greater separation between river and swamp. Inflowing water follows principal channels and has relatively little contact with the swamp environment, the waters of which have accumulated by seepage into the papyrus and *Typha* zones and which may be lost largely by evaporation rather than by return to the flowing channels.

The lower flows which would follow the completion of the Jonglei canal might be expected to reverse this process. A higher proportion of water may spend longer within the sedimenting and reducing environment of the swamps, even though the lower absolute amounts of water involved will lead to a reduction in the total area of inundation. Knowledge critical to a convincing prediction of water movement within the *Sudd* is lacking. Although the pioneering work of Sutcliffe (see Chapter 5) has established the nature of the flow channels south of Bor, the shape and depth of channel cross sections in the main swamps farther north must be largely a matter of conjecture. Would lower flows become increasingly retained within a single main channel or would a relatively large amount of water continue to spill north-eastward due to the possible absence of a substantial eastern river network? Only a far greater understanding of the detailed morphology of the basin would permit more than the most speculative suggestions as to the effect of future river flows on the relationship between swamp processes and the quality of outflowing water.

Notes

1. Conductivity, as it applies to water analysis, is a measure of water's capacity for conveying electrical current and is directly related to the concentration of ionised substances in the water.
2. Estimated from various observations at *c.* 4 km h^{-1} in the main channel (see e.g. Rzoska, 1974).
3. *Adsorption:* the capacity of all solid substances to attract to their surfaces molecules of gases or solutions with which they are in contact. In contrast to the process of absorption the material adsorbed does not penetrate the surface of the solid.
4. The inclusion of Lake Nuong is questionable: although it superficially resembles other lakes of its type, the degree of water exchange with the river may prove to be considerable. Generalisations concerning the *Sudd* are almost inevitably based on insufficient data.
5. For a treatment of the calculation of Euphotic Depth (that part of the water column which receives sufficient light for net photosynthesis to occur, given the presence of chlorophyll) see e.g. Ganf, 1974.

7

Ecology of plants in the swamps and floodplains

Introduction
The ecology of the plant communities of the Jonglei area was studied as part of the Range Ecology Survey (RES, 1983) and the Swamp Ecology Survey (SES, 1983). Plants are clearly important in the ecosystem as food for herbivores, but, for the purposes of this study, the importance of the vegetation communities lies mainly in their role as integrators of environmental conditions. Because plant communities are relatively long lived and immobile, their distribution generally responds to long-term trends rather than to short-term fluctuations in the environment. An understanding of the distribution of the plant communities, and the factors controlling them, can thus be of considerable predictive value. In this account, the vegetation types are presented as a series, beginning with those found in open water and ending with those of the driest land. In each case, a brief description is given of the physiognomy and species composition and, where available, of the productivity and importance to other organisms. Factors controlling the distribution of each type are discussed. In a final section, comparisons are drawn between the vegetation of the Jonglei area and that of other major floodplains in Africa. Changes in the vegetation since the first written records are discussed in Chapter 8.

Throughout this section it should be remembered that most of the studies on which it is based were carried out in the southern part of the area, which may not be typical of the whole. In particular, the belt of land between the river and the 'high land', which remains unflooded by the river, is relatively narrow so that grazing pressure may well be abnormally high. However, checks from the ground and from the air suggest that conclusions drawn in the south can generally be applied elsewhere.

The vegetation of the major land types (see Figure 8.3)
The river and the swamps
Flowing waters

The strong currents and high turbidity of the waters of the main river channels (see Chapter 6) make them unsuitable for submerged macrophytes, except for occasional plants of *Ceratophyllum demersum* trapped in marginal vegetation. There is a floating fringe of water hyacinth (*Eichhornia crassipes*) up to 4 m wide. This may be invaded by trailing plants from the swamps, such as *Commelina diffusa* and *Cayratia ibuensis* which help to stabilise the mat. Free-floating duckweeds such as *Lemna aequinoctialis* sometimes grow among the water hyacinth.

The factors limiting the growth of higher plants also limit the growth of phytoplankton, which is very sparse. It is least sparse in the dry season when turbidity and river flow are at their lowest. *Melosira granulata*, a diatom, dominates at this time, but is replaced by a blue-green alga, *Lyngbya limnetica*, during the wet season. The river phytoplankton is so sparse that it is hard to see how it can be a useful food source for animals.

Lakes

Within the swamps there are a large number of bodies of open water, varying enormously in size and in the degree to which they are isolated from main river channels. At one extreme there are lakes which are virtual expansions of the river, with a short replacement time[1] and often with very turbid water. Deltas often develop very rapidly where rivers flow into such lakes. An intermediate lake type is more isolated, but is large and shallow, and has a long replacement time. Here the water is often clearer. Finally, there are small lakes which are isolated and completely surrounded by swamp vegetation. These often have dark water and may well be oxygen poor.

Lakes support communities of both floating and submerged macrophytes. They also contain phytoplankton, and an abundant algal periphyton[2] growing on the surface of submerged plants.

The most abundant free-floating plant is the water hyacinth. This reached the system Upper Nile in about 1957 (Gay, 1958), and has since spread throughout it, largely replacing the Nile cabbage (*Pistia stratiotes*) which was formerly common (Migahid, 1947).[3] Water hyacinth often completely covers small lakes, mainly those with a weak connection with a river channel. It is itself colonised by the swamp grass *Vossia cuspidata*, which forms circular patches which may themselves be colonised by *Cyperus papyrus*. This may represent a successional series from open water to swamp, but comparison of aerial photographs taken in 1976 and 1980 shows that hyacinth mats may also regress to open water. The ecology, control, and possible uses of hyacinth on the Nile in the Sudan have been the subject of a special study by a German team (Freidel, 1979).[4]

Until recently, control of *Eichhornia crassipes* has relied on periodic spraying with the herbicide 2,4-D,[5] and this continues when security considerations permit. Biological control has been tried, and a weevil and a moth have been introduced and appear to be established (Irving & Beshir, 1982).

Submerged plant communities occur in the clearer lakes. One, in which *Ceratophyllum demersum* is common, is characteristic of more turbid sites. Another, with two species of *Ottellia* abundant in it, is found mainly in the shallow waters of inflow deltas. Open water sites with a firm substrate and clear water are often dominated by *Naias pectinata*, with *Vallisneria aethiopica* and some *Ceratophyllum*. Finally, in very sheltered sites over soft organic muds, *Nymphaea lotus*, *Trapa natans*, *Ottellia* spp., *Potamogeton* spp., *Naias pectinata* and *Ceratophyllum demersum* make up a species-rich community. The smallest and most isolated lakes usually seem to be free of submerged plants, perhaps because of the low oxygen content of the water.

Swamps

Three swamp types can be distinguished in the study area. All are dominated by giant swamp herbs – *Vossia cuspidata*, *Cyperus papyrus* and *Typha domingensis*. All are found on land flooded throughout a normal year.

Extensive *Vossia* swamps cover only about 250 km², mostly in the south and close to flowing water, but the species is frequent as small patches beside the Bahr el Jebel as far north as Adok. *Vossia* forms floating mats over water up to 4 m deep; its only common associate is water hyacinth, which sometimes forms a lower layer in more open stands.

Cyperus papyrus swamps cover about 3,900 km². They form a fringe along the Bahr el Jebel up to 30 km broad in the south, declining to 50 m in the north, and disappearing completely east of Wath Wang Kech. Papyrus declines in stature from south to north, and also declines away from the river (Figure 7.1). The tallest plants, 5–6 m high, thus occur close to the river in the south of the area. Over much of the swamp papyrus probably forms a floating mat. Associated species are few, and most are climbers. They include *Coccinia grandis*, *Cayratia ibuensis*, *Luffa cylindrica*, *Zehneria minutiflora*, *Vigna luteola*, and the fern *Cyclosorus interruptus*. They tend to be commonest at channel margins where there is more light.

Typha domingensis swamps cover about 13,600 km². They are most extensive in the central and northern parts, away from the main river channels. They are very inaccessible and have therefore been inadequately studied. Few species grow with *Typha*, but in the wetter parts there is sparse water hyacinth, and a few duckweeds (*Lemna aequinoctialis* and *Spirodela polyrrhiza*). The more accessible *Typha* swamps form unstable floating mats beside subsidiary river channels. Aerial observations, particularly of the tracks left by oil-prospecting vehicles, suggest that most of the *Typha* swamps are not floating mats, but are rooted in a firm substrate covered by rather shallow water.

Which factors control the distribution of the swamp types? No experimental evidence is available, but observations made in the Swamp Ecology Survey (SES, 1983), and during the work of the JIT (1954; Sutcliffe, 1974) suggest that water depth and its seasonal variability, together with the nutrient status of the water, are the most important controlling factors. The main characteristics of the requirements of each swamp type are summarised in Table 7.1.

Finally, about 1,500 km² are covered by large areas of open water with floating and submerged aquatic plants. All occur within areas otherwise

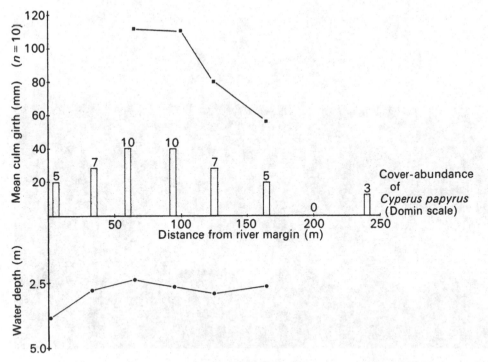

Figure 7.1. The performance of *Cyperus papyrus* in relation to water depth and distance from the main river channel. Culm girth (upper line) is well correlated with culm dry weight (Thomson *et al*, 1979 & SES, 1983). Domin scale of cover abundance runs from 1 (one or two individuals without significant cover) to 10 (more than 90% cover).

occupied by *Typha*; it may be very tentatively suggested that such communities occupy sites which are of low nutrient status, but too deeply flooded to be colonised successfully by *Typha*.

Seasonally river-flooded grasslands

These grasslands, flooded annually to varying extents and depths by the river, form the *toic*[6] which yields the dry season grazing essential to the Nilotic peoples. Two main grassland types can be distinguished, dominated by *Oryza longistaminata* and *Echinochloa pyramidalis*.

Oryza longistaminata grassland[7]

These grasslands cover about 13,100 km^2 – 19% of the study area, mainly in the south. The perennial rhizomatous wild rice-grass, *Oryza longistaminata*, makes up 80–90% of the standing crop. A variety of other species, none common, make up the residue. Many of the associates (e.g. the sedge *Cyperus albomarginatus* and the herb *Melochria corchorifolia*) occur mainly in the

Plate 11. The *Cyperus papyrus* fringe of the main channel of the Bahr el Jebel. This monotonous fringe, some 5 metres high, extends for hundreds of kilometres along the river and blocks all view from it. The water hyacinth which usually forms the outer edge is absent here. Photo: Michael Lock.

Plate 12. Papyrus swamp interior. This is relatively open swamp with a lower layer of water hyacinth. It is, however, exceptionally tall for the area; the man shows the papyrus to be 4–5 metres high. Photo: John Goldsworthy.

Plate 13. Swamp vegetation in the centre of the *sudd*. The more substantial patches are *Typha domingensis*, surrounded by *Vossia cuspidata*. The patches of open water carry dense aquatic vegetation, mostly submerged. A large fire is burning in the background. Photo: Alison Cobb.

Table 7.1.*Swamp types and their associated environments and plants*

	Vossia	C. papyrus	Typha
Water depth	Variable	Deep	Shallow
Depth fluctuation	Large	Small	Small
Nutrient status	High	High	Low
Area (km²)	250	3,900	13,600
Associated species	*Eichhornia crassipes*	*Luffa aegyptiaca*	*Eichhornia crassipes*
	Utricularia inflexa	*Cayratia ibuensis*	*Spirodela polyrrhiza*
		Ipomoea rubens	*Lemna aequinoctialis*
		Vigna nilotica	*Ludwigia suffruticosa*
		Zehneria minutiflora	
		Melanthera scandens	
		Coccinia grandis	

hollows of the *gilgai* micro-relief,[8] at least in years of low flood like 1981 and 1982. The generally very level surface is broken by mounds as well as by hollows; these may be part of the *gilgai* pattern, or the bases of old termite mounds from an earlier period of lower flood levels. Such high spots carry grasses such as *Sporobolus pyramidalis* and *Hyparrhenia rufa*.

Records from the Nyany area (Table 7.2) suggest that *Oryza* grasslands flood when river discharges at Mongalla exceed 3.7 km³ per month. *Oryza* does not reach its maximum standing crop, and does not flower, unless it is deeply flooded for several months. Thus in late 1980 and early 1981 there was a standing crop of *c*.8 t ha⁻¹ and inflorescences were abundant. In 1981 and 1982 the flood was much lower, the standing crop hardly attained 1 t ha⁻¹, and no flowering took place; indeed in 1982 there was barely enough material to support dry season fires. In both these years, *Oryza* flowered in the deeper water of temporary pools, and in 1981 it flowered and reached 7 t ha⁻¹ farther south where the flooding was deeper and more prolonged. Plants of *Oryza* growing in deep water are always more vigorous and a deeper green colour than shallowly flooded or unflooded plants. The JIT (1954) found that *Oryza* occurred in sites flooded for 135–287 days each year to depths of 65–121 cm.

Oryza grasslands are burned in most years, usually early in the dry season, or as soon as the flood water has receded enough to allow burning. If moisture

Table 7.2. Mongalla discharges (millions m³ per month), and flooding of *Oryza* grassland at Nyany, 1980–2

	1980	1981	1982
Jan	3,810	3,538 +	3,007
Feb	3,376	2,990 +	2,352
Mar	3,942	3,194	2,356
Apr	3,350	3,390	2,250
May	4,085 +	3,243	3,038
Jun	3,870 +	2,980	3,069
Jul	4,424 +	3,252	3,007
Aug	4,521 +	3,619	3,720
Sep	4,030 +	3,840 +	3,180 +
Oct	4,052 +	3,957 +	3,813 +
Nov	3,970 +	3,630 +	4,080 +
Dec	3,851 +	3,385	3,689
Total	47,281	41,018	37,561

Notes: A + denotes that *Oryza* grassland near Nyany was flooded for all or part of the month.

remains in the soil, some regrowth occurs, but is very slow until speeded up by the first rains and standing crops remain below 0.2 t ha^{-1}. These rains also stimulate the germination of annual grasses such as *Digitaria debilis*, and small sedges. In years of early flood these may be drowned by the rising water, which stimulates rapid growth of *Oryza* that continues through the flood season. Flowering occurs in August–October, followed by seed shedding. Once the seeds are shed and water levels begin to fall, *Oryza* begins to dry out.

Although *Oryza longistaminata* is a perennial, it produces plenty of viable seed. At Nyany, abundant seed was produced in 1980; seedlings appeared in both 1981 and 1982 although no seed was produced in 1981. There is presumably a dormancy mechanism (as detected by Diarra (1978) in Mali) that prevents all seeds from germinating in the year after shedding.

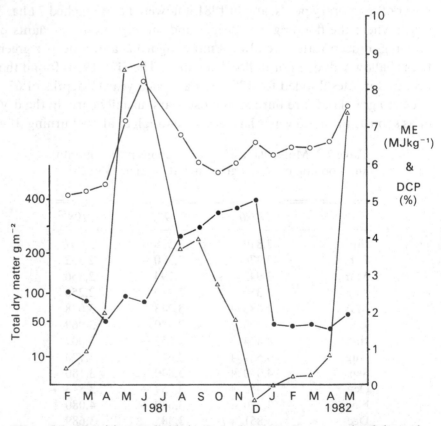

Figure 7.2. Nutrition available in *Oryza longistaminata* grassland through the year. (●) – total dry matter (gm m^{-2}); (○) – metabolisable energy (MJ kg^{-1}); (△) – digestible crude protein as percentage of dry matter. For methods of calculation, and comments on negative digestibilities, see note 9 (after RES, 1983).

Oryza grasslands provide high-quality grazing for much of the year, although in dry years the regrowth after burning is sparse and cannot provide enough bulk though it is rich in protein. Figure 7.2 shows the changes in total dry matter (standing crop), and in its energy and digestible protein content[9] in 1981–2. The figures do not make clear that the protein content of the dead material of *Oryza* was generally higher than that of other grasses, and was often eaten during the dry season.

Echinochloa pyramidalis grassland

These grasslands occupy sites farther from the river, and are thus flooded less frequently and less deeply than the *Oryza* grasslands. This statement is at variance with the conclusions of the JIT (1954), who found that *Echinochloa pyramidalis* occurred closer to the river than *Oryza*. However, *E. pyramidalis* exists in several forms in the area, not sufficiently distinct to rank as species; the great variability of the species is well-known (Clayton & Renvoize, 1982). Within the study area, two forms with distinct ecological preferences are important. A tall vigorous form (referred to for convenience as *E. pyramidalis* (W)) occurs close to the river, extending into the edges of papyrus swamps, and is presumably the form referred to by the JIT (1954). Another, smaller form is found on the landward side of the *Oryza* grasslands; it was referred to as *E. pyramidalis* (D).[10] In the Range Ecology Survey, the importance of *E. pyramidalis* (W) may have been under-estimated because of the low floods in the study years (1981–2). In one site where it was studied, it produced a very high biomass by the middle of the wet season, albeit with a high proportion of stem, and low in usable protein. Figure 7.3 shows the changes in biomass and available nutrients in this grassland during the year.

Echinochloa pyramidalis (D) occupies about 3,100 km². This grassland is flooded in years when river levels are high; as a very rough estimate, it would be flooded when river flows into the swamps at Mongalla exceed 4 km³ per month. On this basis, they would have been flooded for about four months in 1980, but not at all in 1981 or 1982 (see Table 7.2). It occupies land which is much affected by soil movements of the *gilgai* type, and sink-holes up to 40 cm deep and wide were frequent. Associated species include the grasses *Oryza longistaminata*, *Sporobolus pyramidalis*, *Digitaria debilis* and *Echinochloa haploclada*, and legumes such as *Desmodium hirtum* and *Cassia mimosoides*. In 1981 and 1982 the survey site (isolated from the river flood by a dyke since 1980) was flooded for several months by rain water. During the year there was a cycle of change in the composition of the standing crop, which is shown in Figure 7.4. It is tempting to surmise that this reflects this community's intermediate position; given more prolonged flooding, *Oryza* would increase;

given less, *Echinochloa* would probably give way to less flood-resistant species. Because these grasslands are never deeply flooded, and *Echinochloa* usually produces some regrowth in the dry season, they can be important year-round pastures. This is confirmed by Figure 7.5 which shows changes through the year in standing crop and nutritional value.

Seasonal pools

Within the *Oryza* grasslands (and, to a lesser extent, the *Echinochloa* grasslands) there are many small seasonally flooded pools. Most of these are associated with mounds, and can be interpreted as the hollows created when mounds were built for the construction of cattle-byres (*luaks*), abandoned since the floods of the 1960s. Such pools are small (usually less than 400 m²), and usually less than 1 m deep. Larger and sometimes deeper pools occur near present and former cattle camps; they may result from the long-continued

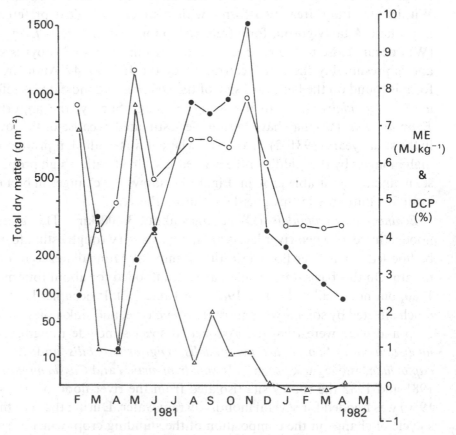

Figure 7.3. Nutrition available in *Echinochloa pyramidalis* (S) grassland (see note 10). Symbols as in Figure 7.2 (after RES, 1983).

removal of mud on the feet of men and animals, and in drinking water, or by deliberate excavation.

In the dry season, the dominant grass *Echinochloa stagnina* persists on the pool bed as stunted remnants, together with two small prostrate herb species, *Coldenia procumbens* and *Glinus lotoides*. In the early part of the wet season, *E. stagnina* makes some growth and tall semi-woody herbs such as *Sesbania rostrata, Aeschynomene indica* and *Hibiscus panduriformis* germinate and may form a dense stand up to 1 m tall. By the middle of the wet season, the pool becomes permanently flooded, either by heavy rain or by advancing river flood water. This kills the tall herbs. *Echinochloa stagnina* then forms a floating mat, often with *Centrotheca aquatica, Vossia cuspidata,*[11] and various submerged aquatic plants. After prolonged flooding, the floating mat declines in vigour, and floating and submerged aquatics such as *Nymphaea* spp., *Utricularia* spp., and duckweeds (*Lemna, Wolffiopsis, Wolffiella*) become abundant.

These pools also support a rich algal flora which, like the higher plants, passes through a series of development stages during the life of the pools (about 200 days in 1982–3) (SES, 1983). Soon after flooding, green colonial algae such as *Pandorina* and *Eudorina* (Volvocales) make up 80% of the volume of free-floating algae. By day 50 these are replaced by desmids (50–60%) and filamentous green and blue-green algae (10–20%). Dense associations of filamentous algae now form on the *Echinochloa* mat. In the final

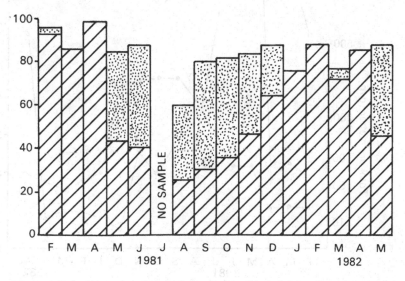

Figure 7.4. Changes in the species composition (by dry weight) in *Echinochloa pyramidalis* (D) (see note 10), grassland through the year. Dotted: *Oryza longistaminata*; hatched: *E. pyramidalis*.

stages, between days 150–200, the pools become concentrated by evaporation and enriched by cattle and birds which feed there at this stage. Euglenophytes (*Euglena, Phacus* and *Trachelomonas*) make up 50–75% of the algal volume at this time, sometimes with up to 20% of dinoflagellates (*Peridinium*). The high concentration of desmids in the mature stages of the pools is noteworthy; they are normally found in nutrient-poor waters, unlike these pools. In the Nile Basin they have been recorded in abundance only from the Bahr el Ghazal (Gronblad, 1962; Gronblad, Prowse & Scott, 1958).

These temporary pools are important to people and their livestock as a source of drinking water. Seasonally, they also produce a crop of fish, and they give good grazing late into the dry season.

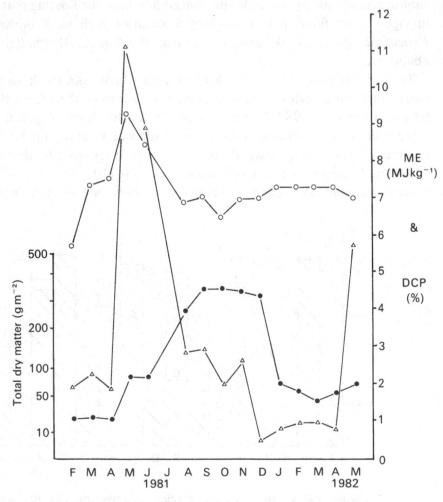

Figure 7.5. Nutrition available in *Echinochloa pyramidalis* (D) (see note 10) grassland. Symbols as in Figure 7.2 (after RES, 1983).

Seasonally rain-flooded grasslands
Heavily used grasslands with Echinochloa haploclada
In the Nyany area, there is a strip of land between the river-flooded and rain-flooded grasslands which is much settled by local people, mainly because it is, at least in parts, sufficiently well-drained for the building of *luaks*, and for cultivation, usually within hand-built dyke systems. The grasslands within this strip are therefore heavily used by any livestock which may stay in the *luaks* throughout the year; during the wet season it is often also grazed by the main herds. This grassland is both varied and patchy. Two sites were studied in detail; one lay in a short grassland with much *Echinochloa haploclada*; the other, discussed below, lay in *Sporobolus pyramidalis* grassland. Flooding of these grasslands is irregular, and is due to heavy rain and creeping flow.

The annual changes, on a dry weight basis, in this grassland show (Figure 7.6) that *Echinochloa haploclada* retains some green growth throughout the dry season, but that in the wet season *Oryza longistaminata* regrows from its rhizomes and becomes the most abundant species. Various species of *Cyperus* are also frequent in these grasslands. Nutritionally, this grassland is of high quality during the wet season because heavy grazing keeps the grass relatively short and in active growth, but during the dry season there is too little material to be useful (Figure 7.7).

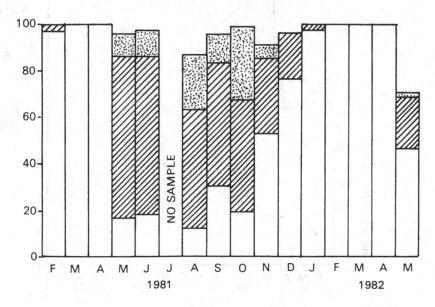

Figure 7.6. Changes in the species composition (by dry weight) in *Echinochloa haploclada* grassland through the year. White: *E. haploclada*; hatched: *Oryza longistaminata*; dotted: *Cyperus albomarginatus*.

Sporobolus pyramidalis grassland

Grasslands dominated by this tussock-forming species are found in a relatively narrow belt to the east of Nyany. Further north, the species is less widespread. Throughout its range in Africa, *S. pyramidalis* is a species characteristic of heavily grazed areas. This may be because it has a very strong root system, so that shoots are not uprooted by grazing animals (Lock, 1972), and also because the young shoot bases are protected inside the tussock base and are therefore not damaged by close grazing (Rabie, 1964).

In these grasslands, *Sporobolus* makes up more than half the biomass for most of the year, but during the dry season it often makes no regrowth, so that the few green shoots of *Echinochloa haploclada* are the only green

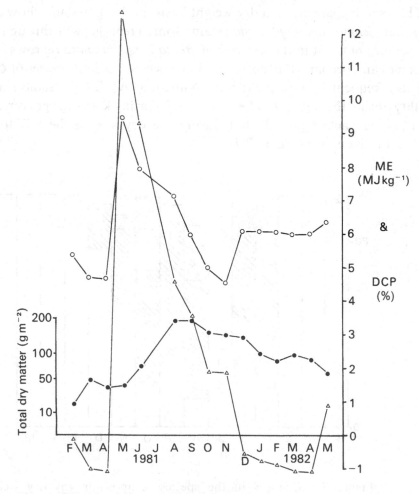

Figure 7.7. Nutrition available in *Echinochloa haploclada* grassland. Symbols as in Figure 7.2 (after RES, 1983).

material. Other common species include *Oryza longistaminata, Cynodon dactylon* and *Cyperus procerus.*

S. pyramidalis is never high in protein, and during the dry season nutrient levels in these grasslands fall well below those needed for maintenance (Figure 7.8). Cattle grazing here in the wet season could select the more nutritious species such as *Oryza. Sporobolus* plays a small but important part in the local economy, as its leaves are plaited into strong string used in hut building.

Hyparrhenia rufa grassland

Grasslands dominated by *Hyparrhenia rufa* cover about 23% of the study area – 15,800 km². They occupy level ground out of reach of river flooding, but are inundated by rain water for varying periods each year. *Hyparrhenia rufa* usually makes up at least 70% of the standing crop, but in some places *Sporobolus pyramidalis, Andropogon gayanus* or *Oryza longistaminata* may also be common. In the north of the study area there is a tendency, noted first by the JIT (1954), for *Hyparrhenia* to be replaced by *Setaria incrassata.* This species is rare in the Nyany area; the JIT worked on it extensively.

Figure 7.9 shows the changes in biomass and nutrients in this grassland

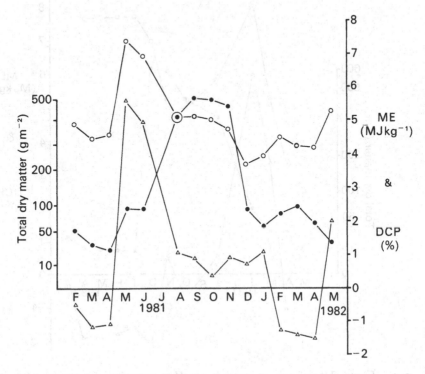

Figure 7.8. Nutrition available in *Sporobolus pyramidalis* grassland. Symbols as in Figure 7.2 (after RES, 1983).

type. Although a biomass of 6–7 t ha⁻¹ is attained by the end of the wet season, up to 90% of this is stem and contains little of value to animals. Only early in the wet season, and again when the grass regrows after an early fire, is there rather sparse grazing of any quality. A high proportion of the *Hyparrhenia* grasslands is burned each year (Table 7.3), mainly with the intention of stimulating regrowth. The quantity of regrowth depends on the soil moisture content at the time of burning; in exceptional cases, when the soil and vegetation are so moist that the fire leaves a large unburned residue, there may be a second fire late in the dry season.

The high unused biomass at the end of the wet season often leads to the impression that these grasslands represent a vast underutilised resource; Figure 7.9 shows, however, that they are very poor in nutrients at this time, most of the standing crop being leafless stem with a mean protein content of 1.85%. Traditionally, these grasslands have been used in the early dry season, and again in the early wet season (see Chapter 12). Traditional use thus depends on using regrowth as and when it is available. Early wet season

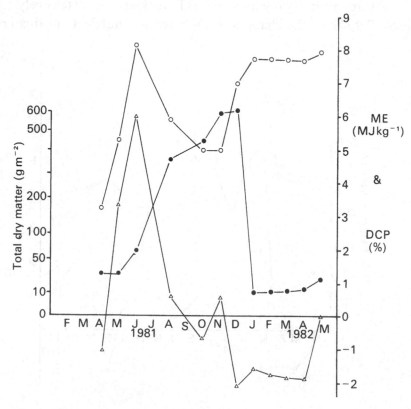

Figure 7.9. Nutrition available in *Hyparrhenia rufa* grassland. Symbols as in Figure 7.2 (after RES, 1983).

regrowth sometimes contains as much as 19% crude protein. Dry season regrowth appears to have less protein (9.4%), and this fell to 4–6.5% as the regrowth dried and reddened.

It is often contended that cattle could be supported on the Eastern Plain for much of the dry season on a diet of *Hyparrhenia* regrowth. Both the Dutch consultants (Ilaco, 1982) and the JIT (1954) investigated this possibility. Both found numerous problems and concluded that it is not a practical proposition. First, the quantity of regrowth is very variable from place to place – in April 1981 it was found that it varied from 2–18 g m² along a 60 km stretch of the Bor–Pibor road (Ilaco, 1982). This is clearly insufficient for maintenance, since although it may be high in protein, there is insufficient bulk. The JIT (1954) noted that cattle being driven home in the evening after grazing would snatch at clumps of tall dry grass, apparently to bulk up their intake. In the

Table 7.3. *Burning and regrowth in Hyparrhenia grassland at the three aerial surveys.*

(a) Burn percentage

	Seasons		
	Mid wet	Early dry	Late Dry
0%	89	42	0
10–25%	0	12	1
25–50%	0	14	16
50–90%	0	12	30
90–100%	0	9	42

Total squares surveyed: 89

(b) Grass greenness

	Seasons		
	Mid wet	Early dry	Late dry
All burnt	0	nd	31
Totally dry	0	nd	58
10% green	0	nd	0
10–25% green	2	nd	0
25–50% green	19	nd	0
> 50% green	68	nd	0

late dry season aerial survey (RES, 1983) not a single 10 km square was scored even as 10% green, and reconnaissance flights at the height of the dry season often showed no trace of regrowth except in a few hollows and dry watercourses.

The *Hyparrhenia* grasslands are the major local source of thatching material. This often conflicts with their use for grazing as, if regrowth is to be grazed, burning should be done early, in October–November, while the grass is not really dry enough to be cut as thatch until late November or December.

Woodlands

Woodlands in the Jonglei area are of several kinds. Most are effectively stands of single species but in the Bor area, and along the Duk Ridge, there are patches of more mixed woodland. None has a sufficiently complex structure to merit being called forest.[12]

Single species woodlands

Of these, the most extensive are dominated by *Acacia seyal* (5,400 km^2) and *Balanites aegyptiaca* (5,300 km^2). Other woodlands are dominated by *Acacia fistula* (mainly in the north), *Acacia polyacantha* (edges of the river-flooded grasslands north of Kongor), *Piliostigma thonningii* (Eastern Plain), and *Acacia drepanolobium* (Eastern Plain). None of these occupies more than 550 km^2 (less than 1% of the study area).

Acacia seyal is a small tree, rarely exceeding 10 m in height, with conspicuous orange bark. It occurs mainly in the northern parts of the study area. Most of the patches seem to be more-or-less even-aged, suggesting episodic mass regeneration. The JIT (1954) attributed this to a series of dry years leading to sparse grass stands which, in turn, by reducing or eliminating fire, allowed tree regeneration. The species is much used locally for firewood, charcoal and building poles, although it is not very durable because of attack by wood-boring beetles (Sudan Government, 1953).

Balanites aegyptiaca is a larger tree than *Acacia seyal*, often attaining 15 m in height. It forms more open woodlands, widespread in the study area, in which the crowns of adjacent trees rarely touch, so that grass growth below is greater. *Balanites* sometimes forms mixed woodlands with *Acacia seyal*. It extends farther into seasonally river-flooded grasslands than other tree species, but clearly cannot withstand prolonged flooding, as trees on land inundated by creeping flow impounded east of the canal were dying in 1982–3. It is, however, widely recorded in Africa from sites subject to seasonal flooding (Eggeling & Dale, 1952; Letouzey, 1963; Suliman & Jackson, 1959).

Balanites is widely used in the area for firewood and for building poles; this

Plate 14. Woodland to the east of Ayod. The large tree is *Balanites aegyptiaca*; the others are *Acacia seyal* and *Acacia polyacantha*. Photo: Michael Lock.

small timber is often obtained by lopping mature trees which regenerate adequately if the lopping is not too severe. The fruit pulp is used to make a drink, or eaten. The plant is much browsed where it is accessible; the leaves are rich in protein and even fallen ones are eaten. Small plants, and shoots low on the trunk, are generally ferociously spiny and this appears to give them some protection. Regeneration seems to be sparse, and many stands appear more-or-less even aged.

Mixed woodlands

These are found in the south, near Bor, and along the Duk Ridge farther north. The southern woodlands are very open and patchy, with their structure often being determined by the *gilgai* hummock-hollow microtopography. A thicket clump, with or without trees, often occupies the top of the mound, and more flood-tolerant species such as *Acacia seyal* and *Zizyphus mauritianus* occur on the pediment. The hollows may be free of trees, or may contain *Balanites*. Other areas, with less well-developed microtopography, carry an open woodland of *Combretum fragrans*. These woodlands can be regarded as the impoverished northern outliers of the much richer savanna woodland which occurs on the well-drained red soils of the ironstone plateau south of Mongalla.

Other mixed woodlands occur along the Duk Ridge but have not been investigated. *Celtis integrifolia* is the most prominent large tree species.

Factors limiting vegetation distribution and productivity
Water
Water and plant distribution

The strongly seasonal and variable rainfall of the region (see Appendix 2) certainly affects the vegetation of the study area, but it seems unlikely that the rather small differences in rainfall regime within it can be responsible for the observed vegetation pattern. However, the way in which the rainwater accumulates on the surface, causing flooding and creeping flow, and the depth and duration of flooding caused by river spill, both vary greatly, and can explain most of the features observed. The flooding regimes tolerated by each vegetation type are discussed in the accounts above.

Water and plant productivity

Plants require water for growth; during the year, four phases of water availability can be discerned in the region. During the dry season, evaporation and transpiration lower the soil water content to wilting point and beyond.

At this time, soils over much of the area crack deeply; such cracks may disrupt root systems. At the end of the dry season, there may be isolated light showers, but these are usually insufficient to penetrate the soil and stimulate plant growth. Furthermore, soil temperatures are high at this time because the soil surface has no protective vegetation cover, and water loss is rapid.

As the wet season asserts itself, rain showers become more frequent and heavier, and rain begins to infiltrate into the soil. Grass growth now begins, but because cover is still low, and soil temperatures high, water loss is rapid and periods of accelerated growth tend to alternate with spells when growth is limited by water scarcity. Small differences in water availability caused by local microtopography can make a big difference to grass growth at this time; *gilgai* depressions and the edges of soil cracks provide relatively favourable environments. As water enters soil cracks, they become sealed by the swelling of the wetted soil. Once this has happened, water begins to accumulate on the soil surface during heavy rains, because it can then only enter the soil by infiltration, which is extremely slow.

In the third phase, water accumulates on the soil surface and creeping flow begins (see Appendix 3). Flooding with river water often also begins at this time. Once the soil surface is flooded, there is little likelihood that plant growth will be limited by water shortage, but, because gases diffuse only very slowly into waterlogged soil, oxygen may be in short supply at the roots. Ions such as iron II (Fe^{2+}) and manganese II (Mn^{2+}) may reach toxic levels in the soil solution. In these circumstances, species such as *Oryza*, which have well-developed aerating tissues, are likely to thrive.

Once the wet season ends, the floodwater disappears by infiltration, evaporation and transpiration, and the soil begins to dry out. Relatively little of the water in the soil can be used by plants (see below), which are also at their maximum size and leafiness so that the vegetation quickly exhausts stored water, wilts, dries and is often burned.

This account of the annual cycle of water availability in the grasslands of the region might lead to the conclusion that added water during the dry season might allow year-long growth and hence greatly increase productivity. The JIT (1954) experimented with dry season irrigation of various grasses, including *Hyparrhenia rufa*, but found that production was disappointingly low. Since they note that there was a marked increase in growth with the start of the rains, it may be that insufficient water was provided. Water demand during the dry season would be very large, as is shown by the evaporation figures (Table 5.1), and such schemes are unlikely to be economic.

Soils
Water availability from sands and clays
The soils of the region are described in Appendix 1. An aspect that may be relevant to plant distribution is the relationship between soil particle size and water availability. Sandy soils hold water less strongly than clay soils, so that more of the water they contain is available to plants. This difference was used by Smith (1949) to explain many features of tree distribution in the drier parts of the Sudan, and may help to explain the presence of relatively dense mixed woodlands on the sandy soils of the Bor area and the Duk Ridge.

Soil nutrient levels
The soils of the region appear to be low in the three most important plant nutrients, with potassium being rather less deficient than phosphorus or nitrogen. Lack of rain is often blamed for the low productivity of tropical grasslands, but a recent detailed study in Mali has demonstrated that only where annual rainfalls are below 200 mm is grassland production limited by water supply; elsewhere, shortages of nitrogen and phosphorus are the main limiting factors (Penning de Vries & Djiteye, 1982). The present low levels of nitrogen may be due in part to long-continued burning, as any nitrogen in herbage when it is burned is lost to the atmosphere. Those authors have, however, demonstrated that sufficient nitrogen returns in rain and dust, and through fixation by micro-organisms, to maintain low but constant soil nitrogen levels. Phosphorus is not lost in fires. It may, however, accumulate in cattle camps, as dung is burned leaving the phosphorus in the ash which is not returned to the grassland. Another essential element, sulphur, is lost in smoke during burning of vegetation; some returns in rain. Any that falls into swamps can be anaerobically reduced to insoluble sulphides which are effectively lost to the system. In Uganda it has been suggested that swamps act as sulphur 'sinks' (Chenery, 1960) and that there is a steady loss of sulphur from grassland to swamps. In Uganda, grass yields can sometimes improve after addition of sulphur (Wendt, 1970).

Fire
Causes and frequency of fires
Perhaps as much as 90% of the dry land vegetation is presently burned each dry season. While occasional fires may be started by lightning, or possibly by spontaneous combustion in the swamps, most are deliberately or accidentally lit by local people. They are usually started to clear tall coarse growth and stimulate regrowth to provide food for cattle in the dry season, but other fires

are started for hunting, by charcoal burners, and by cattle raiders and their pursuers. Local edicts may prohibit early burning in certain areas to preserve thatch supplies, but in most places fires are lit as early in the dry season as possible to get the best regrowth from residual moisture in the soil. As a rule, little lasting damage is done to grasslands by burning; the grasses are well adapted to fire, with the growing points well protected just below the soil, or in the middle of dense tussocks; many have awned seeds which bury themselves in soil cracks (Lock & Milburn, 1971).

Fires and woody vegetation
Because many woody species are vulnerable to fire when young, and need several consecutive years without fire to regenerate, one of the most important aspects of fire is its role in influencing the balance between grassland and woody vegetation. Long-term experiments in Nigeria and Zambia have compared the effects of this balance of regular early burning (at the beginning of the dry season), late burning (at the end of the dry season) and complete protection (Charter & Keay, 1960; Trapnell, 1959). In both sites, not far from the boundary between forest and savanna, both early burning and complete protection tended to encourage woody vegetation to the extent that burning became difficult after 10–15 years. Late burned plots, on the other hand, tended to lose their woody vegetation as mature individuals were killed and not replaced. Neither of these experiments is strictly applicable to the Jonglei area, as both were sited where rainfall was higher and the dry season shorter. A better comparison is with an experiment in northern Ghana in which, after 30 years of protection or early burning, no canopy closure had taken place and no fire-tender species had appeared (Brookman-Amissah *et al.*, 1980). While it is likely, in the climate of the Jonglei area, that exclusion of fire would encourage tree regeneration, such a policy would be impossible to enforce, and would be likely to lead to widespread destructive late burning which would reduce regeneration. Early burning, which experiments suggest gives some help to tree regeneration, and which is the current practice over much of the area anyway, is the most satisfactory policy.

Plant diversity in the Jonglei area
Why are there so few plant species?
An important point to be made about the vegetation of the Jonglei area is that it is poor in species. During the Range and Swamp Ecology Surveys about 350 species of higher plants were definitely identified. There is no doubt that further coverage of the Eastern Plain, the Duk Ridge, and the northern fringes of the area, particularly in the wet season, would add many

species to the list. However, even if the number was doubled to 700, that is not many for a tropical region of 67,900 km². Queen Elizabeth National Park (Uganda) has perhaps 1,400–1,500 species recorded in an area of 1,980 km² (Lock, 1977), and the Parc de Kivu (formerly the Parc National Albert) in Zaire has 1,960 species recorded from an area of 9,080 km² (Robyns, 1947–55).

The Jonglei area is poor in species because it has a low habitat diversity – there are rather few vegetation types, each covering a huge area. It is also virtually devoid of relief – no hills or rocks to support unusual species. It is also a harsh environment, with extremes of drought and flood, intense fires and poor soils. Similar areas elsewhere, however, are richer. Is it possible that the poverty of species points to the whole region being a young one in geological terms, so that there has been insufficient time for a rich flora to develop by immigration or evolution? In support of this speculation, there is only one endemic plant, the remarkable swamp grass *Suddia sagittifolia* (Renvoize *et al.*, 1984). On the other hand, the 10–11 km of sediment reported to exist beneath the area (Adamson & Williams, 1980) would require a lengthy period of deposition and thus a long history of swampy conditions. We know, though, that for an indefinite period ending some 10,600 years ago neither Lake Victoria nor Lake Albert had an outflow, so that inflows to the Sudd must have been greatly reduced (Livingstone, 1980). But did it dry up completely? The answers to this and many other questions of vegetational history may lie in the analyses of bore-hole data – yet to be released by the oil exploration companies.

The Jonglei area compared with other African floodplains

There are a number of other extensive floodplains in Africa. To the north of the Equator, there is the Inland Delta of the Niger, and the basin of Lake Chad. To the south, there is Lake Rukwa with its surrounding grasslands, the Bangweulu Swamps, and the Kafue Flats. The important features of each of these are given in Table 7.4.

Of these African swamplands, the Kafue Flats probably provide the closest environmental parallel to the swamps of the Nile. The Inland Delta of the Niger, and the plains around Lake Chad are both situated in very much more arid regions than the Nile Swamps. Rainfall is less, the dry season longer, and evaporation higher. Lake Rukwa is a rift valley lake with no outlet, and is saline. Much of its surrounding grasslands are thus heavily influenced by salts and some of the grass zones, such as those dominated by *Sporobolus spicatus* and *Psilolemma jaegeri*,[13] are more typical of the rift valley lakes of East Africa than of other African floodplains. The Bangweulu Swamps are fed by

Table 7.4. *Characteristics of floodplain ecosystems in Africa.*

	Jonglei (Sudan)	Bangweulu Basin (Zambia)	Kafue Flats (Zambia)	Rukwa Valley (Tanzania)	Niger Delta (Mali)
Annual rainfall (mm)	850	1220	780	780	200–900
Soil type	Black clay	Alluvium (various textures)	Black clay	Black clay	Mostly clays
Total area of wet season floodplain (km^2)	30,000	730	6,240	520	11,500
Dry season	Dec–May	Jul–Dec	Apr–Sep	May–Sep	Oct–May
Wet season	Jun–Nov	Jan–Jun	Oct–Mar	Nov–Apr	Apr–Nov
Grassland; above-ground biomass	200–2,100	60–240	100–600	nd	500–1,200

Notes: 1. All clips at soil level except Bangweulu; clips there at water level.
2. All five sites are burned regularly, and all support fisheries. All except Bangweulu support transhumant populations who move in search of grazing.

Source: Bangweulu: Grimsdell & Bell, 1975.
      Kafue: Rees, 1978.
           Sheppe & Osborne, 1971.
      Rukwa: Vesey-Fitzgerald, 1970.
      Niger Delta: ILCA 1981a.

16 rivers and have only a single outlet. The Kafue Flats lie along a section of the Kafue River, a tributary of the Zambezi; they are flooded by spillage from the river at high water, like the floodplain of the Bahr el Jebel.

In spite of the rather large differences in the environment, the zonation of grasslands around these African floodplains is remarkably similar. Species such as *Vossia cuspidata*, *Cyperus papyrus* and *Typha* spp. surround the central areas of deep open water. These are in turn surrounded by a zone which is seasonally flooded and which supports swamp grasses such as *Echinochloa stagnina*, *Oryza longistaminata* and *Echinochloa pyramidalis*. The outermost grass zone, flooded for only a short time each year, is very often made up of *Hyparrhenia rufa*.

Records of change in other floodplains of Africa are sparse, but changes have certainly occurred and most probably parallel the very extensive changes that have taken place in the Nile floodplain (see Chapter 8). Lake Chad, for example, is now (1986) rising after dropping to very low levels. Lake Rukwa certainly fluctuates considerably from year to year, and may also show longer-term oscillations of mean level (Vesey-Fitzgerald 1969). Flooding in the Kafue Flats varies with river discharge.

To conclude, the Bahr el Jebel floodplain is an environment whose harshness and relative uniformity is reflected in the small number of vegetation types and few species that are found there. It is, however, not unique, and comparisons with other African floodplains can be useful in understanding its complexities, and in predicting the changes that may take place in the future.

Notes

1. Replacement time – the time needed for the waters of a lake to be completely replaced by inflow and outflow.
2. Periphyton – the plants and animals that adhere to the surface of submerged aquatic plants.
3. It has been suggested (Chadwick & Obeid, 1966) that this replacement was due to the waters of the Nile being more suitable for the growth of *Eichhornia* than of *Pistia*. It seems much more likely that, in most waters, the growth forms of the two species account adequately for the replacement. *Pistia* has no leaf-stalks, and so cannot raise its leaves in crowded conditions. *Eichhornia*, on the other hand, can develop long leaf-stalks in crowded conditions, and would shade out *Pistia*. *Pistia* persists in temporary pools, where it seeds freely and thus survives the dry season. It is also found within the swamps, in sites remote from the river, in 'old' water (see Chapter 5).
4. Freidel (1979) estimated the area covered by water hyacinth in the Sudan as 111.7 km^2. This is in striking contrast to the figure of $3,000$ km^2 given by Obeid (1975) (quoted by Pieterse 1978). Obeid's figure was used by Pieterse to calculate an annual water loss from the Nile of 7.12 km^3 – about a tenth of the annual flow, but in the light of Freidel's estimate this figure is clearly much too high.
5. 2,4-D = 2,4-dichlorophenoxyacetic acid
6. *Toic* (pronounced 'toich'). River-flooded grassland which provides dry season grazing. The term is used by the Dinka to embrace river-flooded grasslands – *Oryza*, *Echinochloa pyramidalis* (S), and *Echinochloa stagnina*, as well as wetter grasslands of *Leersia hexandra* and *Vossia cuspidata*. To some extent, grasslands may be referred to as *toic* early in the dry season, but not later, when they are completely dry. It is also sometimes used of the main swamps, but is not so used here.
7. *Oryza longistaminata* is the species referred to by the JIT (1954) as *Oryza barthii*.
8. The term *gilgai* is fully explained in Appendix 1.
9. As part of the Range Ecology Survey (RES, 1983, 3, 123) the commoner grasses of the Nyany area were collected and dried at monthly intervals for analysis of their nutritive value. The full results of these analyses are given in RES, 1983. The two parameters plotted in the diagrams cannot easily be measured directly, but can be calculated using regression equations. For digestible crude protein (DCP), the equation derived by Milford & Minson (1965), using tropical Australian grasses, was used. It is based on the analytical figure for 'crude protein' (CP), which is itself estimated by multiplying the nitrogen content of the sample by 6.25. The equation is:

$$DCP = 0.899CP - 3.25$$

This equation is very similar to one derived by Glover *et al.* (1975) from less extensive comparisons of the relationship in East African grasses. Where the analysed crude protein figure is less than 3.62%, the DCP percentage is a negative number. Whilst a negative percentage is a mathematical impossibility, such figures indicate that the food concerned contains no digestible protein, so that the animal must inevitably lose weight through drawing on its own reserves of body protein.

The parameter 'metabolisable energy' is a measure of the energy value of the food to the animal. This value depends on two factors – the percentage of the material eaten that can be digested, and the calorific value (energy yield) of the material digested. Fats and oils (lipids) have the highest value. Because the analyses were carried out by two laboratories which used different schemes of analysis, it was necessary to use two equations. For material analysed at Kuku (Sudan), the equation

$$ME = 14.3 \times 0.017\, CP - 0.019\, CF$$

was used. (CP = crude protein; CF = crude fibre). For material analysed at Newcastle (UK), the equation used was

$$ME = 0.15\, IV \text{ (where } IV = in\ vitro \text{ digestibility)}$$

The results from the two methods appear broadly comparable but it appears that the Newcastle method may give higher energy values for dead material and for late wet season living material, perhaps because protein is often much lower, and fibre much higher than in the hays for which this formula was originally derived.

10. *Echinochloa pyramidalis* (D) – representative specimens at Kew – Harrison 1051 and Goldsworthy 81/284.
    Echinochloa pyramidalis (W)– representative specimens at Kew – Goldsworthy 81/97.
        The two forms were probably distinguished by the JIT (1954) using Nuer names; *E. pyramidalis Dutyang* (D), and *E. pyramidalis Reil* (W).

11. *Vossia cuspidata* exists in two forms in the area. The one in the seasonal pools is small, lacks irritant hairs on the leaf sheaths, and has only one or two spikes in the inflorescence (see Lock 82/63 at Kew). The swamp form is much larger, has irritant hairs on the leaf sheaths, and has 7–10 spikes in each inflorescence (see Lock 82/58 at Kew).

12. The definitions of vegetation types used here follow those of White (1983).

13. *Psilolemma jaegeri* was formerly known as *Odyssea jaegeri*.

8

Vegetation change in the Jonglei area

Introduction

There is a tendency to look upon vegetation as unchanging, having developed under conditions which have remained constant for centuries. As we have seen in Chapter 4, however, variations in the discharge of the Nile have been very large even in the present century and there is every reason to suppose that such variations have often occurred in the past. Thus the vegetation of the area has been continually subject to changes in flooding regime, and in this chapter we shall consider the effects of fluctuations in river discharge on vegetation during the present century. The first part deals with changes over the whole area; the second part treats a small area, the Aliab Valley, in more detail, because it was examined thoroughly in 1951–2 and again, albeit briefly, in 1982.

Useful written records of the vegetation of the region begin about the beginning of this century. Garstin (1901, 1904) includes some notes on the vegetation in his massive survey of the river. Broun (1905) travelled to Bor in a sailing boat in 1905, collected specimens, and wrote an account of the vegetation. The maps prepared by the Sudan Survey in 1928–30, by ground survey and aerial photography, include notes on the vegetation and also provide an accurate boundary for the swamps and for lakes and channels within them. The JIT (1954) described and mapped the vegetation. Since the mid-1970s the area has been covered by satellite imagery, conventional vertical aerial photography and by low-level oblique aerial photography.[1] These together were used as the basis of a vegetation map (RES, 1983, 10 Map 3).

The gross pattern of vegetation

Vegetation maps of the area can be drawn for three dates. Figure 8.1 uses the maps prepared for the Sudan Survey in about 1930; Figure 8.2, that of the JIT (1954) for 1952, and Figure 8.3, based on maps prepared by the

Range Ecology Survey (1983) represents the situation in 1980. Of the three, the last is undoubtedly the most accurate, being based on numerous aerial transects as well as ground traverses and satellite imagery. The first is very vague, as it has been based only on the sparse and scattered notes recorded on the map sheets, some of which date from much earlier than the map itself.[2] It

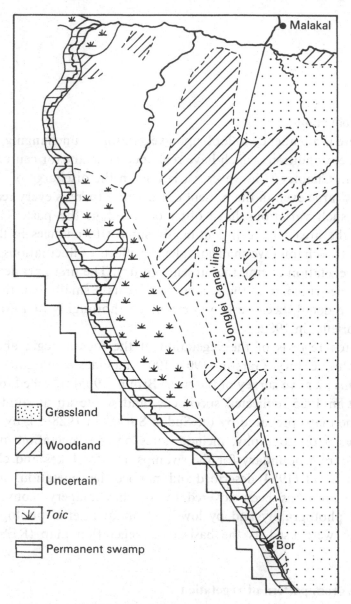

Figure 8.1. Vegetation of the *Sudd* Region in 1930. From information contained in the Sudan Survey maps. All boundaries, except that of the permanent swamp, very approximate (after RES, 1983).

is probable, however, that the boundaries of the swamps are accurate in the earliest map, but they are less certain in the Jonglei Investigation Team map, as this was based mainly on river transects.[3]

The maps show very great changes between 1952 and 1980, but lesser changes between 1930 and 1952. The vegetation types mapped are gross types

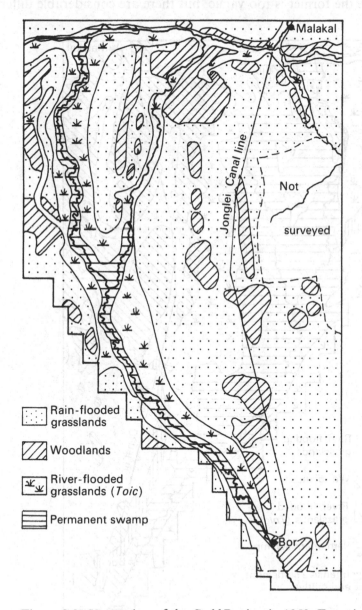

Figure 8.2. Vegetation of the *Sudd* Region in 1952. From information in JIT (1954). The area of permanent swamp is probably under-estimated (see text, p. 175), (after RES, 1983).

(for details see Chapter 7); these, and their components, are now considered in more detail.

Woodlands

No useful comparisons can be made between the maps of 1930 and 1952, because the former is too vague, but there are considerable differences

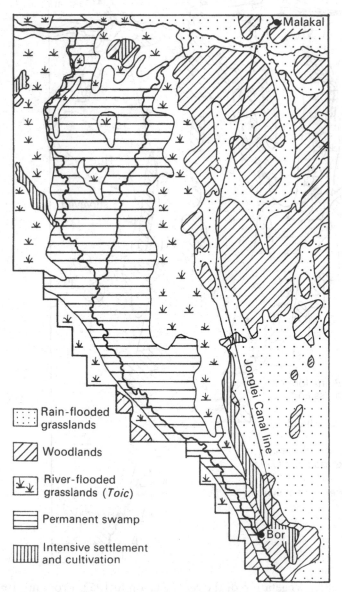

Figure 8.3. Vegetation of the *Sudd* Region in 1983. From ground and air surveys and LANDSAT imagery (see RES, 1983).

in the extent of woodlands shown in the 1952 and 1980 maps. The former shows woodland extending almost continuously from south of Baidit to north of Jonglei village, and the accompanying text states that it extended as far north as Kongor with another line of wooded country farther east. Older residents of this area confirmed in 1980 that the 1952 map was correct, with Jalle, Lilir and Maar all formerly standing in dense *Acacia seyal* woodland, where now there is only open grassland. There are, however, some sites where stumps and coppice shoots of large trees persist, mainly on elevated sites such as termite mounds.

In the north, the 1952 map shows Woi lying within *Acacia/Balanites* woodland. There are now patches of young *Acacia polyacantha* and *A. seyal* woodland nearby but no mature woodland. Local residents reported that many trees had been killed by floods in the early 1960s. In the extreme north the maps give little hint of change, but it is reported that there was formerly much *A. seyal* woodland near Fangak and the Zeraf jebels; this has now disappeared.[4] In the south, the boundary between the woodland and grass-

Plate 15. Aerial view of Jonglei village in 1982. The rise in river levels has all but inundated the site, which stands on an artificial mound within papyrus swamp. The huts are built mainly from papyrus stems. Clumps of sugar cane stand at the edge of the village, and the channel connecting it to the river is visible at top left. Photo: Alison Cobb.

land on the Bor–Pibor road is in the same place in all maps and there is generally little evidence for change.

Why the change? Clearly the floods of the early 1960s must have been responsible for much. Although both *Acacia seyal* and *Balanites* can resist seasonal waterlogging, they both succumb if flooded for longer. This process of deforestation by flooding could be seen by 1982 in places where water was becoming ponded against the east bank of the canal at its northern, partially completed end, within two years of the creation of the embankment. Likewise the lack of change east of Bor, where there is no river spill eastwards, points to the river as a factor. A further cause of woodland retreat may be cutting for firewood and building timber – a resource which must have been scarce after the floods – at a time when extensive resettlement must have increased the demands on it (see Appendix 10). It must be assumed that dead trees would quickly be destroyed by fire in the dry season.

Grasslands

To the east of a line running south along the Duk Ridge, and then through Kongor to Bor, there is little evidence for change. All maps show this as *Hyparrhenia rufa* grassland (although the species is not given in the 1930 map). These areas, being rain-flooded and thus above normal river-flood heights, would not be expected to change in response to alterations in river level. Use of the area by the local people has traditionally been seasonal (see Chapters 10 and 12), with the wet season flooding and the lack of drinking water in the dry season effectively giving the grasslands two annual rests from grazing.

To the west of the line defined above, changes are more obvious. From Jonglei northwards to the Khor Atar, the 1952 map shows a vast area of *Hyparrhenia rufa* grassland, and this is confirmed in the text. In 1980 virtually all of this area was occupied by *Oryza longistaminata* and *Echinochloa pyramidalis*, both species of seasonally river-flooded sites, with *Hyparrhenia* virtually absent except on patches of higher ground. Older local people confirm that *Hyparrhenia* formerly extended very much farther west than it does now. It would seem, therefore, that there has been a real and extensive spread of river-flooded grassland since the early 1960s.

A final change in the grasslands which deserves mention is an apparent increase in *Sporobolus pyramidalis* in a belt extending south from Kongor. JIT (1954) make no mention of the species except on the Zeraf island, where it occurred on rather higher sandier ground with *Hygrophila auriculata*, in areas where wet season grazing was intense.[4] Both *Hygrophila* and *Sporobolus* are often reckoned to be indicators of overgrazing (Lock, 1972), and it may be that the belt south of Kongor has become infested with *Sporobolus* as a result of restriction of the people and their cattle to a narrow belt of land between the river-flooded land to the west and the rain-flooded grasslands to the east.

Swamps

On a gross scale, comparison of the 1930 and 1952 maps suggests a slight contraction of the swamps over the period, but the difference is small and probably not significant. However, between 1952 and 1980 there has been a large increase in the area of the swamps, from 6,700 km^2 to 19,200 km^2. This is undoubtedly linked to the floods of the early 1960s and the subsequent increased discharge volumes. The distribution of each swamp type may now usefully be considered.

Lakes and pools

There is rather more information on the distribution of lakes than for other components of the swamp system, because a detailed map of the river channel and associated lakes exists, made in 1901 (Garstin 1901). Additionally, aerial photography allowed lakes to be mapped accurately in the 1930 surveys. Changes have occurred throughout the system. North of Adok, Garstin's map shows large numbers of lakes which are absent from the 1930 map. There can be little doubt that these lakes disappeared as a consequence of the clearance of *sudd* blockages from the main channel between 1901 and 1904. Garstin (1901) even notes that 'lagoons drained into the river as a result of the removal of blocks.' It is noteworthy also that an earlier map (Watson 1876), made at a time when the channel was clear, shows almost no lakes at all.

In the southern part of the swamps, the number of lakes increased greatly between 1930 and 1980, with the entire Wutchung lake system being new (SES, 1983) (see Figure 8.4). The timing of the appearance of lakes suggests that rises in river levels, caused either by obstructions or by greater flow, are a cause of lake formation. Lakes could be formed either by the drowning of large patches of vegetation, or by fragments of a floating mat being detached and carried away by the current. Fragments of evidence suggest the latter; one pair of aerial photographs (see Figure 8.5d) show a change of this kind, and many cores taken from the Wutchung lakes, which have appeared since 1930, showed no trace of drowned vegetation mats.

It is also to be expected that lakes should be infilled by vegetation and disappear through normal phases of plant succession. Material produced by plants and sediment trapped among their roots both accumulate and raise the lake bed so that it is eventually colonised by swamp vegetation. In the short term, there is little evidence of this happening. Aerial photographs taken in

1975 and 1980 (Figure 8.5a) show that some smaller lakes do not change at all. Others show changes that can be explained by movement of floating

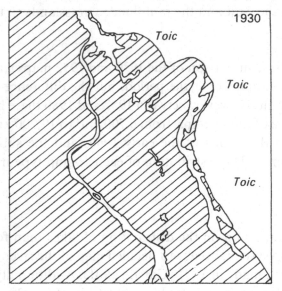

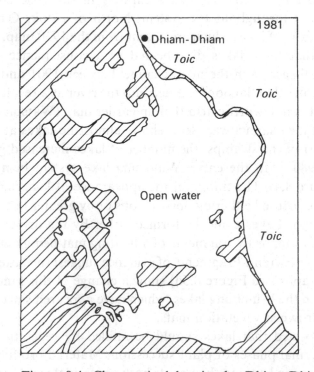

Figure 8.4. Changes in lakes in the Dhiam-Dhiam – Wutchung area, south of Jonglei, between 1930 and 1981. Shaded areas represent swamp vegetation.

vegetation mats (Figure 8.5b, c), while yet others enlarge, or lose a floating vegetation cover. Only where river channels enter lakes is there rapid development through the deposition of sediment in deltas (Figure 8.5e). Small lakes may change slowly because very little nutrient matter can enter

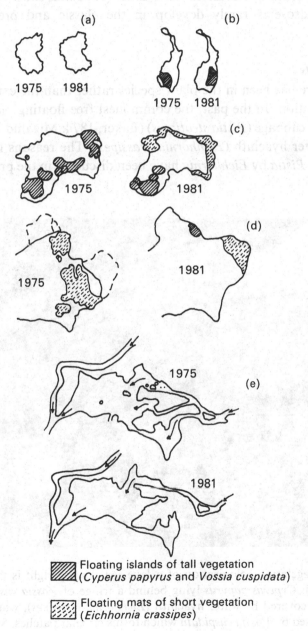

Floating islands of tall vegetation
(*Cyperus papyrus* and *Vossia cuspidata*)

Floating mats of short vegetation
(*Eichhornia crassipes*)

Figure 8.5. Changes in pools, lakes and internal deltas in the southern part of the *Sudd* between 1975 and 1981. For discussion, see text.

them, so that the development of a floating mat of papyrus or *Vossia*, which must contain a substantial nutrient store, is greatly slowed. However, available observations cover only a very short period, and it must be remembered that such a natural succession is likely to be overtaken by changes in the river discharge. Furthermore, Walker (1970) has shown that in the British Isles hydroseres rarely develop in the classic and predicted fashion.[5]

Flowing river channels
The main change here has been in the plant species rather than the extent or distribution of vegetation. In the past, the commonest free-floating plant on the river was the Nile cabbage (*Pistia stratiotes*) (Baker, 1874; Migahid, 1947), but now it is the water hyacinth (*Eichhornia crassipes*). The reasons for the rapid replacement of *Pistia* by *Eichhornia* have been discussed in the previous chapter.

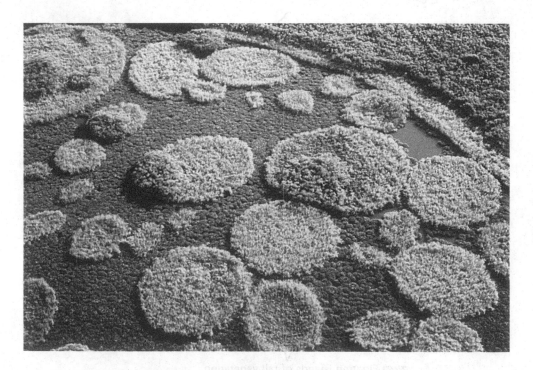

Plate 16. Vegetation colonisation of a lake. At the top right is the lake margin, with *Cyperus papyrus* lying behind a fringe of *Vossia cuspidata*. The lake is covered by water hyacinth (*Eichhornia crassipes*), which has been colonised by *Vossia cuspidata* which forms circular patches. Some of these have in turn been colonised by *Cyperus papyrus*, itself forming darker circles. Photo: Michael Lock.

There are many references to the blockage of the river by vegetation (the word *sudd* itself means 'blockage'). The history of these and the clearances

Table 8.1 *The history of blockage and clearance of the Nile in the Sudd**

Date	Event
pre-1863	Bahr el Jebel open. Baker travels from Khartoum to Gondokoro (near Juba) in 40 days in 1862.
1863–8	Bahr el Jebel blocked south of Lake No.
1870–1	Bahr el Jebel and Bahr el Zeraf still blocked. Baker, however, manages to struggle through the upper Zeraf after great difficulties.
1872	Both channels closed.
1874	Bahr el Jebel cleared by Ismail Ayoub Pasha. Gordon travels from Khartoum to Gondokoro in 25 days.
1878	An unusually high flood once again leads to blockages.
1878–80	Bahr el Jebel cleared by Marno.
1899	After the establishment of Condominium rule, the Bahr el Jebel is found to be blocked south of Lake No. Peake commences cutting in December 1899, and clears Blocks 1–14 (which was near Adok) by April 1900.
1905	Clearance completed with the final removal of Block 15 south of Adok; originally this was an aggregate of blocks some 35 km long in all. Between 1900 and 1905 it was bypassed via a channel to the west through lakes and small waterways (the False Channel).
1905–65	Channel generally open, although the site of Block 10 continued to pose occasional problems; Garstin was briefly trapped there in 1901. Other blocks occurred occasionally; in the autumn of 1926 a blockage 8 km long occurred above Shambe, and in 1926 another blockage in the same region near Lake Papiu, 7 km long, caused the mail boats to divert via the Atem and Awai Rivers.
1969	Blockages between Bor and Shambe cause the abandonment of the western channel past Kenisa, and the establishment of the Atem as the main navigation channel. This is now sometimes known as the Bahr el Jedid (New River). This blockage was probably caused by a combination of high water levels and infrequent steamer passage due to the problems of the civil war.

Source: * Compiled from Garstin (1901, 1904); Hope (1902); Broun (1905); Sculthorpe (1967) and JIT (1947).

Note: No similar history can be reconstructed for the Bahr el Zeraf; it has probably been blocked more frequently. It was traversed by Migahid in 1943 (Migahid, 1947), and was blocked in 1980 although the southern end had been reopened for oil prospecting work.

that removed them are summarised in Table 8.1. It seems likely that the main channel remained open after 1900 because of the passage of steamers, which dislodge fragments from the river banks before they become dangerously large, and also may report minor blockages which can then be cleared before they build up and consolidate. The future of *sudd* blockages is discussed in Chapter 16 below.

Swamp vegetation types

Comparison of the 1930, 1952 and 1980 maps shows very great changes also in the gross area of swamp. The area in 1952 was about 6,700 km^2; in 1980, about 19,200 km^2. It is more difficult, however, to decide how the relative proportions of the major plant communities of the swamps (see Chapter 7) have changed. The main species concerned are *Vossia cuspidata, Cyperus papyrus* and *Typha domingensis*. Their habitat preferences were summarised in Table 7.1

The evidence for change in the distribution of these species is as follows. In 1874 (Watson, 1876) and 1904 (Garstin, 1904) the boundary between papyrus

Plate 17. A *sudd* blockage of the main channel of the Bahr el Jebel. This block appears to be composed largely of water hyacinth, and is unlikely to persist for long. More substantial blocks are formed by *Vossia cuspidata* and *Cyperus papyrus*. Photo: Alison Cobb.

and 'grass swamp' is marked between Shambe and Kanisa; Watson does not mark a boundary, but there is a note that just south of Kanisa 'the ambatch and papyrus have altogether disappeared'. At the present time, continuous papyrus extends almost to Bor, and discontinuous and still abundant papyrus reaches Malek. Garstin (1904) and Grogan (1901) describe the Atem as flowing through grassland. It now flows through papyrus throughout its length and its bank can only be reached from the land at one or two places. Thus papyrus has extended southwards, displacing 'grass swamp', which was composed of *Vossia cuspidata*.

JIT (1954) describes *Typha* as follows: 'Very often forms small areas of pure stands but is not a major constituent; it is often found in pure stands up to a feddan in area in the riverain swamp pastures.' The present distribution is very different. *Typha* swamps cover some 13,600 km². The major areas lie north of Jonglei and east of the Bahr el Jebel, extending north to Buffalo Cape and Fangak, often in a virtually continuous expanse up to 80 km wide, broken only by the channels of the major rivers. They occupy areas mapped by the JIT as 'riverain grassland' and '*Hyparrhenia rufa* grassland'. It seems unlikely that the JIT missed all the huge area of *Typha*, although they did not cover some of it, and they had to estimate the extent of the swamps from the river. However, they surveyed levelled lines across some of the area and overflew parts of it, and it must be concluded that the apparent vast expansion of *Typha* is real.

To summarise: the area occupied by *Vossia* has moved southwards; the area occupied by papyrus has done likewise, and there has been an enormous expansion of *Typha*. The difference in area between the swamps in 1952 (6,700 km²) and 1980 (19,200 km²) is 12,500 km²; the area of *Typha* swamp in 1980 was 13,570 km², so that *Typha* more than accounts for the increase in area. There is in fact no real evidence for any overall increase in papyrus, only for a change in its distribution. Very similar conclusions can be drawn from the comparisons between the vegetation of the Aliab Valley in 1951 and 1982, described in more detail below.

These changes in vegetation distribution can be accounted for in terms of the relationship betwen river level and the consequent patterns of overspill. This was dealt with in more detail in Chapter 5, but is outlined briefly here. The slope of the river surface and of the banks are not parallel, that of the banks being slightly steeper. This is demonstrated in Figure 8.6. As one proceeds downstream, a point is therefore reached where the banks dip below the river surface level, and overspill begins. This overspill point will move upstream and downstream as the height of the river rises and falls during the year, and will also move upstream in response to persistent high discharges and downstream in response to low ones. Thus, as a result of the persistent high discharges that have pertained since the 1960s, spilling now takes place much farther upstream than before, and species such as papyrus, which must be permanently wet, have moved upstream.

To the north of Jonglei, a large amount of water spills eastwards and never returns to the main river, being lost instead by evaporation and transpiration from the swamps. Since the river is now much higher than before 1960, much more water must spill in this way, accounting for the much greater proportion of water now lost completely from the system.

The behaviour of the main swamp constituents can now be considered in relation to the changes outlined above. *Vossia*, due to its ability to withstand strong currents and great fluctuations in water level, and its ability to form floating mats, tends to occur at the head of the swamp where overspill is seasonal. Thus its movement upstream is not unexpected. It also appears to have a high nutrient requirement, which would be fulfilled at the inflow end of the swamps.

Cyperus papyrus is a species of permanently inundated sites. The pattern of vigour of the species, with a marked decline in stature from south to north in the swamp at the river margin, and also a decline in vigour away from the

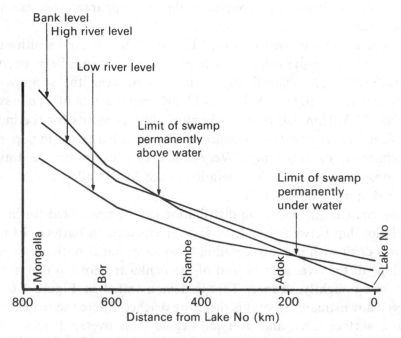

Figure 8.6. The relationship between bank height within the *Sudd*, and high and low river levels (after Migahid 1947).

river, suggests that it is dependent on nutrients carried by the water, which are progressively depleted. Ions showing depletion along the river transect (Chapter 6) are phosphate and nitrate, among others. Both are important to plants. The loss of nitrate is to some extent counterbalanced by a gain in ammonia, but not all plants can absorb the ammonium ion. Papyrus normally grows as a floating mat over water of some depth so that it can adjust to changes in water level if not anchored. Papyrus rhizomes contain few air spaces (Tadros, 1940a, b) and may therefore be susceptible to oxygen deficiency if flooded. This would explain the restriction of the species to sites where there is sufficient water depth for a floating mat to be formed throughout the year; in shallower sites the mat would become attached to the substrate at low water levels and the rhizomes would 'drown' when levels rose. Whatever the underlying cause, the rise in water levels has allowed papyrus to extend its range upstream.

Typha has expanded more than any other species as a result of the floods. It tends to occur remote from the river in sites where it is probably firmly rooted in the substrate, in shallow water which is likely to be of low nutrient concentration. This type of habitat has increased greatly as a result of the rise in river levels and the increase in the proportion of water which leaves the river and fails to return. It is worth mentioning that in other parts of Africa *Typha* swamps are normally associated with sites of internal drainage, with alkaline water and with more dissolved material than the more acidic waters occupied by papyrus (Howard-Williams & Walker, 1974).

Changes in the Aliab Valley

A part of the extreme southern end of the swamps, the Aliab Valley, was surveyed in detail during the work of the JIT (1954). The survey was carried out in 1951 and involved the cutting of 15 transects across the valley. These were surveyed and levelled, the vegetation was recorded along them, and soil samples taken. In the report of the JIT, the cross-sections are presented, together with generalised contour and vegetation maps. In 1982, the area was resurveyed from the air by flying along the original transect lines at a height of 150 m and, once again, recording vegetation boundaries and the positions of lakes and channels. The investigator of the area in 1951 also took part in this 1982 data collection.

Topography and flooding in 1951

The floodplain between Juba and Bor is several metres below the wooded ground on each side, and widens steadily from four to 10 km. The main river channels swing from one side to the other of the margin of the floodplain and enclose a series of basins between the river and the wooded banks. At the lower end of each basin there is a connecting channel where upstream spillage drains back into the main river. The Aliab Valley is one such basin.

The hydrology of these basins and their relationship to the main swamp system north of Bor was discussed in Chapter 5. The floodplain in the Aliab reach is restricted by the banks on each side, thus preventing the total loss of large amounts of water as happens in the main swamps farther north. In spite of this possible problem, the detailed work on the relationships between topography and vegetation carried out there in 1951 makes it an exceptionally valuable site, some of the changes in which will now be described.

Between 1951 and 1982 increased river flows after the rise in Lake Victoria have altered spill channels and the levels and range of flooding. The observed response of the vegetation confirms that the system is sensitive to changes in flooding regime.

A floodplain cross-section (Figure 5.7) presented a complex contrast between the high alluvial banks of past and present main channels and the beds of drainage channels. However, the Aliab topography is simplified by a longitudinal profile of river bank, floodplain and river levels. These levels diverge; the river levels have the shallowest gradient, then the river banks, while the floodplain levels are steepest because of sedimentation.

For a given flow the river level crosses the bank at some point; downstream the bank is submerged while upstream a number of spill channels pierce the bank – 370 in the Aliab in 1952. The increase of spilling with river flow depends on the geometry of the bank and spill channels, and will only change dramatically when the flow regime alters.

At the upper end of the valley, water leaves the river through spill channels and flows north in drainage channels. At the lower end, where the bank is submerged, widespread flooding spreads laterally across the valley. This is complicated by the high banks of the Aliab, itself a former river channel, which inhibited lateral flow and split the valley into isolated basins (see Figure 5.7).

However, because of the topography the flooding has a greater vertical range at the top of the basin than at the lower end, where the range corresponds to the river level. During floods the inundation profile is parallel to the river profile, while at low flows flooding is more nearly horizontal. Thus flooding can be visualised as rotating vertically about a hinge at the lower end of the basin.

Although approximate areas of flooding in the Aliab could be estimated from the elevation at which river levels rose above the river bank, more precise estimates of the volumes and depths of flooding required analysis of inflow and outflow measurements for a given reach. This was not possible for the Aliab as much of the flow bypasses Bor. However, comparison of inflow and outflow for the reach below Mongalla made it possible to relate the volume of flooding to river flow. Because level and vegetation transects were available, and the vegetation revealed the flooding profile, the elevation of flooding throughout the basin could also be related to Mongalla inflow (Sutcliffe 1957, 1974).

Vegetation in 1951

The vegetation surveys along transects were reduced to percentage distribution at different level bands for whole transects and for kilometre segments. The pattern was broadly similar, with *Echinochloa stagnina* occupying the lower levels, with increasing proportions of *Vossia cuspidata* and *Cyperus papyrus* at the downstream ends of the basins. These species dominated the lower end of the Aliab. The higher levels of each transect were occupied by *Phragmites karka* and *Echinochloa pyramidalis*. The boundary between the first group, classified by the JIT as deep-flooded species, and the second group, known as shallow-flooded species, was revealed to be clear-cut in terms of elevation. The level of this boundary could be plotted against distance from the river to show the lateral water slope, or parallel to the river to give the longitudinal slope.

The longitudinal slope of the vegetation boundary is parallel only to the river bank; it was therefore possible to deduce that the maximum depth of flooding controlled the main boundary, and comparison of the boundary surface with the normal volume of flooding revealed that the critical depth was about 120 cm. On the other hand, the limited distribution of papyrus in the area could be most easily explained by the hypothesis that a range of flooding of about 150 cm, rather than the maximum depth, controls its distribution.

These controls, together with the pattern of flooding and the topography, explain the dominance of shallow-flooded species in the top end of the reach and of each basin, and the prevalence of deep-flooded species at lower elevations throughout the sections and their dominance in the lower end of the reach and the basins, including the Aliab. The concentration of papyrus in the lower end of each basin and the Aliab is explained by the rotation of flooding about the lower end of the basin and the increased range resulting from this. In addition, the distribution of *Vossia cuspidata* is distinguished from other deep-flooded species by its ability to withstand high water velocities, so it appears in areas of spilling. The dominance in 1951–2 of *Oryza longistaminata* to the west of the Aliab was distinctive but less easily explained in terms of hydrology; the area was fed by local runoff.

Hydrological changes between 1952 and 1982

As described in Chapters 4 and 5, the Lake Victoria rise in 1961–4 doubled the base flow at Mongalla without affecting the variable component; the mean annual flow at Mongalla rose from 26.8 km^3 in 1905–60 to 50.3 km^3 in 1961–80.

As a result of the rise in Lake Victoria, the river levels near Mongalla rose initially to a profile well above the alluvial banks of the river (Figure 8.7), and this led to new spill channels developing and existing spill channels growing into major river channels. Analysis of Mongalla rating curves showed that the increase in flows caused a rise in levels between 1960 and 1963, but that levels then fell by about 1 metre for a given flow until by 1967 a new rating curve had been reached which had changed little by 1980. This change was associated with a fall in mean bed level of about 1 metre in 1963–4 because of the scouring effect of the floods. Thus the valley adjusted to some extent to the increased flows in this way and by the expansion of spill channels; the change in flooding was more complex than a simple extrapolation of the earlier regime.

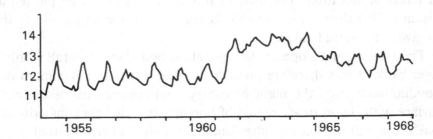

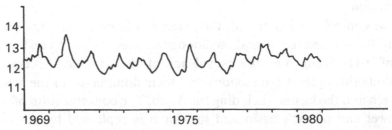

Figure 8.7. Monthly mean levels of the Bahr el Jebel at the Mongalla gauge. The level of the left bank at Mongalla is 13.8 m on the gauge; the right bank farther downstream is almost certainly much lower.

The initial rise in river levels was thus reduced by 1966 by the fall in bed level and increased spillage. Although spill channels below Mongalla which were initially navigable had tended to silt up by 1982, a mass of channels was observed threading through the basin. As a result the outfall channel near Gemmeiza which had been almost stagnant in February 1952 was flowing strongly in February 1982.

Five new channels carry this concentrated flow from below Gemmeiza across to the west, where it spills into, and through, the Aliab basin in a maze of large open channels observed in February 1982 (Figure 8.8). Although the present spill channel exits have grown from cuts in the river bank recorded in 1951–2, the contrast with the flooding of March 1951 is remarkable. In that dry season there was virtually no water in the Aliab valley south of Malek, and although there was standing water farther north, its low level showed that it was fed from the main river near Bor. The river Aliab offtakes had already silted up and carried no spill in 1951–2; it is tempting to date this channel to the previous high flow period of 1875–95.

Because the extra flow from 1961 has been diverted through the Aliab floodplain, the main hydrological contrast between 1951–2 and 1982 is not that the maximum flooding is much higher in this reach but that the minimum each year also spreads through the Aliab valley. The protection provided by the alluvial banks of the main river has been destroyed, and the rotation of the flooding about the lower end of the valley, which exaggerated the range of inundation within it has been replaced by flooding spreading from the upper end. The combined effect of the changes will have been to increase the duration of flooding through the reach from Juba to Bor, to increase the maximum and minimum levels, but above all to decrease the range and to increase the flow velocities down the floodplain. In other words the flooding lasts longer, varies less between seasons, and flows faster than 30 years previously.

Changes in vegetation

The vegetation changes in the Aliab above Bor between 1951–2 and 1982 are illustrated in Figure 8.8. The most obvious change is the spread upstream of *Cyperus papyrus* and *Vossia cuspidata*. Whereas in 1951 papyrus was confined to a narrow belt north of Malek, it has now spread over most of the lower end of the valley and towards the river south of Malek. There are now isolated areas of papyrus at the upper end of the valley near the main spill channels. This area was always to some extent a separate basin, with a surface organic horizon indicating prolonged flooding; this will have

194   *The Jonglei Canal*

increased, but the close connection with the river through the enlarged spill channels will have decreased the range of flooding.

The main extension of *Vossia cuspidata* is to a wider band around the spill

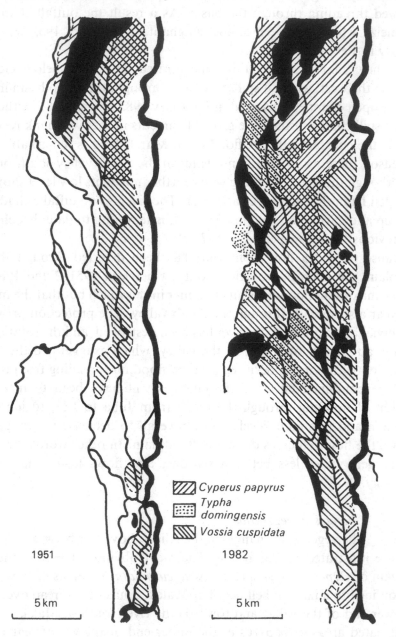

Figure 8.8. River channels, lakes and vegetation in the Aliab Valley in 1951 and 1982. 1951 map derived from JIT (1954); 1982 map after RES (1983).

and drainage channels and especially into the area to the west of the Aliab above Minkaman which was previously protected from flooding by the high banks of the Aliab. The depth and duration of flooding will have increased particularly in this area, while the velocities of flow will have increased throughout the year; its prevalence confirms the tolerance to flow velocities of *Vossia cuspidata*.

Of the shallow-flooded species, *Oryza* has migrated from the sheltered basin above Minkaman to a narrow band near the forest below Tombe. On the other hand, *Typha* has appeared in a number of similar locations remote from the river.

Phragmites appears to have spread from the banks of the main river and the Aliab to cover a wider band near the river. In fact it was present in 1951–2 over 10% of the sections as far north as Malek, though limited to higher elevations (Sutcliffe, 1957, Figure 31). Because it was only then mapped where dominant or near-dominant, the spread may not have been as great as appeared from the air. The decline of *Echinochloa pyramidalis* may also have been exaggerated by the difficulty of recognising it from the air, but where it used to be present over most of the valley, it now appears to be mainly confined to the higher ground near the river.

These changes of the vegetation over a 30-year span are compatible with greater depth and duration of flooding over the valley, an increase in the area whose range of flooding is limited by proximity to spill channels linked directly to the river, and increased velocities of flow down the whole valley.

Similar changes have occurred in the Mongalla basin, where papyrus and *Vossia* have spread upstream from Lake Buri near Gemmeiza, while *Echinochloa pyramidalis* remains present at the upper end of the basin near Mongalla. Above Mongalla, where in 1952 grazing used to be limited by the dominance of *Phragmites*, large numbers of cattle were observed to be grazing *Echinochloa pyramidalis*, with *Oryza, Phragmites* and *Vossia* also present.

The effects of change

These changes, first in the pattern of flooding and secondly in the distribution of plant species, have in turn had their effect on land use. Many traditional dry-season cattle camp sites and grazing areas on the west side of the valley have been flooded and consequently were not used in 1982; nevertheless, those on the high bank of the Bahr el Jebel were still occupied in 1982, in precisely the same locations as they had occupied 30 years previously.

Most of the floodplain species of wild mammal, such as buffalo and tiang common in the Aliab valley in the early 1950s, had been edged out by the increased flooding and changed vegetation. The only animals able to thrive under the new conditions were hippo, and elephant, of which some three hundred were present in the valley throughout the 1982 dry season, evidently more than the small numbers seen in 1951 (Sutcliffe, 1957).

The land use changes may be placed in a wider context. There is evidence that in the period 1875–95 Lake Victoria levels were as high as recently and therefore Bahr el Jebel flows would have been much higher than in 1951–2 and perhaps comparable to 1982. In 1874 the Tombe channel was the main river (Lyons, 1906) and spill conditions could have been similar to the present. In this early period the Bari are reported to have owned large herds of cattle, but the discrepancy between these reports and the lack of grazing was discussed in JIT (1954). Sutcliffe (1957) speculated that the high flows of the late nineteenth century and changes in river profile might have been responsible for a change in vegetation. The recent change in vegetation above Mongalla and the return of cattle, albeit Dinka cattle, for grazing in the area makes this more likely and illustrates the variability of the system.

Conclusion

In fact the main conclusion of this chapter must be that the interaction between upstream hydrology or lake levels, local topography and patterns of flooding, vegetation and land use is dynamic rather than static. The mechanics of the interactions may not have changed but the results of a rise of East African lake levels have demonstrated more clearly than any static description could have done, how topography and hydrology determine the vegetation.

Notes

1. The aerial photographs were taken for the following clients:
   1975 – Chevron Oil Co.
   1977 – Mefit SPA (for Southern Regional Development Plan)
   1981 – Executive Organ of the National Council for Development of the Jonglei Canal Area (by Geosurvey Ltd, Nairobi).
2. Some of the notes on the sheets are clearly copied from the map prepared by Liddell (1904) – *Geographical Journal* 24, 651–5, map p. 708.
3. P. Howell, personal communication, 1985.
4. P. Howell, personal communication, 1985.
5. The classic hydrosere, or succession from open water to dry land, passes through the following stages: open water; submerged plants; plants with floating leaves; plants with emergent leaves; reedswamp; flood-tolerant scrub; wet woodland; dry woodland.

Part III

The people of the Jonglei area

Introduction

The Jonglei Area

By 'Jonglei area' is meant that part of the southern Sudan occupied by people who will in one way or another be directly affected either by the actual physical presence of the canal passing through their territory or by ecological changes along the natural channels of the White Nile system caused by its operation. The extent of this area has never been precisely defined; indeed, it would be difficult to do so with accuracy until the canal is operational and its effects have been monitored for some years.

The basic map (see Figure 10.1), covering an area of 67,000 km², which features here and in various contexts throughout this volume, is taken from aerial surveys carried out on the instructions of the Jonglei Executive Organ in 1981–2. This does not correspond with what is now called Jonglei Province; for administrative purposes that incorporates also Akobo and Pibor Districts which will not be directly affected. Moreover, the area arbitrarily delineated by these aerial surveys does not include all the territory of those Dinka and Nuer living west of the Nile, but whose cattle are driven in the dry season to its floodplain (see Figure 12.1). There also seems good reason to suggest that boundaries of the area should extend farther northwards to incorporate more of the Shilluk and at least some of the northern Dinka (see also Figure 10.2).

Tribes and tribal territories

We cannot concern ourselves here with the complexities surrounding the definition of the word tribe in segmentary societies such as those to be found in the Jonglei area. We use the term simply to mean groups or associations of people having a common name in common use, occupying and claiming rights in identifiable territories.

The boundaries shown in Figure 10.2 are those which, though they had changed substantially over the previous two centuries or more, as Chapter 9 will relate, were eventually stabilised by the Condominium Government and used to define territories for administrative purposes, the limits of the

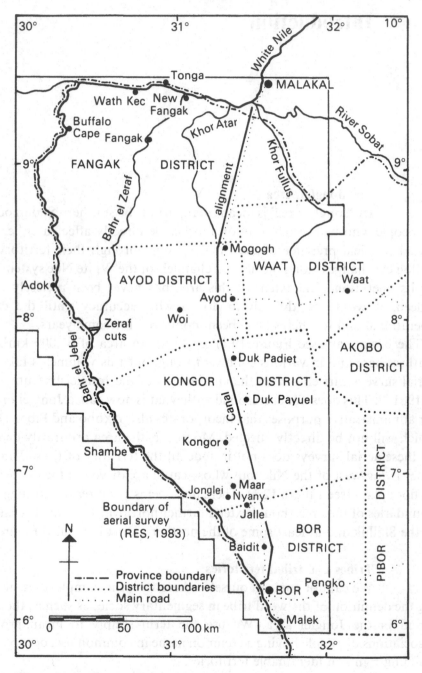

Figure 10.1. The Jonglei area.

jurisdiction of Chiefs' Courts, and later those of the executive responsibilities of local government units. Despite massive population movements away from flooded areas during the 1960s, particularly eastwards from the Zeraf Island, it appears that the original inhabitants would, if challenged, continue to claim collective rights of tenure within them. In theory at least, recent migrants are

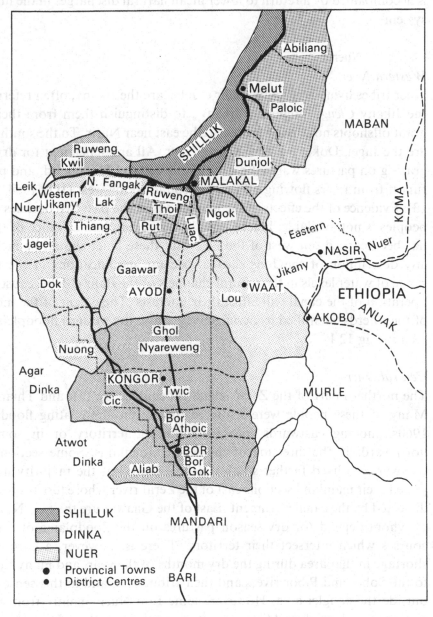

Figure 10.2. Approximate territories of peoples affected by the Jonglei Canal.

there on sufferance and by agreement, though weight of migrant numbers as well as kinship affiliations may have in many cases made their acceptance mandatory. Little information is available on this subject, but it is one that would repay research because similar movements may occur in reverse when flooded areas are reduced by the drainage effects of the canal, especially if this is accompanied by a return to lower mean natural discharges in the main river system.

Nuer

Western Nuer

Nuer tribes living west of the Bahr el Jebel are the Jikany, often referred to as the Jikany *Cieng*, 'homeland Jikany', to distinguish them from their immigrant offshoots now living farther to the east near Nasir. To the south of them are the Jagei, Dok, Aak and Nuong Nuer. All are dependent for dry season grazing on pastures watered partly by the Bahr el Jebel flood, and partly by runoff from rivers flowing off higher ground to the west (JIT, 1954, pp 212–13). Evidence of the effects of the high floods of the past two decades on these peoples is not available; indeed, hardly any research in advance of the canal has been carried out west of the Nile. But these Nuer should be included in any description of the Jonglei area, because modifications in the seasonal range of water levels in the natural channels of the Bahr el Jebel caused by the operation of the canal will affect their interests. The extent of the movement of their herds, identified by aerial survey, on to the western floodplain will be seen in Fig 12.1.

Central Nuer

The northern part of the Zeraf Island is occupied by Lak and Thiang Nuer. Many of these people were scattered during the devastating floods of the 1960s, moving eastwards into neighbouring territory or in some cases northwards in the direction of the Nuba mountains. Some sections of the Gaawar also lived farther south on the Island, but the majority have now joined their mainland sections east of the Zeraf river whose territory is already dissected by the canal alignment. East of the Gaawar are the Lou Nuer, many of whom depend for dry season pastures on the floodplains of the watercourses which intersect their territory. There is, however, a chronic water shortage in that area during the dry months of the year, and many Lou move to the Sobat and Pibor rivers and their tributaries, and in this sense are well outside the Jonglei area. However, some Lou Nuer, mainly from the Gun section, have long shared the *toic* pastures of their southern Dinka neighbours

Plate 18. A young Dinka in traditional dress (beaded corset and colobus-skin kilt) with his 'song-bull'. The horns of the bull are asymmetrical due to artificial distortion of the pattern of growth, regarded as increasing the attractiveness of the animal. Photo: John Goldsworthy.

by agreement, though some conflict over scarce resources or the threat of disease sometimes arises.

Eastern Nuer
The Eastern Nuer will not be directly affected by the canal, but they are frequently mentioned in the text. They are those Nuer, mainly Jikany but joined by other Nuer immigrants as well as Dinka accretions, who moved eastwards during the last century, and now occupy territory, centred on Nasir, up to, and in some cases, across, the Ethiopian border.

Dinka
Dinka immediately north of the Nuer in the Jonglei area are the small Thoi, Rut, Luac and Ruweng tribal groups. Many are bilingual in Dinka and Nuer, and all have close relations with their Nuer neighbours through proximity, intermarriage and the sharing of natural resources. A small pocket of Rut Dinka live in the middle of Gaawar Nuer country, tending the shrine of Luak Deng (see Howell in Lienhardt, 1961). This is of religious importance to Nuer as well as to Dinka farther south, and in the case of the latter may have been a focus for political cohesion among them now largely lacking.

Southern Dinka
The bulk of the Dinka likely to be affected by the canal live to the south of the Nuer, immediately to the east of the Bahr el Jebel, and we refer to them as the Southern Dinka. In Kongor District, bordering the Gaawar Nuer, are the Ghol and Nyareweng Dinka. Their wet season settlements are along the Duk ridge, a sandy, better drained series of outcrops of higher land which runs from Mogogh in Nuer country south to Duk Payuel in Dinka country. South of the Nyareweng are the Twic, a much larger group centred on Kongor itself. Beyond them, in Bor District, are the Dinka who give that district its name, the Bor Athoic and the Bor Gok.

Western Dinka
More than two-thirds of the Dinka people live west of the Nile. The majority will not be affected. However, reports relating to the present Jonglei canal scheme make little mention of the Cic and Aliab Dinka, who live to the west of the Bahr el Jebel in territory south of the Nuong Nuer. Yet some of their cattle move on to dry season *toic* grazing along the western side of the main channels (RES, 1983) which is bound to be affected by changes in the river regime. No mention is made either of the Kwil and Ruweng Dinka living

north and west of Lake No, some of whose cattle used to be brought to riverain pastures dependent on spill from the Bahr el Jebel, though these are areas where ecological conditions may well have changed since the floods of the 1960s.

Northern Dinka

Other Dinka who may be directly, though perhaps not greatly, affected by changes in river levels are the Dunjol, Paloic and Abialang who occupy territory stretching from north of the lower Sobat area as far as Renk. Many make use of riverain pastures along the east bank of the White Nile, but in most years have alternative dry season grazing on the edges of the Machar Marshes and along the Khors Wol and Adar. Their country is, moreover, much better suited to rain grown crop production than their southern neighbours, and is one of the few Nilotic areas where there is usually a substantial surplus of grain for export. Ngok Dinka living both sides of the lower Sobat may also be marginally affected.

Shilluk

Only those Shilluk living in the extreme south of the country and around Malakal are usually included in the Jonglei area, yet the rest, occupying the narrow strip of territory northwards along the west bank of the White Nile, are no less likely to feel some effects from the altered hydrological regime the canal project will bring about. Not only may seasonal flooding take place at a higher level, but in the dry season much of the present floodplain of the river and the numerous islands which are inundated at high discharge levels may never be exposed (see Chapter 16). The precise effects downstream of the canal exit have yet to be determined. Although the Shilluk are sedentary and less dependent upon livestock, the Nile with its seasonal fluctuations is also important for their extensive fisheries and, indeed, their limited but expanding small scale irrigated gardens along the river.

Other peoples

Though they are now in what is called Jonglei Province the Anuak and Murle inhabitants of Akobo and Pibor Districts will not be directly affected and are not included in the Jonglei Area, but it would be unwise to assume that they will not encounter indirect effects. There will, however, be no hydrological and consequent ecological changes in the Pibor river system, and Murle do not have to cross the line of the canal to reach dry season pasture, while the Anuak are in any case sedentary. The Murle might be subject to pressures from their Nuer or Dinka neighbours nearer the Nile, a process which occurs in reverse at present, if the latter suffer from any grave diminution of essential land resources. Already affected by pressures from Bor Dinka in search of dry season grazing are the Mandari, and south of them the Bari. Such pressures have been intensified by the recent increase in mean discharges and changes in ecological conditions in Bor District described earlier. If the operation of the canal reduces dry season grazing resources the pressures will increase.

Human population

Estimates of the population of the *Sudd* region have always presented special difficulties; figures calculated at different times over the last 50 years must be regarded with extreme caution. In particular it would be wildly misleading to attempt any calculation of population trends from comparison of these figures. Various estimates are given in Appendix 6, as well as some explanation of the special difficulties of accurate census work in highly mobile societies such as those of the Nilotes in this area. During air surveys conducted in 1981 dwelling houses (Arabic *tukl)* were observed and mapped according to density. These are shown in Figure 10.3 since they indicate the areas of wet season occupation which also relate to the distribution of higher, better drained land to which the people retreat at that time of year. The map also corresponds broadly with the wet season distribution of population illustrated in the SDIT (1955) report. Although the two maps are not strictly comparable there is evidence of the retreat from villages occupied prior to 1961 but inundated since then in, for example, the southern part of Zeraf Island, as well as indications of greater wet season concentrations east of it.

The bias of research in the social sciences – and indeed in other fields – in the context of the canal and its impact has been predominantly towards Bor and the southern Dinka. Unfortunately less attention was paid to the Nuer farther north, and none at all to people west and north of the Bahr el Jebel who will also be affected by changes in the natural regime of that river system. Unfortunately, too, socio-economic research and survey work has, with certain exceptions (see, for example, El Sammani and Kadouf, and the reports of Dutch Consultants in this field), been lacking in depth and time-scale. An attempt to provide a comprehensive picture of interrelationships between different economic trends, social change, and the movement of peoples was not successfully undertaken when the opportunity was there. There is a particular lack of reliable quantitative data relating to economic processes. This has been a major difficulty in presenting the findings recorded in Chapter 11. The almost complete lack of recent work among the Shilluk has necessitated their omission from that chapter. Space has also necessitated a degree of generalisation in the presentation of this account, which masks not only the complexities of Nilotic society but differences between Dinka and

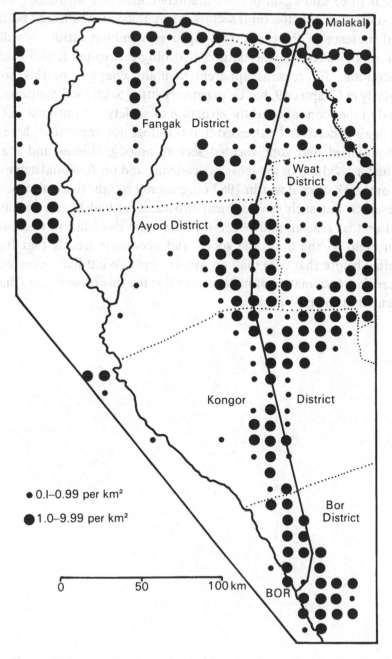

Figure 10.3. Distribution of *tukls* (huts) within the Jonglei area permanent settlements in relation to the canal alignment (RES, 1983).

Nuer and between sections of the same peoples; nevertheless the overall picture should present at least the essentials of the human background in assessing the impact of the canal.

It needs to be said again that we are here concerned with those sectors of Nilotic society, mainly the rural sectors, likely to be most affected both by the physical presence of the canal, the improved communications it will bring, and the modifications to the natural hydrological regime it will cause. The rural economy, for reasons apparent in many chapters of this book but particularly in Chapter 19, has changed very little. Such modifications as have occurred in the economy and the structure of society are outlined in Chapter 11, but again it needs to be stressed that this does not imply that there are not now modernised and sophisticated sectors among Nilotes and many distinguished individuals in the sense of academic and professional achievement.

The outbreak of civil war in 1983 once more brought to an abrupt end not only research but such development processes as had begun to have their effect. For that reason, and because of the total devastation that hostilities have brought to the area, the social and economic trends that had been discernible before that date may no longer apply. What had occurred before may therefore be a matter of history, and it is for this reason that Chapter 11 is presented in the past tense.

9

Environment and the history of the Jonglei area

Introduction

The purpose of this chapter is to present in historical perspective the impact of the environment on the peoples now living in the Jonglei area. The time range employed is long, but it is adopted specifically to emphasise the continuity of human adaptation to environmental change in the region.

It is now possible to write with greater confidence about the pattern of environmental changes, population movements, settlements and economic activities from the early nineteenth century to the present. Anything before that, with the data so far available, must necessarily be speculative. Yet knowledge of the geological, climatological, and linguistic past of the area is growing, and, while there has been no systematic archaeological investigation of the Jonglei area itself, it is possible to bring together these disparate bodies of information in order to suggest some long-term patterns in human settlement. It is important to understand, at least in very general terms, that the occupation of the Jonglei area by Nilotic-speaking pastoralists, their differentiation from each other, their movements and their economic activities have taken place against the background of a progressive drying out of the region over millennia. In any specific historical period there have been erratic shifts in the availability of water, vegetation and dry land which were bound to affect human settlement. The most recent movements of Nilotic societies can therefore be recognised as modern representations of a far older pattern.

The pre-historic and early historical reconstruction proposed here is presented as a hypothesis, to be tested against future archaeological, climatological and linguistic data. All dates prior to the late eighteenth century must be regarded as very tentative. 'Ethnogenesis' is not an issue here, nor is any attempt made to retrace in detail the itinerary of mass migrations. Instead, the main concern is with the origin of a way of life, with the development of the economic and social systems which are characteristic of the region today.

Prehistoric and early historic settlement

East Africa, the Nile basin and the Sahara were far wetter some 5,000 years ago than they are today (Grove, 1977; Schove, 1977; Harvey, 1982), and the onset of drier conditions at about 4,000 BP may have been a factor which accelerated the differentiation of language groups in that vast region, also fostering the development of agriculture and pastoralism (Sutton, 1974; Harvey, 1982). The process that began then can be seen continuing in the Jonglei area today with the constant search for reliable sources of water, the interdependence of cultivation and herding, and the traces of different population alignments produced by periodic short-term movement and occasional long-term migrations.

The emergence of the ancestral Nilotes from the larger Eastern Sudanic language family, centred around the confluence of the Blue and White Niles, may have begun some 5,000–4,000 years ago at a time when there was a substantial southward movement of the cattle-keeping complex into the southern Sahara, the Sahel and the southern Sudan (David, 1982*a* & *b*; Ehret, 1982 & 1983; Smith, 1980; Stemler, 1980). The gradual drop in the level of the Nile (Livingstone, 1980), and the southward retreat of the vegetation belt at this time were probably factors in the adoption of herding by the Nilotes and their movement south (Ehret, 1982 & 1983; David, 1982*a* & *b*).

The southern Sudan remained particularly wet, and when Lake Turkana was connected to the White Nile by the Sobat–Pibor system about 7,000 years ago, and briefly again 3,500 years BP (Harvey, 1982; Adamson & Williams, 1980; Livingstone, 1980) there might have been greater inundations in the eastern plains. Such conditions appear to have inhibited the spread of herders and their cattle from their homeland between the White Nile and Sobat. The Nilotes seem to have penetrated the Jonglei area only late in the first millennium BC (David, 1982*a* & *b*; Ehret, 1983).

The western Nilotic settlement

Extensive human occupation of the area south of the Sobat may have been made possible by the severance of Lake Turkana from the Sobat–Pibor system during the general decline in East African lake levels, and by the increasing restriction of the Nile waters to the fault-bounded trough which contains the Bahr el Jebel (Livingstone, 1980; Adamson & Williams, 1980). This opened up a vast swamp-free area between the western and eastern river systems flanking the plain where many pastoralists are now found. It is an area that now becomes progressively drier towards the east, with a corresponding deepening of the water table level (JIT, 1954; RES, 1983). The Sobat river became the main link between the western and eastern edges of the Jonglei area, and it was used as such a link in recent times during Anuak and Nuer movements into lands east of the Nile.

The area west of the Nile has been for long the most favourable for settlement and although there is as yet insufficient archaeological evidence, it is possible that cattle-keeping Nilotes began to occupy the far western edge of the floodplain of the Bahr el Jebel system by the middle of the first millenium AD. Humped cattle appeared in this same region some time during the fourteenth to fifteenth centuries AD, and this seems to have corresponded with a change in settlement patterns from nucleated villages on artificially raised mounds to the more scattered and temporary settlements associated with modern patterns of transhumance (David *et al.*, 1981; David, 1982*a* & *b*).[1]

It has recently been suggested that the southward movement of the Western Nilotes, which culminated in the Lwo migrations in East Africa, was in some way initiated by Arab nomadic incursions into the White Nile valley following the decline of the northern Christian Nubian kingdoms in the thirteenth to fourteenth centuries AD (David, 1982*a* & *b*; Oliver, 1982). This may be a plausible chronological explanation but, quite apart from the fact that such Arab incursions might not even have happened (Spaulding, 1985), it is inconsistent with what is known about the vigorous, almost militant northern movement of peoples out of the southern Gezira by the end of the fifteenth century (Hasan, 1967; O'Fahey & Spaulding, 1974; Adams, 1977; Hrbek, 1977). It also entirely overlooks environmental factors.

While there are many uncertainties, it is possible to infer from early Nile records in Egypt that there were considerable periods of relatively low rainfall in the headwaters of the Nile during the first half of this millennum (Schove, 1977; Herring, 1979*a*; Harvey, 1982). There seem to have been a number of extensive low rainfall periods in the southern Sudan, some lasting for decades, between AD 750 and AD 1460 (Schove, 1977; Harvey, 1982). This could account for a period of continual movement by small groups searching for water and pastures within and beyond the Jonglei area. The swamps would have contracted during these periods (JIT, 1954), allowing for the multiplication of scattered settlements closer to the river than was possible during periods of higher discharge. It would seem that this succession of low years was followed by a relatively stable period roughly from AD 1450 to 1800 (Schove, 1977; Harvey, 1982).[2] Movements during these three centuries could have been more consistent and less extreme, conditions for both cultivation and herding would have improved (subject to local variations in annual flooding), and there could have been a sustained population growth and concentration in many areas, especially along the ridges of elevated ground. If this was so, then we can better understand the expansion of the Lwo and Dinka societies during this period.

The Shilluk, Anuak and Dinka

It is suggested, on linguistic evidence, that Western Nilotic speakers lived between the White Nile and Sobat earlier than previously assumed, and that their penetration of the *Sudd* region was facilitated, and to some extent encouraged, first by the general desiccation of the area and then by a succession of relatively dry periods. This corresponds with the general pattern now being suggested in recent historical studies of the environment of related peoples in East Africa (Herring, 1979*a* & *b*; Webster, 1979), a region affected by many of the same climatic changes as the southern Sudan (Schove, 1977; Grove, 1977).

The modern East African Lwo communities grew, in part, by absorbing earlier Nilotic settlers during their migrations (Herring, 1979*b*; Webster, 1979). The Anuak and the Shilluk, too, appear to have grown by ingestion rather than expansion. Clan myths of both these peoples refer to the frequent incorporation of foreigners and also note constant movements up and down the White Nile and Sobat rivers (Pumphrey, 1941; Howell, 1945*b*; Crazzolara, 1951, JIT, 1954, and Evans-Pritchard, 1940*b*). Since Shilluk myths about Nyikang, their first *reth*, or king, who led them to their present territory, acknowledge the prior presence of other peoples on the Nile (Pumphrey, 1941, Crazzolara, 1951), it is possible that the Shilluk kingdom was not founded by the sudden incursion of a new people, but through the reinforcement of an existing Nilotic presence on the west bank of the White Nile. A combination of environmental factors, which facilitated communication along the river and concentrated the population on long continuous ridges, and political competition with the neighbouring kingdoms of Sennar and Tegali, created the conditions for the development of a kingdom in the sixteenth and seventeenth centuries (Wall, 1976).

The Dinka conquest of the east bank of the White Nile, which is supposed to date from the last quarter of the 18th century (Wilson, 1903; Westerman, 1912; O'Fahey & Spaulding, 1974), can be seen in the same terms. There are Shilluk traditions of Dinka on the east bank nearly a century earlier (Hofmayr, 1925). Dinka may have been among those who moved south of the Sobat throughout the succession of drier periods in the seventh to fifteenth centuries AD. Population increased during the more stable period which followed, not only in the east, but along the edges of the ironstone plateau in the Bahr el Ghazal where greater concentrations of people and cattle are possible. Modern traditions of present-day Bor district being the home of the western and northern Dinka prior to their expansion may well refer only to movements in this later period, which is still part of the historical memory. The last half of the eighteenth century does seem to have been a time when the Dinka, bolstered by an influx of settlers from the south, consolidated their hold on the ridges west of the Bahr el Zeraf, north of the Bahr el Ghazal, and from the khors Adar and Fullus down the east bank of the White Nile to Melut, Paloich and Renk (Hofmayr, 1925; Beavan 1930, Chatterton, 1933 & 1934; Bedri, 1939 & 1948; Howell, 1945*b*). This movement to existing Dinka settlements on the White Nile and Bahr el Ghazal could have been encouraged by the series of famines and low Niles which took place in the early 1720s and 1780s (Schove, 1977), as well as from localised inundations and political confrontations following any competition for reduced resources (Wyld, 1930; Maclaglan, 1931). The northern and western Dinka regions do allow relatively high population concentrations. By contrast the Dinka living south of the Sobat at the turn of the eighteenth to nineteenth centuries seem to have been scattered into smaller, more autonomous groups.[3] To understand how this came about we must look at the process of Dinka fission and amalgamation.

The western Dinka claim an eastern origin. According to their current political theory, which is based on both historical and recent experience, political groups spread over larger territory as they increase in size, segmenting as they grow. Segments move away from the parent body as they become larger, claiming their own wet-season camps to satisfy their expanded needs. These new settlements in turn attract later settlers, but the original settlers generally occupy the best land, leaving the more marginal areas to newcomers. The newcomers are prone to move off in search of better sites once their own numbers increase to the point where they are capable of surviving on their own. This process is encouraged by the fact that it is usually the largest lineages that dominate the dry season pastures as well as the wet season camps, thus forcing smaller groups to search for better areas on their own. In the process of migration the composition of each tribal group changes, for each is merely a temporary collection of individual groups within a well-defined territory (Lienhardt, 1958).

By the beginning of the nineteenth century a number of smaller western Dinka groups had attached themselves to eastern Dinka communities, and other Dinka in the east were spread over a vast territory from the Bahr el Jebel to the Pibor (Wyld, 1930), indicating a greater freedom of movement in the eastern than in the western plains at that time. This was to prove an important factor in the subsequent Nuer occupation and domination of the central Jonglei area.

Nuer expansion

It can now be seen that Nuer expansion of the nineteenth century conforms in many respects to the general pattern of Nilotic movements in the Jonglei area.[4] The Nuer who originally settled on the sandy ridges west of the Bahr el Jebel developed compact communities. The western Nuer were frequently restricted in their movements, with the *Sudd* in the east and with the western swamps regularly merging into one large swamp during years of extensive flooding. The ridges themselves were (and are) generally free from floods, while the water table is shallow and water near at hand during the dry season (JIT I, 1954). Shorter seasonal movements are required than in the east; there can be a greater concentration of population in a compact area, and the Nuer tribes in the west are territorially restricted (Evans-Pritchard, 1940*a*). During the centuries prior to 1800 the Nuer grew not only through natural increase, but through the assimilation of incomers, a process represented constantly in clan myths.

Nuer myths also refer to the origin of feuds, and feuds in a restricted area can be self-perpetuating, especially when fuelled by competition for irregularly available resources. Most Nuer testimony claims that the Nuer were suffering from population pressure at the beginning of the nineteenth century (Jackson, 1923; Coriat, 1931; Evans-Pritchard, 1940*a*; Crazzolara, 1953; Jal, 1987). This pressure may have resulted from an increase in population, but it seems mainly to have been brought about by the decrease of land available for grazing, cultivation and settlement. The hydrology of the area is such that prolonged flooding periodically makes land unsuitable for human use. The first quarter of the nineteenth century was indeed a period of low Niles, drought and famine throughout the Sudan and in adjacent territories (Schove, 1977; Herring, 1979*a*; Harvey, 1982). The worst period seems to have been the early 1820s, which is the date of the beginning of the Nuer migration eastward, as calculated from age-sets and corroborated by the Shilluk king-list (Johnson, 1980).[5] Nuer accounts of that period mention an exceedingly variable Nile, ranging from excessive localised flooding to an abnormally shallow river. This is consistent with what we know about the hydrology of the Bahr el Jebel. A series of years of low river builds up vegetation blockages. With water backing up behind the blockage the area immediately up-river becomes flooded, while the water level immediately down-river drops further. We can infer from the admittedly incomplete Nile records we do have that this is more likely to have happened in the 1820s than in the preceding or succeeding decades.

The Gaawar and Lou Nuer were, according to aural tradition, the first to cross the Bahr el Jebel and settle on the periphery of the higher ground occupied by Dinka on the Zeraf island (Jal, 1987). The process of migration seems to have been similar to that of the Dinka described above, and the earliest Nuer settlers to the east were those who had been confined to the least attractive and most vulnerable land in the west (Johnson, 1980; Jal, 1987). The movement began in the first quarter of the nineteenth century for three reasons: 1) prior to this the west bank of the Bahr el Jebel had been able to contain the bulk of the Nuer population; 2) the prolonged dry-period of the early nineteenth century imposed harsher conditions on life in the west while reducing the flood level in the east, making large scale eastward movement possible, 3) the eastern Dinka population was by this time scattered and sparse, at least relative to the Nuer, while the western and northern Dinka presented much denser and more compact populations capable of deterring a Nuer advance.

The Lak and Thiang followed on to the Zeraf island and pressure from these newcomers and renewed flooding of the lower ground following the rise of the river in the late 1840s (Grove, 1977) forced the Lou and Gaawar farther east, displacing the Ngok and Luac Dinka from the Zeraf ridges. In about 1830 part of the Jikany moved east, too, following a different route across the White Nile and along the Sobat (Crazzolara, 1951, JIT I, 1954, Johnson, 1980 & 1982, Jal, 1983). The last half of the nineteenth century experienced a general trend of high Niles (Grove, 1977), and these had a particular effect on the population of the Zeraf valley. The high flood of 1878 contributed to a major internal conflict among the Gaawar, and the continuous high levels of 1892–6 encouraged the Bar Gaawar to occupy the northern half of the Duk ridge.

Demographic shifts within the area east of the Bahr el Jebel as a result of Nuer expansion can only be guessed at. Those Dinka who fled the approach of the Nuer tended to amalgamate in settlements around the confluence of the White Nile and Sobat, or to join the Twic Dinka to the South (Johnson, 1980 & 1982). Nuer communities retained their links with the west. Cattle flowed back to the western settlements (Crazzolara, 1953), there were continuing marriage contacts between some western and eastern Nuer groups (Coriat, 1923), and a steady stream of individuals came from the west to take advantage of the larger eastern pastures and the herds they supported (Johnson, 1980). By far the greatest single source of Nuer increase came from the assimilation of the original Dinka and Anuak inhabitants (Johnson, 1982). Throughout the nineteenth century the Dinka population shrank and their territory contracted as the Nuer communities grew and expanded.

The growth of modern Nuer society in the Jonglei area during the nineteenth century was not achieved through a 'major population replacement' (David, 1982*a* & *b*), but through assimilation. This created changes in the social and political structures of Nuer society. Eastern Nuer lineages ceased to be territorially defined (Evans-Pritchard, 1940*a*), new lineages of *diel* ('original settlers') were founded and new *kuar muon* ('land priests') were created by common consent (Lewis, 1951; Howell, 1954; Evans-Pritchard, 1956; Johnson, 1980). The most significant development was the emergence of the Lou and Gaawar prophets, who espoused a social ideology which transcended kin-affiliation and embraced the Dinka and Anuak neighbours of the Nuer (Johnson, 1980 & 1982).

Political and administrative factors affecting modern settlement

Political as well as environmental events influenced the direction, timing and extent of the Nuer occupation of parts of the Jonglei area. The Jikany Nuer took advantage of a Shilluk civil war early in the reign of the *reth* Awin to begin their eastward movement (Crazzolara, 1951, Johnson, 1980 & 1982, Jal, 1983). Slavers based in armed camps in the Zeraf valley in the 1860s and 1870s chased most of the Dinka away from the Duk ridge, allowing later Nuer occupation there (Johnson, 1982). The Jikany settled on the right bank of the Baro only after a Mahdist expedition in *c*.1890 dispersed the Anuak (Michel, 1901, Evans-Pritchard, 1940*a*). Gaawar movement southward, after the 1916–18 flood, was halted by the intervention of the Anglo-Egyptian government. Throughout the 1930s, 1940s and 1950s, successive administrations tried to retard the steady movement of Nyareweng Dinka (escaping floods or unpopular chiefs) into Lou country and Lou (seeking refuge from exessive flooding in the rains and extreme aridity in the dry season) into Anuak and Jikany territory.

Alien governments began to make an impact on the Jonglei area as early as the 1820s when the first Turco-Egyptian armies entered the White Nile and Sobat valleys before finally penetrating the *Sudd* in the 1840s. But it was the Anglo-Egyptian government of 1898–1955 which attempted the first comprehensive administration of the Jonglei area, thus affecting life in the rural areas in a sustained and systematic way. From 1902–30 many parts of the Upper Nile were subjected to a succession of pacification campaigns. It was this recurring demonstration of the government's superior force which was the foundation of later administrative regulation in both its restrictive and beneficent effects on local societies.

The 1920s saw the beginning of rural administration in the real sense, with the building of motor roads linking administrative centres in the interior, public health measures such as mass smallpox vaccination campaigns, the registration of taxpayers, and the organisation of local courts administering customary law. All these activities were expanded and consolidated through-out the 1930s and 1940s, especially in the appointment of headmen, sub-chiefs and chiefs in the administrative hierarchies, the expansion of court work, and the government sponsorship of inter-tribal and inter-district meetings to settle matters and problems of mutual concern through discussion and negotiation (Johnson, 1986a).

Activities of this kind influenced population movements and distribution. The Anglo-Egyptian government was especially concerned with regulating contacts and relations between groups. The appointment of a hierarchy of territorial and kin-group chiefs, the allocation of local labour duties (road work, wood cutting, etc.) and the listing of taxpayers meant that changes of residence across district boundaries had to be approved and recorded by the administration. Court centres became the focus of political, social and economic interest just as the main centres, as we shall see, also became the sites of dispensaries, cattle auctions and grain markets.

Economic interdependence in the Jonglei area: floods, famines and kinship

Droughts and floods are constant features of life in the Jonglei area which define the type of pastoralism and crop production most of its people can follow. Nuer and Dinka of the Jonglei area may regard them-selves as predominantly pastoral but cultivation is essential for any society there to survive (Evans-Pritchard, 1940a; JIT I, 1954; see Chapter 10). An examination of the last hundred years in the Jonglei area reveals the wider importance of crop production, especially in the face of recurring high floods such as those of 1878, the 1890s, 1916–18, the early 1930s, the late 1940s, and the 1960s. We can suggest that there is a regional economic net-work which frequently cuts across ethnic and political boundaries, and which has often been able to distribute the products of both pastoralism and cultivation, evening out surpluses and deficits in different places at different times.

As we have seen, village settlement patterns of both Nuer and Dinka are greatly influenced by the topography of the region (Chapter 10). Nuer villages on the ridges of the Zeraf island and west of the Bahr el Jebel tend to be more compact than the straggling settlements of Dinka and Nuer living between the Sobat and Bor (Liddell, 1904; Struvé, 1907; JIT, 1954). The dispersal of dwellings in the lower-lying areas does not necessarily increase the area under cultivation during the rains, since homesteads are built to take advantage of even mild elevations and are not always separated from each other by flood-free land. There are only a few areas where pastoralists, such as the northern Dinka and the eastern Jikany, have been able to produce regular grain surpluses during much of this century (JIT, 1954). Throughout the region as a whole the Nilotes have had to develop a variety of strategies for confronting the environment. These have included deliberate short-term concentrations of economic resources, and the long-term exploitation of specific areas where crop production was more frequently stable.

In the late nineteenth century the prophets Deng Lakka and Ngundeng Bong succeeded in concentrating some economic resources in small centres. Deng Lakka moved to the Duk ridge, in part because its well-drained sandy soil was good for cultivation. Once there his concentration of cattle, accumulated through his activities as a prophet, produced a surplus of animals, milk and meat that attracted and provided for the destitute over a wide area, Dinka as well as Nuer. The Lou prophet, Ngundeng, created a more regular supply of food around his 'Mound' north of Waat. This mound was built in part from mud excavated from two large *hafirs* on either side of it, providing a new reserve of water for part of the year in this notoriously dry land. Surrounding the mound for a mile or more was a plantation of maize and sorghum which was cultivated by Ngundeng's followers and visiting supplicants. Enough was harvested to provide for the needs of visitors who came year round from various Nilotic communities. These deliberate accumulations of animal and agricultural wealth declined after the death of both prophets early this century.

Because conditions vary throughout the Jonglei area, shortages are usually localised and relief can come from distant kinsmen (JIT, 1954: see Chapter 10). The main anti-famine measure developed by the Nilotic peoples in the nineteenth and twentieth centuries was the expansion of the kin-network, enabling a number of communities to take advantage of local grain reserves, particularly from the Twic and Ngok Dinka areas and from the eastern Jikany.

The Jalle and Kongor areas of the Twic Dinka are noted for artificial settlement mounds built on intermediate (rain-flooded) land with protective embankments surrounding the cultivations (JSERT, 1976a). These embankments appear to be used more extensively by the Twic than by any of the other peoples in the Jonglei area. The inhabitants of present-day Bor District have usually been unable to produce all their grain needs, but the Twic area, before the floods of the 1960s, used to produce enough grain to supply some food and seed to other Dinka from Bor and the Duk ridge, as well as to more distant Nuer. During the late nineteenth century the Twic provided refuge for a number of Dinka fleeing from slavers and the Nuer advance, and Twic wealth in cattle began to increase at that time. But the Twic began to marry their own daughters to the Lou Nuer, not only to increase their herds through bridewealth (see Chapter 10), but also to establish kin-links with the Lou to enable Twic to get grain and food when their own crops failed, their cattle died, or their land was flooded. The Lou reciprocated by coming to the Twic for relief when Lou country suffered shortages during times of Twic prosperity (Johnson, in press).

The Ngok Dinka now living along the Sobat used their position in much the same way as the Twic. They settled along the Sobat after being expelled from the Zeraf island by the Lou in the mid-nineteenth century. Their new land was near permanent water and good for cultivation. Following the Nuer conquests there were a number of Ngok and Ngok descendants living among the Lou and Gaawar. Through these kin-ties the Nuer frequently came to the Ngok Dinka in dry years to exchange livestock for grain, or even to cultivate in Ngok territory. Occasionally the Ngok, too, would send their cattle to graze on the Khor Fullus near Nuer camps, and a number of Ngok youths were initiated alongside Nuer youths (Evans-Pritchard, 1934 & 1936, Anon. 1932). Mutual co-operation between the Ngok and Lou was well developed by the beginning of this century (Wilson, 1905).

This inter-ethnic kinship network expanded in part as a response to persistent dislocation through widespread flooding and cattle epidemics (Johnson 1988). During all the periods of major flooding of this century (1916–18, 1931–4, 1946–9, 1961–4), each of the peoples living in the imme-diate Jonglei area were forced to seek assistance by crossing political and ethnic boundaries. The Nuer and Dinka of the Duk ridge have had to exchange grain and cattle with each other and with peoples farther south and east; the southern Shilluk have frequently crossed the White Nile to obtain grain from the Lak Nuer; the Twic and Bor Dinka have received assistance from kin across the Bahr el Jebel but have also turned increasingly eastward to the Lou. With each catastrophe, new kin-links were established through inter-marriage (often described as a deliberate strategy for rebuilding depleted herds by exchanging girls for cattle with more favoured peoples); with each subsequent catastrophe kinship obligations were activated. And so it has continued, with kin-ties following and strengthening the lines of feeding (Johnson, in press).

Exchanges based on kinship and reciprocity are not the only ways in which livestock and grain have circulated, and the more favoured or stable areas of the immediate Jonglei area are not the only ones which have been tapped. The country to the north and north-east of the Machar Marshes, inhabited by the Meban and Koma, has been another source of grain frequently exploited by peoples in the Jonglei area. The Meban, living between the Daga, Yabus and Tombak rivers, are separated from the northern Dinka by a wide belt of dry land, and from the Nuer by swamp and savannah. Throughout much of this century they have been visited by small groups of Nuer and Dinka exchanging livestock for grain. These relations were often interrupted by raiding, but during most of the first 30 years of this century an extensive trade network flourished throughout the extended Jonglei area when the eastern Jikany Nuer controlled the Ethiopian firearms and ivory trade. The main items of exchange were firearms and ivory but also included cattle, grain and tobacco. For a while the grain producing areas of the Meban and eastern Jikany were linked to the grazing lands of the eastern plain and the Zeraf valley, until the government intervened for reasons of security, and suppressed the trade (Johnson, 1986*b*).

Conclusion
This historical survey has attempted to show the effects of environmental fluctuations on population distribution and social and economic activity in the Jonglei area. We have seen that just as the peoples of this area live between an annual alternation of extreme aridity with excessive inundation, so historically a past period of long-term desiccation is contrasted with a more recent period of repeated flooding. The first period allowed for the occupation of the Jonglei area, the second accounts for the current distribution of the peoples within the area.

The peoples of the Jonglei area have sometimes been described as living in equilibrium with nature. In so far as this conjures up a picture of stasis, this is clearly not the case; nature in this region is too erratic to allow a stable balance. What does happen is that environmental fluctuations in one direction are compensated for by movement in other directions. The restrictions of the environment do not encourage a static existence in any area of life. The harshness of the environment encourages inter-dependence at the same time that it fosters competition, not only within, but between communities. There have also been considerable and constant changes in social and political configurations in response to political as well as environmental changes. It is not sufficient to explain Nilotic history solely in terms of the internal dynamics of Nilotic societies. Changes in the environment are a prominent feature in the direction of other kinds of changes in this region, and the importance of this fact has been largely ignored in most historical studies and many recent anthropological commentaries. If future historical research about the Jonglei area is to provide us with a clearer understanding of the development and functioning of the societies contained within it, then the inter-play of the peoples with their environment must become the context for any chronology and analysis of social and political history.

Notes

1. The date of the arrival of humped cattle in the Bahr el Ghazal area is fairly certain, but the archaeology of the White Nile is as yet too tentative to establish a date for the introduction of humped cattle there (Kleppe, 1982). At the same time, the date of the oldest known traces of human occupation at Nyany (AD 1000) indicate that pastoralists may have already been exploiting the very heartland of what is now the Jonglei area well before the arrival of humped cattle.
2. There are great difficulties in using the Nile data for this period (Harvey, 1982), but other climatic evidence can be used to help construct a general picture of conditions at this time (Schove, 1977).
3. It has been suggested that the eastern Dinka population had been thinned out by migrations to the north (Johnson, 1980). Gabriel Jal (personal communication) has concluded from his own collection of traditions about Nuer migrations that a succession of small feuds had scattered the Dinka into small, mutually antagonistic communities throughout the central part of the Jonglei area.
4. Anthropological discussions of Nuer expansion which are not based on field research, or on a thorough analysis of historical sources, can be found in Sahlins, 1961; Newcomer, 1972; Glickman, 1972; Southall, 1976; Sacks, 1979 and Kelly, 1985. None have had access to sufficient data which would enable them to construct their theories around either an accurate reconstruction of Nuer and Nilotic movements, or a detailed knowledge of the local variations and historical changes in the Jonglei environment.
5. Kelly (1985) proposes the suspiciously precise date of 1818 as the beginning of the Nuer eastward movement. This date is computed through a re-working of the dates of the early-nineteenth century Shilluk king list, but without sufficient reference to the internal evidence of the king list itself. There is also no attempt to confirm this computation by comparison with Nuer historical traditions, Nuer age-sets, or external environmental evidence. Kelly also makes the error of reversing the order of Nuer migrations, placing the Jikany before, rather than after, the Gaawar.

10

Society and rural economy in the Jonglei area

Introduction

Dinka, Nuer and Shilluk live in an environment that is unpredictable and capricious. The plains of the Jonglei area contain four main land types, described here in necessarily very generalised terms as high land, that is areas of marginally higher, better drained and sometimes sandier, ground; intermediate land or rain-flooded grassland; *toic*, a common Nilotic word for the floodplains or river-flooded grasslands bordering the main channels and tributary watercourses; and permanent swamp. Each of the first three land types is seasonally essential for the maintenance of the rural subsistence economy, while the fourth is of marginal economic value, except for the small and widely scattered fishing communities of the central and southern part of the *Sudd*.

These topographical features combine with climate to restrict the range of agricultural opportunity. For Nilotes the most significant aspect is the extreme contrast between the wet and dry months of the year. Rain generally begins to fall from about mid-April, and with intermittent periods of dry weather, rises to a peak from July to September. By contrast, December to the following April is a period of almost total drought (see Appendix 2). Rainfall is extremely variable, not only from year to year but month to month and place to place. Such variations influence the availability of rain-fed grazing grasses and the success or failure of crop production.

There are, of course, different combinations of climatic and topographical factors throughout the area. The central Nuer, for example, have on the whole access to greater expanses of river-flooded grassland in the dry season and animal densities are lower than among their Dinka neighbours to the south. Some parts of the Jonglei area are subject to greater and more frequent rain flooding than others. There is, however, sufficient homogeneity in almost all aspects of the environment to justify the degree of generalisation which is inevitable in the brief description of land use and the rural economy that follows.

Annual rainfall and its distribution is not the only climatic factor of importance. Those parts of the Jonglei area subject to river spill from the White Nile system are also affected by fluctuations in rainfall in the upstream catchment far outside the region (see Chapters 4, 9 and 11). Seasonal climatic variations cause the rise and fall of the river. Maximum discharges determine the extent of the inundation of the floodplain and the natural production of river-flooded grasses; minimum discharges affect the accessibility of these pastures to the people and their herds. This requires emphasis because it is the main feature of the hydrological and hence ecological regime which will be most affected by the operation of the canal. Moreover, these variations do not occur only from season to season or year to year; as we have seen, there have also been major periodic climatic changes. The sustained high discharges into the *Sudd* beginning in 1961 and the consequent massive increases in permanent swamp and river-flooded grasslands is an outstanding example. Reference to Figures 8.1–8.3 and Table 1.2 will illustrate their magnitude. The significance of the consequent changes in the ecology and economy of the Jonglei area are discussed in other chapters.

Nilotic society, despite substantial changes that have occurred and which will be discussed in Chapter 11, is still very largely dependent on a rural economy. Given the climatic and topographical features outlined above, it is an essential subsistence strategy for the Dinka, Nuer and Shilluk to exploit a wide range of resources through a mixed economy. To differing degrees the economy of the Dinka and Nuer involves transhumant animal husbandry, moving according to the seasons over high land, intermediate land and *toic*, in association with crop production restricted to the marginally better drained land during the rains, supplemented by fishing in the rising and receding waters of the rivers, and hunting the abundant wildlife during the dry season. There are, however, greater similarities between the Dinka and Nuer than between both these people and the Shilluk, who are largely sedentary. Accordingly the Dinka and Nuer are treated together in what follows; we shall return to the Shilluk later.

Dinka and Nuer
The rural economy: seasonal variations in land use
High land
Flooding, which reaches a peak in the late wet season, necessitates the concentration of people and their livestock in the comparatively limited areas of high

ground. To speak of these as 'high land' is a relative and imprecise description, since such areas, though usually better drained, are rarely more than a metre or so above the surrounding country, often less. It is here that permanent homesteads are located; permanent in that they can last for several years compared with the flimsy grass shelters or screens of the cattle camps. Among the Dinka and Nuer, villages are not close clustered in any ordered pattern. Individual huts and homesteads are sited to take advantage of patches of better drained relatively flood-free ground. A homestead will usually comprise a man, his wife or wives, and their children. Homesteads, or clusters of homesteads occupied by kinsmen, each with a cattle byre (*luak*, Plate 19), are linked by networks of paths and form villages whose members are for the most part closely related and share a sentiment of common affiliation. Village boundaries are often well defined by topographical features which may only be significant in the wet season when water courses and swamps can create barriers to communication, but it must be stressed that kin-links commonly extend beyond the boundaries of villages and even larger politically allied groups. It is around the homesteads that older, married people plant their crops, cultivation being the concern of each household independent of its neighbours, though the heavier tasks of clearing the land require collective action dependent on kinship obligations and good neighbourly relations as described in Chapter 13.

For most of the rainy season, particularly from July to October, community life within the framework of the village is at a peak of economic and social activity. Fields have to be cultivated, weeded and protected from pests, and cattle have to be herded in ever-increasing proximity as the surrounding countryside becomes waterlogged or flooded. Wet season camps are a particular feature of the Dinka of the southern part of the area, where herds continue to be tended by youths and unmarried girls well into the rainy season. In most other areas, however, and in the case of the southern Dinka towards the height of the rains, livestock are pegged at night and for part of the day in cattle byres adjacent to the homesteads, in an atmosphere of pungent dung smoke to protect them from mosquitoes and biting flies (see Chapter 12).

Intermediate land or rain-flooded grasslands
By far the greater part of the Jonglei area consists of hugh expanses of intermediate or rain-flooded grassland in which *Hyparrhenia rufa* predominates (see JIT, 1954 and Chapter 7). Soils are heavy cracking clays which do not allow deep percolation when the rain falls, and although intersected by networks of drainage channels, some shallow some well defined, these plains

are subject to variable levels of flooding during the wet months of the year. Such land, rarely with a slope of more than 10 cm/km (see Figure 10.4), is transitional between the marginally high ground of the villages and the seasonally inundated floodplains along the rivers and their tributaries. It is characterised by open grassland with scattered patches of acacia woodland or more sparsely distributed *heglig* trees (*Balanites aegyptiaca*). Timber from these woodlands is another essential resource for the people; it is from this that the frames of their buildings are constructed, while the mature dried grasses from intermediate land provide material for thatching.

Rain-flooded grasslands are of less value to livestock husbandry than might be supposed from a superficial view of the waving masses of tall green grasses to be seen at the height of the rainy season. Whatever their potential value as grazing might be at this stage of growth if they could be kept heavily cropped, they are inaccessible to livestock owing to flooding and muddy conditions. It is not until the land begins to dry out, usually towards the end of November, that the cattle are driven away from the villages, herded by the younger people, leaving the older generation to reap the second harvest. By then these

Plate 19. A Dinka *luak* (cattle-byre). The grass thatch (of *Hyparrhenia rufa*) covers a wooden frame and is held in place by string woven from local grass (*Sporobolus pyramidalis*). Three pied crows sit on the roof. Photo: John Goldsworthy.

grasses are tall, coarse, and unpalatable, but the old growth dries out rapidly and is fired to produce green and nutritive regrowth so long as moisture is retained in the soil. To begin with there is a daily exodus from the villages to these regrowth pastures, but it is soon necessary to take the herds farther afield and to set up small transient cattle camps.

The use of these grasslands in the grazing cycle is of great importance because grasses on the verges of the higher ground and in the vicinity of the villages have already been heavily cropped during the rains, and the river-flooded grasslands are still under water and inaccessible. In most places the period of usage of these intermediate lands is a relatively short one. Except in depressions which are flooded for longer periods and produce quite different grass species, the clay soils crack and dry out fairly rapidly, grass growth ceases, and the pools that provide drinking water are soon exhausted. Intermediate land pastures are, however, needed again at the end of the dry

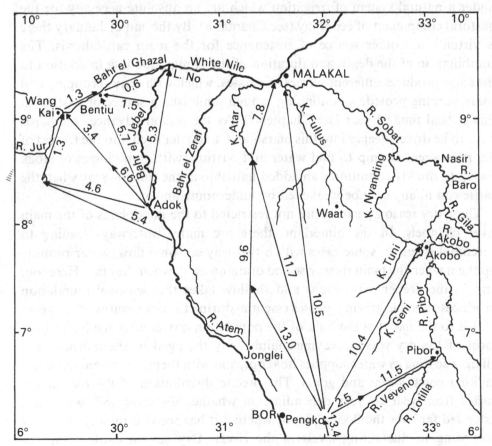

Figure 10.4. Ground slopes in the Jonglei area. Figures denote centimetres per kilometre (*source*: JIT, 1954).

season, for those areas that have been burnt and cropped earlier soon regenerate after the first fall of rain in April and May, when the rivers begin to rise once more to inundate the floodplain pastures. These rain-flooded grassland plains are therefore intermediate in two senses: they are intermediate in level, a factor which combined with rain and intermittent sheet flooding or 'creeping flow' (see Appendix 3) produces the grass species to be found there; and intermediate in the timing of their usage in the seasonal grazing cycle – at the end of the rains and once more when the rain begins to fall and the rivers to rise again.

River-flooded grassland or Toic

The low-lying land on either side of the main channels of the rivers and their tributaries is subject to spill from excess water that cannot be carried by the channels themselves. This is *toic*, where seasonal inundation produces grasses under a natural system of irrigation which are an absolute necessity for the pastoral component of economy (see Chapter 8). By the end of January there is virtually no other source of sustenance for the main cattle herds. The combination of the depth and duration of the annual flooding in relation to the soils, produces different species of grasses, which with heavy cropping and some burning provide a continuing though diminishing source of pasture at this critical time of year (see Chapter 7). As the season advances the herds have to be driven deeper into this marshland, and later even into the fringes of the permanent swamp to find water and pasture, with some losses of stock owing to muddy conditions, an added hardship at the time of year when the cattle can in any case be weakened by undernourishment.

These dry season pastures are not restricted to the floodplains of the main river channels. In the hinterland there are many waterways leading to perennial rivers, in some cases with a two-way seasonal flow, water running up them when the main rivers rise and draining away when they fall. Here too, and in the deeper depressions and shallow lakes this seasonal inundation produces valuable grazing grasses essential during the dry months of the year.

It is to the *toic* that the bulk of the population moves after the harvest for most of the dry season, leaving mainly only the aged or the infirm in the villages so long as water supplies hold out, and with them for sustenance a few milking cows, sheep and goats. The precise distribution of the *toic* camps varies from year to year depending on whether the river spill water has extended far over the floodplain, or whether it has receded rapidly or slowly according to fluctuating levels in the rivers. Dry season cattle camps in February and March are usually large, and it is at this time of year that there is the greatest concentration of population. It is a time of year, too, when for younger people there is the greatest opportunity for social intercourse, an opportunity to renew acquaintance with kin and friends from different villages and sections, to court and begin to negotiate marriages. It is regarded by all Nilotes as the best time of the year.

Wildlife tends to concentrate in these regions for the same reason that the cattle are driven there, for grazing and water, so it is also a time for hunting. Although frequently pursued as a social, sporting activity, in recent years hunting appears to have become a more important source of food (RES, 1983), perhaps first made more necessary by shortages of meat from other sources following the reduction in live-stock during the warfare and heavy floods of the 1960s. Fishing reaches a peak when the floods begin to recede at the end of the rains, again when the people descend to the floodplains in the dry season, and then when the river floods start to rise once more (Plate 26).

During the dry season people tend to move between camp and village, visiting the more sedentary elements among their kin for example, or replenishing their supplies of grain or other commodities. Distance undoubtedly dictates the frequency of these movements. Except for the Lou Nuer, some of whom move into the territory of Ghol, Nyareweng and Twic Dinka, Nuer have more extensive and productive riverain pastures than their southern Dinka neighbours, but they have greater distances to travel and are less well provided with permanent water supplies near their villages. Most of the Nuer population therefore tend to accompany the herds to the cattle camps and many villages are totally abandoned at the height of the dry season. The southern Dinka have been better supplied with wells and *hafirs* (excavated tanks), though there has been much recent deterioration from lack of maintenance (see Chapter 11). Dinka villages therefore tend to be occupied for longer periods, often right through the dry season, by a larger proportion of the community.

There are also differences between Dinka and Nuer in the pattern of movement, formation and structure of their cattle camps. Smaller and more scattered wet season camps are common among some of the southern Dinka, and when in the dry season they regroup, their camps are larger, more organised in layout, and almost always set up on sites with long established usage (Plate 20). The Nuer, on the other hand, usually spend rather longer in smaller camps moving over wider riverain grazing areas – partly because these are available to them (Plate 24). When they do assemble in larger numbers it is in more fragmented form, and they always avoid the use of the same sites (see Chapter 12).

Distances between villages on higher ground and the cattle camps in the *toic* vary greatly. In some cases, especially where higher ground for permanent

Figure 10.5. Seasonal variation in flooding (from Air Survey, RES, 1980).

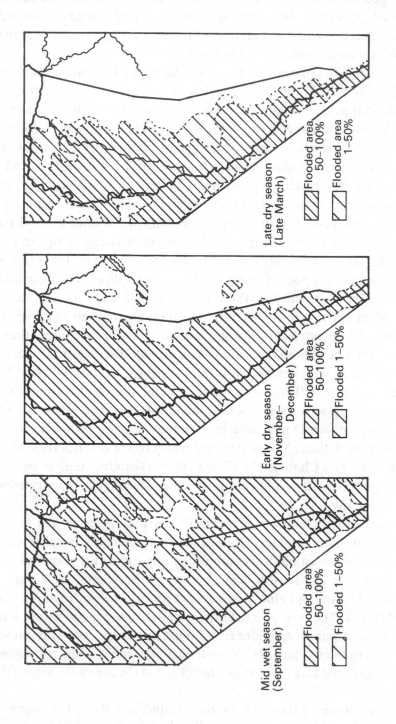

habitation is on the alluvial banks of the rivers, they are close enough to allow almost daily communication. In others people must migrate over anything up to a hundred kilometres, though movements vary from year to year according to the accessibility of river-flooded pastures, and sometimes depending on prevailing political relations between tribes and tribal segments who traditionally share, or by force of ecological circumstances may be compelled to share, dry season grazing resources.

Permanent swamp

Beyond the *toic*, fringing the main river channels (Plate 27), are expanses of permanently flooded or heavily waterlogged swamp, characterised by papyrus and bulrushes (see Chapter 7), as well as open water. Such areas are only of marginal use to the bulk of the population and their livestock. There are, however, specialist fisher-people, *monythany*, who live in the heart of the *Sudd*, their villages sited on the alluvial banks of the main river channels, higher than the surrounding country, or on mounds built up from detritus over the years. These people subsist very largely on fish and what other foodstuffs they can trade for fish. Many *monythany* have been forced to this form of livelihood from economic necessity, particularly owing to the decimation of their herds by flooding and disease (see Chapter 11).

The ambition of many is to return to cattle-keeping as a way of life, and since they are able to sell their fish, usually in sun-dried form, for cash either directly to their pastoral neighbours or in markets farther afield, they are able to buy cattle and build up their herds again. Most have kinsmen among the pastoral communities and leave cattle in their care, until they feel able to join them. The social composition of these fishing communities therefore tends to be fluid, people joining them following natural disasters, or leaving them as a result of successful marketing and investment in livestock. This has probably always been so, but particularly in recent years following losses in livestock during the 1960s on the one hand and rising prices and expanding markets for fish on the other.

Seasonality and variations in diet

The subsistence economy is maintained by the seasonal uses of the land types described above. Milk, clarified butter and meat from cattle and smaller stock, cereals and vegetables from the cultivations, fish, the meat of wild animals, edible roots and wild fruits, are all seasonal constituents of the diet. This seasonal variation, and the fact that an abundance of one source of food may balance shortages in others, together with the way in which people move from one type of land to another to make the best use of available resources according to the time of year, has been referred to as the 'balanced utilisation of land types' (JIT, 1954, p. 232). It would, however, be more correct to say that the livelihood of these people is dependent upon the success of such a balance and that this is by no means always achieved.

There are formidable obstacles to the success of crop production (see Chapter 13). Rainfall is extremely erratic in its distribution, expecially in the early months of the wet season. Early clearing of the land is crucial, and the return from the cattle camps of much needed labour may be delayed by lack of rain to fill pools for water supply. Even when the crop has been successfully sown, the seedlings may shrivel and die for want of rain in the ensuing weeks. The second crop may fail owing to accumulations of rain combined with poor drainage and hence excessive inundation; torrential storms may flatten the crops when they are most vulnerable, and those on higher, sandier soils that escape such damage from standing water may not mature for want of moisture if the rains cease early.

While certain varieties of sorghum have a remarkable capacity to withstand these extremes, and it is for this reason that it is the staple crop throughout the area, these hazards combine with generally heavy impermeable soils low in nutrients and organic matter, plant diseases and pests, to make crop production a discouraging undertaking, especially in the central and southern parts of the area. Crop failures, however, are rarely total over the region and frequently fairly localised, so that people in times of want can draw upon the resources of their more fortunate kin or barter livestock for grain with others. The Jonglei area has nevertheless long been a net importer of grain, though such figures as are available should not be taken as precise indicators of total needs but of supplementary requirements.

If cereals, particularly sorghum, are a major constituent of diet, animal husbandry is equally important though open to other hazards. Most Nilotes regard themselves as essentially pastoralists and say that they would prefer to live on the products of their cattle alone. But because of the threat of major losses from disease, particularly rinderpest, and the reduction in numbers of cattle and smaller stock sometimes caused by excessive flooding, this aspiration cannot be realised. A mixed economy is essential. Cattle, sheep and goats produce animal protein through meat and milk which, despite seasonal variations in quantity, provides an element of consistency. Average daily milk yields are always relatively low, though there are considerable differences between Dinka, Nuer and Shilluk cattle in this respect (see Chapter 12). There are also wide seasonal variations. According to the most recent assessment average daily milk yields reach a peak in the early months of the rains when the herds have returned to the intermediate plains to feed on the nutritive growth of rain-fed grasses, and by contrast show a marked depression when the cattle are on the riverain pastures in the dry season (see Chapter 12). Earlier assessments suggest a greater fall in milk yields towards the end of the rains when the more restricted pastures at the edges of the higher ground have been extensively cropped (Payne & El Amin, 1977; Watson *et al.*, 1977). This agrees more closely with the view of the Jonglei Investigation Team which, while recognising a diminishing return from riverain pastures as the dry season advances, suggested a similar or sometimes greater drop in yields in November and December (JIT, 1954, p. 247, p. 314). This discrepancy probably merely reflects the great variations from year to year and place to place. There is also a much greater though again seasonably variable consumption of meat from cattle, sheep and goats than is generally supposed, which represents a substantial offtake for local use.

Sheep, though valued for their meat and as an easily marketable product, are particularly vulnerable to flooded conditions. Moreover, being grazers they have to be driven to the riverain pastures in the dry season and do not flourish in the muddy conditions often found there. Goats on the other hand are browsers, and can thrive on shrubs, bushes, seed pods and dried leaves on the higher ground round the villages, and their average daily milk yields are double those of the wet season, the reverse of the general pattern of milk yields among cattle (see Chapter 12). They are therefore an exceptionally valuable source of food, especially for those elements of the population who remain in the villages through the dry season.

Fish provide a substantial input to the diet, though here again there are not only seasonal variations, but years when fish are more abundant than in others. Edible roots and wild fruits also augment the diet (JIT, 1954, p. 246), though many of these are not sufficiently palatable to be readily consumed except in time of real hunger. Nevertheless, in one form or another different indigenous vegetation yields edible products throughout the year.

Comprehensive and detailed nutritional studies among Nilotic peoples are conspicuously lacking.[1] Indeed, such studies among transhumant, highly mobile communities, whose sources of food vary so widely from year to year, would be difficult to conduct. They are, in the case of the Dinka and Nuer, further complicated by the fact that elements of the population, according to age and division of labour, tend to disperse and foregather again at different times of the year. Quantitative data are not available and such abstractions of the fluctuating nature of nutritional levels as do exist (see, for example, JIT, 1954, pp. 244–9 and Figure E.12) are based on generalised observation rather than long-term statistical sampling. What is apparent is that for the people there are very considerable variations in nutritional levels, and that at certain times of the year there is almost always a narrow margin between sufficiency and real hunger. The extreme seasonality of nutrition is characteristic of the area, a classic example of variable dietary stresses to which different elements of the population are subjected at different times of year, a subject which is a matter of current debate (Longhurst, Chambers & Swift, 1986).

Grain consumption in the form of porridge and beer reaches a peak after the harvest; it falls for most of the population when they move to the cattle camps. Nilotes do not, unlike their north-western neighbours the Baggara Arabs, use their oxen as beasts of burden in order to carry large quantities with them. The consequent drop in calories consumed daily is, however, partly offset by the consumption of animal products, including the meat of wild animals, and fish. Life in the cattle camps is fortunately relatively more leisurely compared with the exertions of the rainy season. Grain may also be in short supply when they first return to the villages; the previous year's crop may have been poor, storage is difficult and subject to pests, and much may have been sold earlier to meet the cost of other requirements (see Chapter 11). The period before the first crop matures is the 'hungry gap' for this reason, and though there is a substantial increase in milk supply if the herds are by then close to the villages, it can be a critical time of year if shortages coincide with maximum labour input in the processes of cultivation.

The supply of cereals will be replenished if the weather favours an early crop. In years when this fails people must rely more heavily on their livestock, and despite a high mortality rate, pastoral production provides a reasonably reliable means of survival when other sources of food fail. Barring disaster from disease or exposure, cattle and smaller livestock are on the whole a safer and more consistent source of food than any other. In a countryside predominantly of swamp and grassland, livestock also continue to be vital sources of material necessities, e.g. sleeping skins, tethering cords, spoons, tobacco pouches, etc., although consumer goods bought in local markets have steadily increased in importance (see Chapter 11). Even cow dung has its uses, especially as fuel and for plastering walls, floors and the exterior of huts; in cattle camps, its smoke and ash are a deterrent against biting insects for man and beast.

Since a combination of shortages does not often happen over very wide areas in the same year, deficiencies in one place will be made good from the resources of another by barter, gifts in pursuance of the obligations of kinship, or by purchase of grain from the marketplace. The subsistence economy is thus maintained, even though the diversity of climatic conditions throughout the year, and from year to year, rarely allows total adjustment to variations in exploitable resources. It is only in years in which all sources of food are much reduced over a wide region, such as happened in the great floods of 1916–18 and again, combined with widespread hostilities, in those of the 1960s, that crises reaching disaster proportions have occurred. It is therefore not only an essential subsistence strategy to make use of the wide range of resources which each land type offers according to the season, but it is a wise strategy that human relationships are organised through a series of extended kinship links which draw the individual and his or her immediate family into a wider network of social and economic mutual obligations. By this means those in want in one year can seek help from more fortunate kin and reciprocate in another.

Kinship and the social significance of cattle

The rural subsistence economy is far more than just the range of crops and animals people nurture. It also embodies a system of distribution and access to economic resources, organised in large measure as social relations. Bonds between age-mates or between neighbours are social relations, but the relations which are most wide-spreading and invested with complex perceptions of reciprocal obligation are those between kin. Kinship relations are created in marriage, which is bound by the rules of exogamy[2], and takes place in a rotation of seniority within a group of men who have common rights in a herd. Marriage involves the transfer of a bridewealth of cattle which are collected by the bridegroom and his kin according to accepted patterns of obligation, and apportioned among the kin of the bride in the same way. The system of distribution is complex, with some variation from tribe to tribe. Those kin who receive a portion of the bridewealth on the marriage of one of their kinswomen are also those who will be called upon to contribute on the marriage of one of their kinsmen. Such obligations on the one hand and rights on the other extend to both the paternal and maternal kin of the bridegroom or the bride. A web of kinship is thus created, centred on the relationship of a man and a woman, but bringing together many different people (see Appendix 7).

Kinship relations have an existence, logic and value beyond the dictates of the economy, but inherent within the kinship system is an important degree of economic security. Relatives help one another in the daily round of economic tasks, herding some of each other's cattle in different camps in the district, thus minimising the risk of cattle loss from disease and raiding, and helping with the major tasks of cultivation.

Disasters caused by climatic fluctuations and disease epidemics have from time to time caused massive diminution in numbers of livestock. Such episodes in the past have been followed by periods of replenishment by natural processes and the redistribution of cattle around the area through both marriage and raiding. If the herds are seriously reduced by disease and other disasters, Dinka and Nuer adjust the numbers of cattle acceptable in bridewealth payments, as it is recognised that the legitimacy of a union can be established without immediately meeting all the claims which they would say should ideally be honoured in this regard. Significantly, in the past at any rate, the need to accept a lower bridewealth did not mean that the cattle-rights of the bride's immediate family took precedence over those of more distant kin. Indeed, the immediate family would normally bear the greater proportional loss, which underlines the importance attached by Dinka and Nuer to the web of kinship (see Appendix 7 and Chapters 9 and 11).

Cattle are also transferred between groups of people to right a wrong: a method of settling disputes that in the distant past was only successful if a whole range of sanctions came into play, which varied according to different circumstances. Disputes over the infringement of rights in women, for example adultery or the seduction of unmarried girls, were composed by the transfer of smaller but conventionally recognised numbers of cattle, as were disputes over bodily harm for which there were scales according to circumstances and the severity of the injury, serving as models to which references would be made in negotiation. Such transfers later came to be enforced by the Chiefs' Courts established throughout by the Condominium government, and perhaps in the process gained a measure of untypical inflexibility. An extreme example is bloodwealth (Nuer: *thung*, Dinka: *apuk*) which is paid by the kinsmen of a killer to the relatives of his victim, without which a state of feud might continue for generations, for it is incumbent upon the kinsmen of the deceased to avenge his death unless restitution is made by these means. Such payments are broadly similar to bridewealth collection and distribution, though, when there is a widely shared will to avoid conflict, people not normally called upon to contribute to bridewealth may co-operate in providing cattle for bloodwealth. The numbers handed over in the past varied according to the circumstances, including the state of the herds, the nature of the act of homicide, and a complex combination of sanctions, including extreme dread of spiritual contamination through physical contact, or fear of violent retaliation. The Condominium administration, however, tended to standardise the level of compensation and added a punitive element in the form of a fine imposed by the Chiefs' Courts they had created (Howell, 1954).

The purpose of bloodwealth is to restore the equilibrium disrupted by a man's death and allow the kinsmen of the deceased to marry a wife in his name with cattle that can be used as bridewealth. In this way kin links are created with other people as if the deceased was still alive. 'A man who dies without issue is truly dead' (Deng, 1972). Children assure both the social and cultural continuity of the deceased's lineage and the memory of his name. The transfer of cattle as bloodwealth is therefore a fundamental factor in re-establishing peaceful relations between groups by enabling a vital obligation to be fulfilled within the more restricted family. Failure to marry on behalf of a dead unmarried agnate in this way, whatever the cause of death, can occasion divine intervention, causing illness or death within the family (Evans-Pritchard, 1940a; Howell, 1954; Hutchinson, 1988).

For Dinka and Nuer kinship and its associated values are fundamental in the conduct of everyday life. An individual's sense of well-being is linked to the perception of being part of a community of kin providing economic and social security. It is through the exchange of cattle that extended kinship is created, and, because of this, and because they are a relatively consistent source of sustenance, cattle are regarded as the source of life. This is a symbolically rich notion, which is expressed in oral poetry, the very personal identification of young men with their 'song oxen' and the use of their colours as familiar names. It is a perception which is also elaborated in religion. In another context life comes from the Divine. In the case of the Dinka, religious representations of divine power have been explained by Lienhardt (1961) as their interpretation of experience, much of which concerns the values and obligations of kinship. Thus it is no coincidence that cattle, with their role in creating and articulating relations between people, are also by their sacrifice a most important means through which people mediate their relations with the Divine. Sacrifice is performed for propitiation of the many manifestations of the Divine in order to avoid or mitigate the natural calamities that beset life, not least sickness and mortality among people and animals. It is called for in expiation of breaches of the moral order that can be the cause of such disasters. Sacrifice is an essential feature of the principal transitional stages of a man's life: maturity and the ritual that accompanies entry into an age-set and the age status it bestows; betrothal, wedding ceremonies and the final stages that confirm the legality of marriage by the birth of children; then death and burial.

After an act of homicide, steps are taken through the sacrifice of cattle to avoid the spiritual contagion of the killer, and, by extension, his kin (see Hutchinson, 1988). Sacrifice is thus essential in the resolution of disputes, particularly those involving violence and bloodshed, in which it is an essential component of the ritual of reconciliation between opposing parties. The social and religious importance of cattle is therefore a most conspicuous feature of human activities and relationships. An assessment of the economic significance of cattle merely in terms of their contribution in milk and meat to the diet would thus only provide a partial indication of their supreme importance to Dinka and Nuer.

Political organisation: the control of territory

Each economic resource is important and valued in its own time and place. But, as we have seen, cattle are more than an economic resource; their exchange is part of the socio-cultural fabric of pastoral Nilotic society (see Appendix 7). People unite to herd cattle within the constraints of the ecology of their area through a system which provides for the ordering of their association. This system is the political system.

Nuer

Since the publication in 1940 of Evans-Pritchard's work on Nuer political organisation an anthropological industry has developed, reinterpreting his analyses of Nuer political institutions. The theoretical impact of his first book, *The Nuer*, within the field of social anthropological theory gave the Nuer the dubious privilege of becoming known as the 'type-site' of the segmentary lineage model of political organisation (Sahlins, 1961). The many reinterpretations of this model since are wide-ranging in their theoretical stances, in the anthropological issues they have raised, and in the insights into Nuer society they have provided (see Karp & Maynard, 1983). However, our knowledge of the Nuer remains primarily rooted in the work of Evans-Pritchard, as virtually all reinterpretations have been based on his published material and none on extended fieldwork more recent than Howell. Most of the very important results of Hutchinson's fieldwork and research (1979–83), cited in Chapter 11, have yet to be published.[3] It is not possible, however, to describe the political system of the Nuer, or indeed of any other of the Nilotic peoples of the Jonglei area, without taking into account some of the implications of subsequent commentaries on *The Nuer*. However, it is also not feasible in the context of a single chapter in this book to enter fully the debate that has emerged, and it must be borne in mind that this debate has been more concerned with theory than with the Nuer as people.

'In describing any Nuer institution we have . . . always to consider it in relation to their pastoral pursuits' (Evans-Pritchard, 1937). Each political section is a territorial unit, with boundaries demarcated by natural features such as swamp, wide stretches of bush, or rivers. Of particular importance for our discussion is that each political section will contain expanses of the three main land types necessary for the conduct of Nilotic pastoralism and other agricultural activity, as well as common fishing grounds. People unite to articulate their rights in their territory through usage: common herding, fishing activity, and collective assistance in the major tasks of cultivation. They also unite in defending it from the encroachments of other sections by force, and combine to negotiate with others who may wish to make use of their resources. Generally speaking the smaller the unit the more real and intense the sense of corporate unity and mutual obligation to combine over political issues or to collaborate in economic activities. The larger a group grows, the more likely sub-sections are to develop within it, each incorporating the economic and social features of the larger unit of which it is a part.

Territorial sections among the Nuer are known as *cieng*, a relative term which can be used to mean a range of associations such as homesteads, hamlets, and villages through to tribal segments. Each political section has a name, which is usually derived from that of the founder of the lineage which acts as its nucleus. Thus a section which has the lineage Kerfeil as its nucleus may be called *cieng* Kerfeil. Not all the members of this section will be members of the Kerfeil dominant lineage, or indeed true descendants of Kerfeil, but most would claim some form of relationship to it.

The term *cieng* thus has both territorial and kinship connotations. What is common to the various associations of people known as *cieng* is the notion of a group of people being gathered together around an agnatic nucleus. At the very lowest level, which provides the conceptual model for the integration of larger associations of which each smaller association is a part, is a man and his wife or wives and their children, together with any affinal kin or sister's sons who may have joined the household. Such a household may join in common herding activities with the households of other agnates, commonly brothers who have rights in a family herd, and who also have in common the expectation of marriage cattle from other sources.

The kin and lineage systems and the idioms in which they are expressed thus provide at least a theoretical framework on which wider political associations rest and can be explained. An important point about descent groups is that they are also formally segmented by reference to their different lines of origin from a single ancestor. The segmentation of tribes and tribal sections is expressed as a value of the formal segmentation of dominant descent groups (*diel*) – clans and their lineages and sub-lineages. In other words the ordering of the segmentation of political units can be expressed as a function of the segmentation of their central agnatic group. Evans-Pritchard illustrated the relationship between tribal segments which are also territorial, political (and as the result of action by the Condominium government often administrative) units with lineage systems.

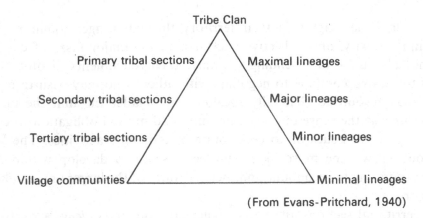

(From Evans-Pritchard, 1940)

Other Nuer lineages (*rul*) attach themselves to the dominant sub-clan or lineage through marriage, or in the case of Dinka (*jaang*) living with Nuer, by adoption or by marriage, thus becoming *maar*, kin, in the widest sense.

'When a Nuer speaks of his *cieng* ... he is conceptualising his feelings of structural distance, identifying himself with a local community, and by so doing cutting himself off from other communities of the same kind' (Evans-Pritchard, 1940*a*). Herein are complementary processes of fission and fusion, processes through which groups can merge in some contexts and stand in contrast or opposition to each other in others. The lines of descent of dominant lineages provide the framework through which such processes of fission and fusion are conceptualised. Kin relations extended across political boundaries through migration and intermarriage are like 'elastic bands' that enable political groups to fall apart and be in opposition, and yet in another context to be held together. In other words ... 'the kinship system bridges the gaps in political structure by a chain of links which unite members of opposed segments' (Evans-Pritchard, 1940*a*).

In various situations, such as the need to protect herds and pastures from outsiders, there is an obligation on members of a territorial section to come together to protect each other's rights and their common property. However, the extent of the mobilisation of the bonds existing between members of a particular *cieng*, reinforced or perhaps even created by common residence and common economic interests, is relative to the context of the situation, as is, indeed, the precise definition of who belongs to and who is outside a group. Thus different situations will involve different scales of action and reaction. Action often involves schism, but the particular divisions relevant to one situation, such as a dispute over an elopement, may give way to unity in a different situation, such as in a fight with another section to avenge an act of homicide.

Among Nuer, therefore, the tradition of descent from a common ancestor provides the guidelines, reinforced by ties of affinity and cognation, for the conceptualisation of the integration of different orders of territorial groups. This conceptualisation is a model which often describes the way people actually combine in pastoral activities in any particular situation. But it is not an unbending rule; it is more a way of putting sense and order into the world at a cultural level, order which allows extreme flexibility of response to different ecological, economic and social situations.

This becomes apparent when considering the extreme mobility characteristic of Nuer and, indeed Dinka, society. This mobility ranges from the seasonal cycle of transhumance, which for the individual or family may well involve pasturing cattle in different camps from year to year according to the availability of resources and the social ties that allow their use, to the large-scale migrations from west to east of the nineteenth century (see Chapter 9), and, more recently, for example, population movements away from the flooded parts of the Zeraf valley, as well as migration in search of work (see Chapter 11). Referring to this mobility, Evans-Pritchard observed that:

> ... it is the clear, consistent, and deeply rooted lineage structure of the Nuer which permits persons and families to move about and attach themselves so freely, for shorter or longer periods, to whatever community they choose by whatever cognatic or affinal tie that they find convenient to emphasize; and it is on account of the firm values of the structure that this flux does not cause confusion or bring about social disintegration (1951, p 28).

It can thus be seen why it is important for Nuer to establish their status by reference to a recognised lineage. A man will wish in particular to father many children who will continue his line, as well as in his lifetime bringing him influence and respect in his village or section. In time, a large lineage may become dominant within a territorial section, thus assuring the name of the lineage founder in posterity. Further, since the legitimacy of children and their position in the structure is determined by the legality of marriage, and that by the transfer of cattle, the social and political as well as the material and economic importance of cattle is firmly embedded in the minds of all Nuer. Membership of a dominant lineage implies prestige rather than privilege, political influence rather than anything approaching social superiority.

Dinka

The political organisation of the Dinka of the Jonglei area is based on a variation of this principle of lineage segmentation, though Dinka tribes are made up of rather looser and less structured associations of smaller

groups. These are known as *wut*. Like the Nuer word *cieng*, *wut* is a relative
term. It has a range of meanings clustered around the notion of a group of
people gathered together to herd cattle. Lienhardt, writing in this case about
the Dinka west of the Nile, has outlined the range of meanings:

> A *wut* is any cattle camp, any site upon which cattle are usually
> tethered or which is associated with camping in the past, any group
> of cattle with or without their herdsmen, or any group of herdsmen
> even without their cattle if it is understood that the purpose of their
> association is the tending and protecting of a herd. By extension
> from this *wut* denotes the section (the smallest formal territorial
> unit of the tribe), the sub-tribe and the tribe (1958).

The conceptual focus of the term *wut* is more narrowly concentrated on
cattle pastoralism than is the Nuer term *cieng* (the Nuer have a separate word,
wec, to denote cattle camp). However, each *wut* will also have areas of high
land containing villages and homesteads (*bai*) with their cultivations associated with it, as well as wet and dry season pastures – the full range of
resources needed for a mixed economy.

Among the Dinka the relationship between territorial segments and lineage
segments is more complex and ambiguous than among the Nuer, because
each Dinka *wut* may have more than one nuclear descent group through
which members of other descent groups explain their residential rights or their
claim to belong to that *wut*. It seems that among the Southern Dinka it is the
most powerful, wealthiest and largest descent groups that become socially
dominant within territorial sections (Ilaco, 1979*a*). It is these descent groups
that have come to provide Court Presidents at a sub-tribal level (Kadouf,
1977).

Indeed, since the early days of the Condominium government, succeeding
administrations have used territorial associations for administrative, organi-
sational and court purposes, and have tended to create *de facto* a greater
degree of political cohesion than may have been the case, especially among
the Dinka, in the more distant past. The Dinka at both ends of the Jonglei
Canal had been much disturbed, and in some cases decimated, by Nuer raids
and incursions during the nineteenth and early twentieth centuries. At one
stage Condominium government policy was to strengthen and even enlarge
Dinka political groupings as protection against further Nuer inroads into
their territory. From an early time during the Condominium period Dinka
Chiefs' Courts created by the government appear to have acquired greater
judicial and effective executive authority than was ever the case among the
Nuer.

Shilluk

The rural economy: seasonal variations in land use

Unlike the mobile, transhumant Dinka and Nuer described above, the Shilluk are sedentary. Whereas Dinka and Nuer habitations are widely scattered to make best use of outcrops of higher, better drained land, and communities reach the peak of concentration in the cattle camps at the height of the dry season, Shilluk villages are built in circular plan and concentrated form. Nevertheless, Shilluk territorial organisation and their economy are in large degree dependent upon land types similar, but not everywhere identical, to those found in areas occupied by Dinka and Nuer. In Shilluk country these land types run in lines parallel and close to the river. To contain all three land types, the boundaries of each territorial unit must therefore run at right angles to the river. This applies the length of their country on the left bank of the White Nile from Tonga to Kaka, but villages in the less densely populated and relatively recent colonies east of the Nile are more widely distributed, though their inter-settlement boundaries are simply extensions laterally across the river from their parent villages to the west.[4]

The nature of Shilluk land usage is a little different from that of the Dinka and Nuer to the south of them. Cattle and other livestock are held in *luaks* in the villages for the whole of the rainy season, but when this is over, they are herded by the youths in cattle camps by the river. Around the villages there are wet season gardens in clearly defined, tenured segments. Here a quick-maturing variety of sorghum, some maize, groundnuts, beans, cucurbits, gourds and tobacco are grown in the rains. Away from the river there are large expanses of rainfed grassland with some grass species similar to those found farther south. The annual rainfall in the more northerly parts of Shilluk country is, however, lower, the flooding less, and the composition of the grassland species somewhat different, mainly *Setaria incressata* and *Hyparrhenia pseudocymbaria* rather than *rufa* (JIT, 1954, pp. 222, 257). This land is relatively more nutrient rich and fertile and substantial crops of sorghum are grown as a second crop in well demarcated fields often some considerable distance from the villages. Fishing is an especially important and popular activity among the Shilluk, both by large crowds spearing fish in the lagoons and pools close to the river, and individual fishermen using harpoons, fish-spears, long-lines, cast and gill nets from dugout canoes or flimsy raft-like craft made from ambatch.

The main difference between the Shilluk rural economy and that of Dinka and Nuer farther south is therefore that they possess far fewer cattle in relation to the human population, depend less on cattle products, are not obliged to migrate with the seasons and, to make use of the three essential land-types, do not have to move over great distances. The only mobile elements in the population are a relatively small number of youthful herds-men who move with the cattle in the dry season, and in most cases milk is easily carried to the permanent villages each day. There is in consequence less seasonal variation in the main staples: cereals, vegetables, fish, milk and milk products. The Shilluk diet is not only more consistently varied but is less subject to seasonal deficiencies, although they too can suffer years in which climatic conditions lead to real hardship (JIT, 1954), particularly in the southern part of their homeland which is the area most likely to be affected by the operation of the canal.

Political organisation: the control of territory

More clearly defined territorial boundaries running at right angles to the Nile, greater population densities, more compactly built villages, a greater dependence on crop production than on animal husbandry, and the absence of the need to migrate over long distances, represent important differences from the Dinka and the Nuer. Attention is also frequently drawn to the more hierarchical nature of Shilluk society – the existence of a divine ruler, the *reth* or king, and the prolonged and elaborate ritual surrounding his death, burial and the installation of his successor (Howell & Thompson, 1946; Evans-Pritchard, 1948). A seemingly less egalitarian system of social and political organisation appears to contain within it a more authoritarian notion of personal leadership than exists among the Dinka and Nuer.

Spread through Shilluk-land there are four main groups of clans. These are the *kwareth*, the descendants of the original tribal leader, Nyikang, and from whose numbers each *reth* is elected from among the sons of a previous *reth*. There are then the *ororo*, a dispossessed branch of the royal house, who have no special distinction other than their ritual functions, particularly on the occasion of the complex ceremonial surrounding the death and installation of a *reth*. There are also the *bangreth*, literally 'the retainers of a *reth*', or their descendants, made up of people of various origins, in the past recruited by enslavement or voluntary enlistment, who serve the king or tend the shrines of his ancestors. Finally, there is the main body of exogamous patrilineal descent groups (*kwa*), the 'commoner' clans referred to collectively as *collo*, which are segmented and scattered throughout the country in localised lineages.

Yet closer analysis reveals that Shilluk territorial organisation and social structure is less far removed from that of the Dinka and Nuer than it might seem. Territorially the Shilluk are divided into two 'provinces' which have no everyday significance, though there is a tradition of north–south rivalry that features large during the ceremonial installation of a *reth*. Within these are a number of smaller territorial, and now administrative divisions (*luak*), which in turn are divided into a number of settlements, termed *podh* (Pumphrey, 1941; Howell, 1941). A settlement occupies a common and clearly defined territory, and has common fishing, cultivation and grazing areas. Within each settlement are a number of villages and hamlets, each of which, with its offshoots, is occupied by a distinct kinship group, the local lineage of one of the main Shilluk clans. One of these lineages will be dominant (*diel*) within the settlement, acting as an agnatic nucleus in much the same way as dominant clans act among the Dinka and Nuer. Often, but not necessarily, this dominant lineage will be from the *kwareth* (Howell, 1952). New villages spring up when young men move off to found their own settlements, in the same way as they do, for example, among the Dinka (Lienhardt, 1958; El Sammani, 1984). Such extensions, following the pattern of lineage cleavage, are the origins of the colonies of Shilluk settlements to be found on the right bank of the White Nile to a point some 35 kilometres below Malakal.

Conclusions

Among all three peoples each political section contains territory made up of the land resources vital for a mixed economy, which is a necessity in the environmental circumstances of the Jonglei area. However, it must also be stressed that kin-links across group boundaries are economically of great significance (see Chapter 9). Such links are established not just between Nuer and Nuer or Dinka and Dinka, but also between Dinka and Nuer and on occasion between either of these peoples and the Shilluk. Links between different families among different peoples are long established, despite the disruption caused by the attempts of the Condominium government in the early 1930s to separate the Dinka and Nuer for security reasons, and later by the floods and civil war of the 1960s. Should rainfall distribution be detrimental to successful crop production or should flooding adversely affect the well being of livestock or epidemics reduce their numbers, kin from other areas can generally be drawn on for food and for the rebuilding of cattle herds. This is a fundamental feature of Nilotic society that recurs throughout this book (see Chapters 9, 11 and 20).

Among the Nilotes in general, historical events outlined in the preceding chapter demonstrate that violent excursions into adjacent territory, cattle raids, forceful occupation and, in the case of the Nuer, resettlement and colonisation, were, in part at least, the social and political repercussions of periodic massive variations in the annual discharges of the White Nile, and consequent major ecological changes resulting in pressure on essential natural resources. Even short-term changes can compel people to turn to the grazing grounds of their neighbours, whether by negotiated agreement or force. Resources amicably shared in times of plenty can become the focus of conflict in less favourable conditions. This dimension of human relationships is an important one in considering the impact of the Jonglei canal, particularly if human and animal populations multiply while natural pasture resources diminish, and agricultural alternatives are not adequate or even feasible.

What has been said here of the structure of Nilotic societies in the Jonglei area, and of the nature of the finely tuned adjustment of their economy to the variable conditions of their environment, is necessarily very generalised. The purpose of this chapter has been to outline some of the basic principles of contemporary rural Nilotic society in order to provide a socio-economic perspective of their extremely flexible reaction to both environmental events and political interventions. This is not meant to imply that Nilotic society has somehow denied history and remained the same over the last century. Both Chapters 9 and 11 outline a series of events and processes which people in this area are part of and have reacted to. It is something of a paradox that these Nilotes are frequently regarded as conservative pastoralists, deeply resistant to change. Such a view is misleading, because Nilotic society, as will be seen, is by no means unchanging. The recent history of Nilotic involvement in Sudanese politics, for example, bears testimony to their vigorous participation in both national and regional economic and political issues.

Any apparent lack of change is a phenomenon that has at least two facets. The first is a problem of the perception of an external observer orientated towards modernisation: change that does not bear the surface trappings of 'westernisation' is frequently ignored or overlooked. Change may proceed through a continual process of subtle redefinition and reformulation of social relations, economic practices and cultural values that may appear to be a continuation of 'tradition'. The second is that the hard-headed mixed subsistence farmer, in a context of high environmental risk which can only be minimised by drawing on a wide range of economic resources, each complementary and each important, is unlikely to experiment with practices that are first seen as of dubious benefit and reliability. When, however, the utility of a practice is proven, the alacrity with which Nilotes will seize upon it is clear. 'It would be wrong to ascribe a conservative attitude to the Nilotic peoples in the sense that they are negative towards introducing new elements into their culture' (Hoek, B.v.d. *et al.*, 1978).

The next chapter will examine the way in which the people of the Jonglei area have responded not only to natural environmental upheavals and the shattering effects of civil disturbance, but to extraneous influences brought about by the penetration of their country by outsiders as well as their own increasing excursions into other parts of the country.

Notes

1. But see Ogilvy, S., 1977.
2. A man and a woman from the same patrilineal descent group may not marry, at any rate if the ancestral link is not too remote to be ignored or annulled by appropriate ritual and sacrifice. Marriage is not possible between those closely related by previous marriages, or between persons related by blood outside legal kinship. Those standing in a close relationship through adoption cannot marry, and a man may not marry the daughter of an age-mate with whom he underwent the ritual and ceremonies of initiation. Inter-marriage is also barred between groups who are involved in an unsettled bloodfeud (see Evans-Pritchard, 1951; Hutchinson, 1988).
3. The work of Sharon Hutchinson is based on fieldwork among the Eastern and Western Nuer between 1979 and 1983. The results of her research will add a new and profoundly important dimension to the anthropological debate.
4. Except for the northern Dinka, other Nilotes in the Jonglei area live in widely scattered settlements as described, but the boundaries of the territories they occupy must encompass all land types and thus run at right angles to the river or tributary whose floodplain supplies the *toic* (see Figure 10.2), usually, in the Jonglei area, on an east-west axis. In the so-called 'Canal Zone' this is, of course, important because it is for this reason that so many people and livestock will have to cross the canal.

11

Recent change among the Nuer and Dinka peoples of the Jonglei area

A mutiny of troops at Bor in May 1983 marked the start of civil war in the Jonglei area, rendering the subject of this chapter a matter of history. The only reliable information available is drawn from the period before hostilities began. It is not yet possible to assess the extent to which fighting and administrative collapse have disrupted life or to judge what the consequences for the future will be. This chapter, in presenting a picture of growing economic diversification before 1983, describes the impact of the previous civil war in the 1960s, which overlapped with the effects of severe flooding.[1]

Economic trends before 1961

The economies of the Dinka and Nuer started to take on their more recent characteristics in response to the limited opportunities for diversification available during the Condominium period. Four main processes have been identified by Hutchinson (1988) in this connection: the growth of the cattle trade; the development of the grain trade; the gradual emergence of the use of cash as a medium of exchange; and the expansion of migrant labour. It is worth describing these processes in more detail as they provide the context for the impact of later events, and underline the fact that though the area began to enter the trading and cash system of the rest of the country long ago, the process for the majority of the population has since then been slow.

The primary concern of the Condominium administration in the early years was to maintain law and order, first by direct intervention and later through Chiefs' Courts established to administer customary law (Howell, 1954). These courts also had limited executive functions concerned with tax collection and the upkeep of government buildings, as well as the sparse network of dirt roads. Although a co-ordinated, systematic approach to development was never implemented during the Condominium period, and, indeed, was not even seriously considered until 1953 (Badal, 1983; Collins, 1983; SDIT, 1955) when it was too late, from an early stage the authorities were well aware that economic stability was one means of promoting political security. For this, and humanitarian, reasons attempts were made to reinforce the rural economy by imports of grain in times of shortage and relief measures when shortage approached famine. Efforts were also made to control epidemic diseases among cattle. In addition the government successfully implemented a number of water supply programmes to further crop production and animal husbandry. Where water was available in the late dry and early wet season, people could begin clearing their land for cultivation earlier and cattle could be brought back from cattle camps sooner to feed off the new growth of grass, thus providing milk for those engaged in the heavy tasks of cultivating, and thereby releasing the labour of some of those involved with herding, as has been described in Chapter 10. This allowed a greater area to be cleared, and early weeding to be carried out more efficiently (see BADA, 1984*b*).

Government revenues accruing from the southern Sudan were, however, extremely meagre throughout the period of Condominium rule. In the pastoral areas the administration at first attempted to offset its budgetary deficit by simply appropriating cattle as tribute. Once Chiefs' Courts became firmly established during the 1930s and 1940s, considerable revenue was also gained through the collection of cattle in court fines. Initially, these were sold almost exclusively to merchants for export to the north. By the mid-1940s, however, an increasing number of Dinka and Nuer were also bidding at government cattle auctions for the best beasts (Hutchinson, 1988). Most bidders had acquired the necessary currency either through the sale of less desirable livestock, such as goats, sheep, aged oxen and unproductive cows or from service as chiefs, chiefs' police, and other forms of government activity. Wages earned on government-sponsored road construction and wood-cutting projects were generally too low to permit direct investment in cattle. Eventually the government succeeded in reducing its vulnerability to major price fluctuations in the cattle export trade by introducing individual taxation for men, though, as Hutchinson (1988) has indicated, this took many years to complete.

Meanwhile improvements in public security and communications allowed the expansion of northern Sudanese trade networks in the Jonglei area. One of the earliest of these was in cattle hides. This trade had its origins in the great numbers of hides available during the rinderpest epidemics of the early 1930s. Northern merchants availed themselves of the opportunity, and the government took active steps to promote the trade, particularly encouraging frame-drying techniques to improve quality. By the early 1950s the trade had reached substantial proportions. Between June 1951 and June 1952, for example, 71,198 hides were exported from Upper Nile Province (JIT, 1954, p. 314; see also SDIT, Table 48). During the 1930s in Nuer areas and somewhat earlier in Bor and Kongor Districts, northern merchants, within the constraints imposed by government policy on the presence of northern Sudanese in the southern provinces, also began to establish a cattle export/grain import trade. Initially this was based on barter, but in due course cash became a medium through which trade was conducted. Dinka and Nuer would also sell or barter grain when it was plentiful after harvest as a way of meeting taxation requirements or acquiring trade goods such as salt, beads, and cloth. Merchants generally sold local grain to the urban populations in administrative centres, exported it or stored it in anticipation of the 'hungry gap' between May and July when necessity would frequently force Dinka and Nuer to buy grain back at higher prices (Hutchinson, 1988). Cash by then had usually been spent on earlier requirements and the high cost of grain at that time of year often had to be met by sales of cattle or small stock.

The introduction of wage labour had an impact on almost all aspects of the economy. In the 1940s Nilotes from the Jonglei area began to travel more widely in search of work in response to rising wage rates and a growing demand for consumer goods – mosquito nets, blankets and clothing, for example – as well as the need to buy grain. An additional stimulus to the acceptance of currency and wage labour by Dinka and Nuer was, as noted above, the creation of government supervised auctions in which female cattle and oxen collected as fines in Chiefs' Courts were available for purchase with cash. As Hutchinson explains, the sale of cattle obtained in this way was a very important factor in attracting local people into both the cattle and labour markets since it provided them with the first real opportunity to turn their cash into what they desired most: young and fertile heifers to increase their herds. During the late 1940s commercial auctions were established in the Jonglei area to which Dinka and Nuer were able to bring their own stock. A factor which hindered the growth of cattle markets, however, was that people themselves were disinclined to bring heifers or fertile cows to market, which in their eyes would have defeated the very object of trading. This limited the number of female stock available for purchase (Hutchinson, 1988). Nevertheless this was the beginning of the cattle trade, and one which provided a real incentive for young men to earn money, even though wages were still very low. As will be seen, it was a trade that gained impetus in the 1950s when earnings from outside the Jonglei area provided much higher returns for labour.

After 1947 these developments coincided with the relaxation of travel and trade restrictions on northern merchants in the south, which was a part of the new policy of integration between the northern and southern Sudan. Because of this, though also due to the increasing supply of imported goods after the shortages and rationing of the war years, the number of traders in the Jonglei area increased (SDIT, 1955, pp. 131–4). In the early 1950s, the resulting growth of trade networks coincided with the rapid expansion of northern Sudanese-owned cotton schemes in the vicinity of Geiger and Renk. By 1954 there were nine pump schemes irrigating some 8,500 feddans (3,570 ha) of cotton, and in that year a further ten schemes were approved (SDIT, 1955, p. 118). For young men interested in earning money to invest in cattle for marriage, work in Renk District was now an attractive option. By 1954 approximately 500 Nuer a month were moving north to these schemes during the dry season. Others were also involved; in some years, for example, around 200 Shilluk a month came north seeking work from July onwards (Chatterton, 1954).

These pump schemes initially proved so lucrative, owing to high world cotton prices at the time, that their expansion soon outstripped the supply of available labour, so much so that after independence Nilotic labour was actively recruited by the government. During the immediate post-independence period an estimated 15,000 seasonal workers were required. Nuer migrants appear to have been among the most sought-after by scheme owners, primarily because they were willing to accept wages on the basis of piece work at a time when local labourers were striking for the introduction of a daily wage (Hutchinson, 1988).

The Paloic Dinka inland from Melut, who by the late 1940s were producing grain for export to both the north and other parts of the southern Sudan, began planting rain-grown American cotton in the 1950s. They, too, were faced by a shortage of labour and also began to employ Nuer migrants (SDIT, 1955, p. 194).

It was not long before Nuer began to go beyond Renk in search of better wages on agricultural schemes and in towns farther north. By the early 1960s many Nuer were employed as unskilled casual labourers in the private sector construction industry in Khartoum. The typical migrant tended to be a young unmarried man who saved as much money as he could, sometimes to the point of extreme self-denial, in the hope of investing in heifers from local auctions on his return south. According to Hutchinson, wages in the pump-schemes rose so rapidly during the later 1950s that it was possible for some migrants to purchase a calf at open market upon returning home with as little as 15 to 30 days' wages. To earn a comparable amount at contemporary pay rates within the Jonglei area would have required about five months (Hutchinson, 1988).

Although local cattle prices soon began to rise as well, employment outside the region continued to be an increasingly effective way of acquiring or expanding a marriage herd, as well as raising money to meet expenses at home, including taxes, school fees, consumer goods and particularly to buy grain in times of shortage.

Dinka were also venturing northwards, especially those from the northern parts of the Jonglei area and Bahr el Ghazal. Some were initially employed on building sites, but many more became, for example, domestic servants, tailors or joined the army (Kameir, 1980). Unfortunately, less is known about the migrant labour patterns of the Dinka than those of the Nuer. The Southern Dinka were much farther away from northern labour markets, and nearby opportunities for employment were more restricted because Juba and other parts of the south were not undergoing the same measure of economic expansion.

The Bahr el Jebel floods: 1961 onwards

The 1960s were years of great hardship for the people of the Jonglei area. Flooding, which reached its maximum extent in the decade following 1961 (see Chapters 4 and 5), led to a contraction of lands available for both agriculture and cattle grazing in the areas closest to the Bahr el Jebel system. On the east side of the main river channels the inundation of pastures and village sites caused a shifting of settlements and cultivations eastwards. In the north-west the major part of Zeraf Island, the home of Lak, Thiang and the westernmost sections of the Gaawar Nuer, was especially heavily flooded and became virtually inaccessible. Many of the inhabitants of this area lost large numbers of livestock (El Sammani, 1984) and moved into the territory of the Gaawar and Lou Nuer, and of the Dinka living east of the Zeraf river in the vicinity of Khor Atar and Khor Fullus, with consequent pressures on available wet season pasture and agricultural land in those areas. Some Nuer even settled in Shilluk territory around Malakal.

In the central part of the Jonglei area Lou Nuer continued to negotiate entry into Gaawar Nuer grazing, but conflict frequently arose, particularly from fear of disease. A little farther south, long-standing good relations between the Lou Nuer and the Ghol, Nyareweng and Twic Dinka (see Johnson, 1982) were important in avoiding conflict since some Lou also continued to bring their herds south into Dinka dry season pastures, while many Dinka moved north-eastwards to settle more permanently with relatives among the Lou. Shortages in natural resources caused by the floods, however, as well as the fear of cattle disease, caused some friction and occasional violence. This is significant because given any diminution of river-flooded pastures under the new hydrological regime created by the operation of the canal, reverse population movements, political realignments and possibly friction can be expected in many parts of the Jonglei area.

The Dinka of Bor and Kongor Districts were beset by river flooding from the west, which temporarily inundated a large part of the territory of the Twic and Bor Dinka. Moreover, the simultaneous and growing threat of Murle raiding from the east resulted in the abandonment of many villages and cultivation areas on the eastern periphery of their high land, and limited the use of eastern intermediate grasslands at the end of the rains (see Figure 11.1), an important interim seasonal resource used for grazing before the herds were

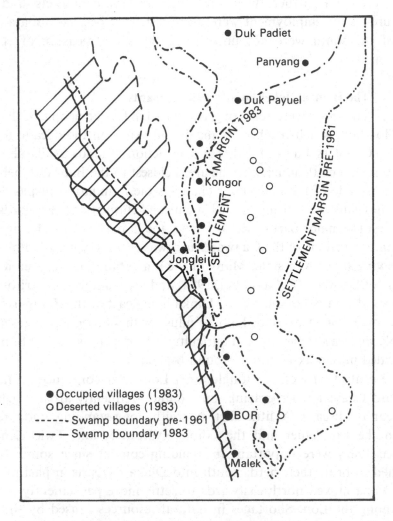

Figure 11.1. Bor-Kongor Districts: Diminution in permanently occupied land caused by floods and raids (Source: S. Zanen).

driven to the *toic* in the west (El Sammani & El Amin, 1978; McDermott & Ngor, 1984; Hoek, B.v.d. *et al.*, 1978). As a result many open wells and tanks which used to provide water at the start of the dry season were no longer maintained and fell into disrepair. Livestock were kept longer in the vicinity of villages, and were taken to the riverain pastures earlier than before, with markedly detrimental consequences, being driven to the floodplain pastures in muddy and waterlogged conditions, with consequent low productivity, lower milk yields and a higher mortality rate (RES, 1983). In this southern part of the Jonglei area *toic* pastures along the eastern edges of the Bahr el Jebel were thus grazed by too many cattle for much of the year (see Chapter 12). Before this period of flooding large numbers of Bor Dinka cattle were annually swum across the Nile to the Aliab *toic*, but high minimum discharges meant that this floodplain remained under water in the dry season, further restricting their grazing resources. Bor Dinka had always driven some of their cattle south into Mandari and Bari pastures (JIT, 1954, pp. 237–40), and increasing numbers of cattle were taken in that direction, causing mounting pressures there also.

Both Dinka and Nuer report that throughout the Jonglei area much of their cultivable land was inundated, and that large numbers of cattle and small stock were lost from exposure, lack of wet season pasture and greater susceptibility to disease. The reduction in territory available for cultivation added land shortages to the usual difficulties of crop production. Yet risks and limitations in crop husbandry were not new (see Evans-Pritchard, 1940; JIT, 1954, p. 210), and the import of grain to meet shortfalls in local production had been a fluctuating but almost constant feature of most of the Jonglei area since administrative records began. During the 1930s and 1940s, for example, crop failures and livestock epidemics had frequently forced the government to introduce famine-relief projects which provided the Bor Dinka with food in return for work. The JIT report concluded that 'Bor District as a whole is always short of grain' (JIT, 1954, p. 367). Nuer areas were also only irregularly self-sufficient in grain before the flooding of the 1960s; for example, in the 21 years before 1953, eight were years of shortage among the Lak, two due to drought and six to floods; and eleven among the Gaawar, three due to drought and eight to floods (JIT, 1954, p. 369).

It would thus be wrong to give the impression that flooding, both localised and more widespread, is an unusual phenomenon in the Jonglei area, or that economic conditions before the 1960s could ever be regarded as stable. Nevertheless, the river floods of the 1960s were far greater and longer sustained at high levels than any other experienced during this century (see Chapter 4). As described in previous chapters, in less extreme circumstances the extended kinship system, established through the pattern of cattle exchange in marriage, stands out as a way of organising the distribution of economic resources, being thus an insurance against food shortages and famine.

The Nilotic kinship system, however, is not an abstraction existing as a set of unchanging rules of conduct, involving both obligations and expectations of assistance and help, divorced from people's actions. People have to invest in the system to maintain it. Thus any seasonal abundance of grain tends to be used, especially in the form of beer, at social events, dances and marriages, or collective activities such as building, as well as in generosity to less fortunate kin and neighbours. No household lives in economic isolation; grain and other food is frequently transferred between the members of different households, particularly in times of localised shortage. The widespread effects of the floods, however, began to push the economic safeguards inherent in the kinship system beyond the limits of viability. Too many people over too wide an area were affected by shortages of grain and heavy losses among their livestock. This distressing situation was one of the main factors in increasing the number of young Dinka and Nuer men who migrated outside the area in search of work (see El Sammani, 1984; Ilaco, 1981*b*).

The impact of civil war: 1966–72

From 1966 to 1972 the hardships caused by excessive flooding were made worse by the disruption of civil war. Some people, like the Nuer to the west of the main river channels, may actually have been protected from the worst excesses of the war by the very floods that beset them (Hutchinson, 1985, p. 627), but those in more accessible places were not so fortunate; flooding rendered large areas uninhabitable and war made habitable areas unsafe. The effects were disastrous; both administrative controls and such rudimentary infrastructure as existed to serve the rural areas collapsed. Villages were burnt down by the army, local leaders executed, and herds were raided by both sides. The rural grain trade fell apart as northern merchants withdrew into towns or left the Jonglei area altogether. Large numbers of young men left their homes to join the guerrilla forces and were either absent for many years or never returned.

Among the Southern Dinka in particular the deterioration in rural water facilities during the civil war, paradoxical in an area suffering from excessive floods, may have been as great a limiting factor to grain production as the reductions in cultivable land from the flooding itself. The collapse and pollution of wells and the resulting lack of water near homesteads during the dry season led to many older, married people who might otherwise have spent that time of year in the villages being forced to move to the cattle camp before completing the post harvest processes of cultivation. Lack of water also prevented their return to begin field clearance and preparation until enough rain had fallen to fill the pools. This was a significant cause of delay in the start of cultivation in areas previously supplied with perennial water.

Some people affected by grain shortages and the breakdown in law and order took refuge in towns. There they hoped to gain access to governmental grain supplies and some measure of protective security. Others moved as far away from urban centres as they could, taking refuge in inaccessibility, while many fled as refugees to the northern Sudan, Ethiopia and even Uganda, to avoid both the army and guerrillas. Many other migrants, however, returned home to join the fight. Large quantities of automatic weapons were brought into the area, many of which remained in people's possession after the war. By 1972 some parts of the Jonglei area had suffered seven years of warfare and ten years of floods.

The local economy after the war: 1972–83

The war ended in 1972 with the signature of the Addis Ababa Agreement. Economic circumstances in rural areas, however, remained difficult. In many parts there had been changes in the location and availability of both grazing and agricultural land. In some districts, particularly Bor and Kongor, villages and fields were still concentrated on land that was frequently waterlogged at the height of the wet season, causing crop failures and reduced productivity (BADA, 1984a; El Sammani, 1984), and the risks of seasonal shortfalls in sorghum remained high. The grave reduction in cattle populations, combined with the dispersal of people from their homes to escape both the floods and the fighting, not only contributed to difficulties in the operation of economic safeguards inherent in the kinship system, but also led to problems for people trying to establish and maintain kinship links through the exchange of cattle in marriage. Steep reductions in bridewealth levels had come to be accepted, allowing the legality of marriages to be established, but the important objective of creating a web of links and reciprocal obligations between large numbers of people was hampered by shortage of cattle to meet the normally recognised claims of all (see Chapter 10 and Appendix 7). This was a major social problem which carried over into the 1970s, and particularly affected young unmarried people.

The restoration of peace in 1972, combined with relief and rehabilitation measures in the Southern Region and the beginning of canal-related projects in the Jonglei area re-opened a number of avenues through which people could acquire cattle, as well as relieving problems of food shortage in the countryside. The economic and social trends which followed were really continuations of processes that had begun before the civil war, reflecting the tendency to renew access to as wide a range of resources and opportunities as were available, an essential response, as we have seen, to a physical environment, the unpredictability of which had been so acutely demonstrated to the people in the preceding period of disastrous floods. These processes now centred, first on the growth of urban populations in Bor and Malakal, as well as smaller administrative and market centres in the area, providing employment opportunities that did not exist before, and allowing the expansion of the educated section of society; secondly, on the growth in the numbers of young men migrating to the north and to Juba in search of work; thirdly, on the increase of trade in fish; fourthly, on the greater involvement of young men in small shop-keeping enterprises; and fifthly, on the renewal and further expansion of the cattle trade.

The growth of urban centres

Bor, Malakal and, to a lesser extent, the smaller towns in the area, became active administrative centres, not just because of the restoration of peace, but also as a consequence of the granting of regional autonomy to the Southern Region and a measure of decentralisation of government responsibilities. In the early 1970s trade was renewed, with northern merchants gradually returning to import and distribute grain and consumer goods as well as commodities such as fuel and building materials needed by the authorities. There were also markets for local produce: people, particularly women, living nearby brought in thatching grass, grain, fish and other foodstuffs for sale. For women living in larger centres brewing millet beer, both illegally and in licensed drinking houses, and distilling illegal grain spirit were important sources of cash (Gruenbaum, 1978). Brewing was a significant factor in the economy in other ways too, because virtually all commercially produced beer was brewed from imported sorghum bought from merchants (Ilaco, 1981b). In general, there were increased sales to the urban populations of sheep, goats, fish, tobacco and other local products to meet the cost of taxes, school fees and the growing demand for medicines and cattle vaccines. The earlier substantial trade in hides did not, however, re-emerge (RES, 1983).

Increasing numbers of people began to come into Bor and Malakal in search of work. Bor was estimated to have been growing at around 8% per annum during the 1970s, much of which was accounted for by immigration (Ilaco, 1981b). By 1981 Bor and Malakal had wet season populations of approximately 13,500 and 50,000 compared with 600 and 12,000 respectively in 1954 (SDIT, 1955, pp. 77–80). Malakal's catchment included much territory outside the Jonglei area; Bor, however, attracted most of its population from within it.

Educated people were reluctant to return to live in the countryside, but there was considerable interaction between people in each town and its vicinity. To make too sharp a distinction between townspeople and country-dwellers, and between the economy of the town and that of the countryside would be misleading. Bor, for example, received an annual temporary influx of approximately 4,000 people in the dry season, partly because water resources were restricted in the villages, and partly because young people enjoyed coming into town at this time of year as a way of meeting others, an extension of the social attractions of the dry season cattle camp. A proportion of many urban incomes was remitted to rural kin not only in cash but in kind.[2] The uncertain economic climate of the Sudan meant that government employees – the government being by far the largest employer – rarely wished to sever links with their rural relatives.[3] Salaries could remain unpaid for months, and in that time town-dwellers were frequently dependent upon what they received from relatives in the countryside. Most men living and working in town also remained socially dependent upon their kin owing to their continued need for marriage cattle from that source. Even educated towns-men retained an active interest in cattle, and tried to keep a herd in the cattle camp tended by kinsmen. Marriages for such people still involved a bride-wealth of cattle, although other items, often from friends rather than kin, sometimes played a part.

Migration in search of work outside the Jonglei area

After 1972, with economic conditions still difficult in the country-side, migration in search of work outside the Jonglei area again increased substantially. Many Dinka and Shilluk, and even greater numbers of Nuer, went north to work on agricultural schemes, or in Khartoum and other towns (Abdel Bagi *et al.*, 1976; Hutchinson, 1988). Nuer continued to work as relatively low-paid casual labourers in the construction industry, where recruitment by building contractors of new migrants tended to be through relatives who had preceded them (Kameir, 1980). Like their predecessors, migrants were mainly young unmarried men, who generally worked away from home for between 6 to 18 months, depending on their motives and success in finding work (Bagi *et al.*, 1976; Kameir, 1980). Increasingly, however, married men began to join the body of temporary migrants. Another development was that Juba, the new regional capital, growing rapidly with the administrative and developmental input that had become disproportionately concentrated there, became an attraction for people from the southern part of the Jonglei area, particularly Bor and Twic Dinka.

For the bulk of the rural population the same economic incentives that led to local urbanisation encouraged migration farther afield (Bagi *et al*, 1976; El Sammani, 1984). The most significant motive behind this labour migration, however, was still the strong desire to acquire cattle for marriage. Even those who went north ostensibly to earn money for other requirements frequently did so in order to avoid having to sell cattle they wished to use in marriage. This was underlined by the way the initiative to migrate would come from the young man concerned. It was not so much a family decision about household economics, although kin would generally be consulted, but more an individual's decision about the development of the family herd and his access to cattle within it (El Sammani, 1984).[4] Clearly the possibility of wages higher than those available in Bor, Malakal and smaller administrative centres was the primary incentive for working so far afield, but there was always the added desire to work a good distance away from rural kin who might feel that they had some claim on a young man's earnings.

According to Hutchinson (1985; 1988), Nuer to the west of the Bahr el Jebel tended to invest their earnings in bridewealth cattle on their return, following the pattern of migrants in the 1950s. Many Dinka did the same, as did Nuer from the central parts of the Jonglei area (Daly 1984). Our knowledge of these central Nuer, however, is less extensive than of Nuer to both the east and west of them. Among the Eastern Jikany, for example, who live close to the Ethiopian frontier along the Baro and Sobat rivers, migrants would frequently buy consumer goods rather than cattle. There was an increasing demand for mosquito nets, mattresses, blankets and colourful clothing, not only for personal use, but as 'courting paraphernalia' (Hutchinson, 1985), a significant deviation from the single-minded pursuit of cattle suitable for marriage, but associated with marriage none the less.

Trade in fish

Trading in sun-dried fish also expanded after 1972. Traditionally fishing was largely a collective subsistence activity, as described in Chapter 10, taking place in March/April and November/December (see also Platenkamp, 1978, p. 66 and Chapter 14). Fishing waters were an important part of each section's territory, and decisions over grazing were usually made with access to good fishing in mind. In the fishing camps and nearby cattle camps large amounts of fresh fish were eaten, and much was sun dried for consumption during the rest of the year. Commercial fishing began to develop in the late 1940s and early 1950s, under the stimulus of big profits to be made from exporting sun-dried salted fish to what was then the Belgian Congo (SES, 1983; SDIT, 1955, p. 225). By 1954 commercial fishing existed in three reaches of the Upper White Nile: Geigar–Melut, Malakal, and Gemmeiza–Juba (SDIT, 1955, p. 111). Commercial fishing in those days was carried out almost exclusively by northern Sudanese. However Dinka and Nuer, particularly the *monythany* fishermen of the southern part of the Jonglei area, were beginning to produce unsalted sun-dried fish for local markets. During the middle 1960s the civil war disrupted this trade, but after 1972 the export market picked up again, some of it conducted by Dinka and Nuer organised by the Fisheries Directorate, but mostly by northern Sudanese fishermen who had returned to the area. In the early 1980s sun-dried salted fish, better preserved and trans-portable over longer distances, commanded between £S1,300 and £S2,300 per tonne to the producer, depending on quality and whether it was retailed direct or through middlemen. As before the civil war, much was sold through Juba to Zaïre (Lako, 1985); some was taken north to the Gezira (SES, 1983).

Many people had become full-time fishermen during the floods of the 1960s, after herds had been decimated. This was particularly true of those Nuer who remained on the Zeraf island (Jal, 1985; SES, 1983, p. 20) and those living west of the Bahr el Jebel near Adok (John Ruac: pers.comm). With the end of the war, sun-dried fish began to appear again in local markets throughout the Jonglei area, caught and processed by Zeraf Nuer, *monythany* Dinka and, with the realisation that fishing was a profitable cash-earning activity, by men who entered the fishing industry with the aim of earning money to invest in cattle (SES, 1983). Unsalted sun-dried fish was also exported to Juba, where in the early 1980s a tonne could fetch up to £S900 (El Sammani, 1984) as well as to other towns in Equatoria. Around Bor and Malakal a limited trade in fresh fish also existed.

Trade in grain and consumer goods

During the 1970s growing numbers of Dinka and Nuer, mainly younger men who had had some education, began to establish themselves as traders in grain and consumer goods. They raised their initial capital from a variety of economic activities – the sale of dried fish, crocodile skins, the products of local crafts such as simple metal working, for example – or from accumulated earnings in the north, and as time went on from wages earned in Bor and Malakal. As might be expected capital was also raised through the sale of livestock, particularly sheep and later cattle (Abdel Ghaffar, 1976; Hutchinson, 1988; Kadouf, 1977; RES, 1983). Despite the initial sale of livestock, however, the longer-term aim of such trading was often to accumulate capital to reinvest in cows for marriage.

Dinka and Nuer traders were to be found in both towns and small bush shops in the countryside. One of the most important factors which inhibited their success was that control of road transport and wholesaling remained predominantly in the hands of northern Sudanese merchants (Ilaco, 1981). This had been the situation before the civil war, when the few Dinka and Nuer who became traders were generally agents for northern merchants (SDIT, 1955; Tosh, 1981). Even the government was largely dependent on northern Sudanese motor transport and distribution networks (Ilaco, 1981), and, partly as a response to this, the southern regional authorities instituted a capital loans scheme in 1975 to encourage southerners to take up trading. Hutchinson (1988) reports that this later failed among Dinka and Nuer in a particularly revealing way. Upon the recommendation of local chiefs and government officers, would-be merchants were granted 50 sacks of sorghum with which to start their businesses. Most recipients, however, simply sold their grain allotments, bought cattle and married. A second, more successful loans programme was then attempted in which trucks rather than grain were granted. Nevertheless, the failure of the first programme would seem to suggest that marriage, not profit-making, was still the primary objective of these early Nuer entrepreneurs.

The links in the wholesaling and transport chain were commonly very involved; even northern merchants operating in the south were often the agents of larger merchants in Omdurman or other northern cities. It was only the particular circumstances of the commercial vacuum created by the departure of northern merchants from the Jonglei area during the civil war that allowed a few Nuer and Dinka to break into the long distance grain import/cattle export trade (Hutchinson, 1988). It was usually those with personal links with merchants in the northern Sudan who were most successful. The vast majority of Nuer and Dinka traders remained economically dependent upon northern merchants who throughout the 1970s, particularly from 1975 onwards, were re-establishing themselves in the south. There was a marked decline in the numbers of Dinka who went into trade in Bor from 1978 onwards, suggesting that it was by then a much less attractive enterprise (Ilaco, 1981). This was perhaps because the small-scale trading sector was saturated and had become less profitable owing to increasing competition from northern Sudanese.

Cattle trade
The increase in migrant labour was, as we have seen, encouraged by the new opportunities to purchase cattle, as was local involvement in entrepreneurial trade. The genesis of cattle trading has been described earlier,

but after the hiatus of the civil war, there was rapid expansion, with regular auctions at Malakal, Bor and smaller centres such as Kongor, Duk Payuel, Duk Padiet, Ayod and New Fangak. Besides sales made locally, many Nuer drove cattle to northern markets, attracted by the higher prices pertaining there, while Bor Dinka drove appreciable numbers of cattle southwards for sale in Juba and beyond.

As among all pastoralists, cattle of different sexes and different stages of growth had differing values for both Dinka and Nuer, some categories of beast being more in demand or more readily disposed of than others. People were naturally inclined to sell old or infertile stock rather than heifers or fertile cows, on which the viability of a herd for both milk and marriage depended. Oxen continued to be used regularly in sacrifice, but were also exchanged for other cattle or sold if the situation demanded it. The majority of beasts brought to auction were in fact oxen (Ilaco, 1981), although the finer beasts, subject to a quite different perception of value, continued to be retained as 'song bulls' (see Chapter 10).

Most oxen brought for sale would be bought by traders for meat sales locally or for export on the hoof to Juba or by steamer to the northern Sudan. In general there were two sorts of traders involved in this relatively large-scale cattle trading: professional traders, often northerners, sometimes operating through Nilotic agents; and Dinka and Nuer trading in their own right. Both kinds of traders purchased oxen at local auctions, and having amassed sufficient numbers drove them for sale to Juba or Malakal. On their return Dinka and Nuer often used the considerable profits made in this way to invest in heifers, only trading with the particular goal of marriage in mind (see Burton, 1978; El Sammani, 1984).

In a similar but more irregular way, some people made use of cattle auctions as part of herd management, selling unproductive stock in order to buy younger fertile beasts, following the pattern begun in the 1940s (Hutchinson, 1988; Kadouf, 1977; McDermott, J., 1984). These people would be bidding, of course, against returned migrants, traders, as well as others who wanted to acquire heifers for marriage, as described above. Just as the growth of the cattle market was a stimulus to the migrant labour movement, so the determination of the migrants to acquire cattle for marriage was a major influence on the development of the cattle trade. If people needed to raise cash for immediate needs, they generally preferred to sell small stock, or grain when available. However, as we have seen, it had long been the case that early sales of grain in times of relative plenty after the harvest could often lead people into a 'poverty trap', which might later in the year necessitate the sale of cattle. In extreme cases this could mean the sale of heifers and fertile cows, their most valued animals.

For marriage, men needed heifers or fertile cows, and the fact that such beasts were brought to market at all is indicative that at least some people were still subject to economic pressures of the kind experienced during the floods and civil disturbances. Through the 1970s there was a gradual upward trend in the numbers of all kinds of cattle brought to auction. It is difficult to assess the proportion of cattle sold in relation to the total cattle population, because detailed information is only available from two auction markets within the area, Kongor and Bor, neither of which are of direct relevance far outside their respective localities. There were no data from markets outside the Jonglei area, such as Juba, where large numbers of cattle from the area were also sold. At Kongor the total number of cattle sold in the auction market was not in any case high. It was very low when expressed as a percentage of the total livestock population of the district; sales in 1982, for example, were only 1.5% and 0.6% of the estimated wet and dry season cattle populations respectively (for seasonal changes in cattle populations see Chapter 12). Cattle from Kongor District were also marketed outside the district, although they were not recorded in great numbers in auctions at Bor (Ilaco, 1981). The percentage of cattle herded in Bor District brought to market in Bor town in 1979 was much higher; between 9.7% (wet season) and 5.7% (dry season) of the estimated total population, though this did not take into account any Bor Dinka cattle exported for direct sale in Juba. This compares unfavourably with the figure of approximately 8% per annum which Dahl & Hjort (1976) suggest is safely available for slaughter from a well-balanced pastoralist herd without prejudicing its long term reproductive viability.[5] This is discussed more fully in Appendix 8.

Murle raids

Pressures on Southern Dinka from Murle raiding in the east became a very serious threat through the 1970s. From interviews with local people McDermott & Ngor (1984) record 112 deaths (of which 28 were Murle) during the raids that took place between 1969 and 1983, and there must have been even higher numbers of injured. Around 14,000 cattle were stolen, 13,000 of them in two particularly large raids. The most significant feature, however, was the increase: 8 out of 14 raids occurred in 1983, the same year that according to Gurdon (1986) the Sudanese forces had again armed the Murle and encouraged them to harrass the SPLA. If earlier Murle plundering had been sufficient to discourage Dinka from the interim seasonal use of grazing to the east, abandon areas with good water supplies, and withdraw from some areas fit for wet season habitation and cultivation – with the result that cultivable land and intermediate grazing were much restricted (El Sammani & Deng, 1978) – the events of 1983 must have confirmed the wisdom of these moves in the minds of the people (see Figure 11.1, which illustrates the contraction of southern Dinka land usage in Bor and Kongor Districts, first by river flooding from the west, then by the threat of Murle raids from the east).

Economic conditions in rural areas

Owing to local grazing shortages during the 1970s some 20% of Bor Dinka cattle were annually taken south into Bari and Mandari lands, a great increase in numbers from pre-flood years (El Sammani, 1984). Even in Kongor District, where there was less evidence of economic hardship, a survey in 1983 concluded that there was a shortfall between production and consumption of grain (McDermott, B., 1984), whereas in the past this had been an area to which people from Bor District could sometimes turn for supplies in times of need. Grain imports through Bor port increased in the latter part of the 1970s, and although there are no data to indicate what proportion of imported grain was consumed in the households and beer shops of Bor town, it is likely that much was purchased by people from rural areas, particularly within Bor District.[6] Given that the flooding of the mid-1960s had also caused a reduction of well-drained cultivable land among the Nuer, where self-sufficiency in cereals had never been assured, it is likely that there was a considerable shortage in that area too.

The consequences of changing economic strategies
The diminishing scope of kinship

One of the effects of labour migration and movement into the towns was to remove much needed labour, however temporarily, from the rural supply. This was sometimes crucial if it continued through the very late dry season, or into the early to middle wet season, the period of intensive field preparation and weeding. It could affect the long-term ability of the rural population to produce its own grain supplies, and it has been argued that there was a decline in the quantity of grain produced in Bor District during the 1970s for this reason (Ilaco, 1981), as well as among Nuer both east and west of the Bahr el Jebel (Hutchinson, pers. comm.). Even a small reduction in the area of land under cultivation would have led to the greater dependence of some people on imported grain. As we have already seen, increased cattle sales gave greater opportunities for young men to build up a herd, and an added incentive to leave the district in search of higher wages, thus diminishing their contribution to agricultural production. In the long run, this could lead to a vicious circle of declining production, perpetuating the need for men to leave the area in search of work (Ilaco, 1981).

The effect, however, of the departure of large numbers of younger men on crop production should not be overrated, because their contribution to this component of the household economy was relatively small in the first place. They seldom took much part in all the processes of cultivation with field preparation or the harvest. Even cattle herding, their main occupation, was not a labour intensive activity, and for both these reasons young men could leave their homes without necessarily causing critical economic disruption. Movement from the countryside was rarely permanent, and although a high proportion of younger men moved for a period into town, took up trade, or went north in search of work, these trends on their own should not be taken as indicative of an unprecedented collapse of the economy. But the effect of married men leaving home for longer periods was more serious, and it would seem that in the recent past the Jonglei area had seen a progressive decline in the amount of grain produced.

Research among the Bor (Ilaco, 1979) and Twic Dinka (El Sammani, 1984; Kadouf, 1977) suggests that changes were taking place in the social system, particularly in the scope of kinship ties, the relative significance of different categories of kin and – especially in the towns – attitudes to conventional values and patterns of behaviour. No such information is available from the Nuer inhabitants of the Jonglei area farther north. We have to return to the observations of Hutchinson for clues at least to the kind of change which had been taking place among these people. Such clues might provide further indications of the nature of the social transformation that may be expected if the canal is completed and there is greater interchange between this and other parts of the Sudan, although more abrupt changes are now more likely to take place as a result of the renewed hostilities. Hutchinson provides a penetrating analysis of these processes among the Eastern Nuer, but observes that similar changes were not apparent to the same degree among the Nuer west of the Nile who, it will be remembered, had been shielded from the full impact of civil war by the barrier of river floods which did so much damage in other respects. Similar harmful physical conditions, on the one hand, and relative freedom from the ravages of war, on the other, had prevailed in the territories of most other Nilotes in the Jonglei area. The Eastern Nuer, though they escaped the floods, had borne the full brunt of the war, and the extrapolation of their experience to peoples living south and west of them cannot be made without considerable reservation. Moreover, they had been subjected to rather different and more direct extraneous influences and interventions since the early part of this century. The frontier with Ethiopia had always been an area of potential insecurity, and among other things the passage of firearms from that country had long been a feature of the region.

One factor that had some bearing on the contraction of kinship ties among the Eastern Nuer was the increasing acceptability of firearms, money and, to a lesser extent, radios and other consumer items, as substitutes for a portion of bridewealth cattle required in marriage. The transfer of firearms in bride-wealth appears to have been particularly significant in this regard. On the one hand, conventional patterns of distribution were weakened by the fact that the value of several head of bridewealth cattle was sometimes condensed in the form of a single rifle. On the other hand, firearms were usually needed for the security of the bride's immediate family and were therefore rarely passed on to more distant kinsmen. Firearms were held by many people in the Jonglei area, but there is no evidence that they featured significantly in the social system in this way. But among the Nuer inhabitants of the Zeraf Island in the 1960s, devastated by flood rather than civil disturbance, canoes became an instrument of equal importance for survival, had a recognised exchange value in cattle, and might sometimes, at least at the time, have been acceptable as a substitute in bridewealth (Ruac, pers. comm). The social implications of the extreme shortage of cattle in this area have not been investigated.

There were influences affecting Eastern Nuer society reported by Hutchinson which were, however, common to the whole Nilotic region. During the Condominium period the establishment of the court system whereby disputes were composed before they could lead to hostilities had already reduced the need for wider political associations for common defence. This, and the greater certainty of grain supplies through imports, had also somewhat diminished the need for extensive social relationships as security against hunger and famine, leading to a tendency for people to split into smaller units of segmentation (Howell, 1954; Hutchinson, 1988). In the Jonglei area, economic hardships and the need for security in times of flood and violence should have recreated a situation where extended kinship obligations were needed more than ever. Yet, as we have seen, they were incapable of fulfilment because a reduction in the herds, itself a cause of hardship, meant that there were insufficient cattle to meet them or to promote extended social links.

In the case of the Eastern Nuer the acquisition of cattle for marriage through the market eventually contributed to the contraction of the effective range of kinship ties. Cattle purchased with money earned as wages were increasingly subject to differing perceptions of ownership than those acquired through bridewealth transactions or the natural increase of herds. There appears to have been some uncertainty and dispute among Nuer about who had rights in cattle bought in this manner. Generally speaking, however, it was deemed less obligatory to subscribe beasts acquired in this way to help meet the marriage needs of others in the extended family as a contribution to the communal family herd. With the gradual regeneration of herds during the 1970s the levels of bridewealth had risen. The risks experienced during the civil war had radically altered people's perception of cattle as a secure form of wealth. Men were therefore keen to marry as quickly as possible on receipt of bridewealth cattle, adding to them beasts bought at auction. It was also prudent for the immediate agnatic kin to maintain control over as many cattle received in bridewealth as possible. One way was to reduce the number of cattle distributed to the bride's mother's side, or indeed to attempt to ignore their cattle rights entirely, though this amounted to a rejection of the essence of kinship.

This strategy had drawbacks for other members of a family lower in the order of marriage rotation, especially if they did not have close marriageable kinswomen, for they could expect nothing from those who had been denied a share in a previous marriage. The action of elder brothers could therefore create difficulties for younger brothers who, when their own turn came, were unable to call upon a wider circle of kin. In consequence there was a weakening of mutual obligations between close agnates, as well as between maternal and paternal kin. Many migrants would be young men who had brothers at home with a prior claim to marry, or men without close marriageable kinswomen and expectations of bridewealth cattle from that source. Thus individual ownership stemming from the purchase of cattle served to promote a spiral of diminishing kinship links, a contradiction of the aims of purchasing cattle in the first place (Hutchinson, 1988).

There is evidence of similar processes of thought and action among the Southern Dinka. The situation tended to encourage a disregard for the customary procedures of marriage, and there was a consequent increase in runaway unions in the vicinity of towns and administrative centres, and their subsequent settlement and legitimisation through the courts. Similar deviations from conventional practice have also been recorded among Eastern Nuer (Hutchinson, 1988). Urban centres provided sanctuary, because the presence of police there reduced the threat of violent reprisal, which in the past often led to extended hostilities and bloodfeuds. Once the case reached the courts, the relatives of the offender would usually be ordered to transfer bridewealth cattle, generally below the norm, irrespective of his place in the marriage order, to the disadvantage of the families of both parties to the misdemeanour. To proceed semi-independently of one's relatives in this manner did not allow the wider social objectives of marriage to be realised (see Appendix 7), and contributed further to the contraction of kinship ties.

Circumstances in the towns allowed other forms of liberation from the more restrictive conventions of the countryside. Money is a less visibly disposable form of property than cattle, and could be managed and appropriated individually. For some young men even temporary residence in the towns was a way to escape the more direct influence of their elders. Widows sometimes moved into town to evade the convention that they should co-habit with one of their dead husband's kinsmen. For some it was refuge from the pressures of other irksome kinds of kinship obligation. This was never totally successful; a common complaint in Bor and Malakal, and even from Nuer and Dinka working as far afield as Juba and Khartoum, was of the hordes of kin who descended on them under the illusion that they were wealthy enough to provide for all.

There was a great deal of disagreement and ambivalence among Dinka and Nuer about the nature and desirability of the changes taking place in their society. This is well illustrated by some of the attitudes that uneducated rural people held towards education and their educated kin. It was considered an asset to have male relatives who had been educated and who, through working for the government, had access to the resources that it controlled. Once part of the administration, educated people were expected to act in the interests of their kin by facilitating the allocation of government resources to their own areas, or by supporting their rural relatives with remittances. They were also expected to find positions for those relatives that came into town seeking work, and to protect their kin from the actions of other members of the administration, who were meanwhile taking care of their own.

However, the different economic and social environment of town-based educated people frequently led them to view their obligations to their rural relatives quite differently. This could arouse resentment on the part of rural people, leading to feelings of suspicion and dissatisfaction with the behaviour of their urban kin. Sometimes this could lead to a brother away working in the town being ignored when it came to his turn in the marriage order.

Perceptions of government

Such attitudes were an expression of the general mistrust of government to be found in rural areas, hardly surprising after the experiences of the civil war between 1966 and 1972, and reinforced by much of what followed during the 1970s and early 1980s. Despite the fact that throughout this period the administration of the Jonglei area was largely conducted by sons of the region, local people perceived an acute separation between government and themselves. Government was seen as a source of power which was alien and potentially dangerous.[7] The government demanded taxes, it could imprison, it had a police force and an army. It also decided the allocation of resources, such as health clinics or rural water supplies, but it was felt that in doing so it rarely seemed to take into account the wishes of rural people themselves. For example, villagers in Bor district claimed that between 1972 and 1983 they were frequently not consulted about matters directly affecting them. In one instance much resentment was expressed about the occupation of higher ground around Bor and Malek by government institutions, since this conflicted with wet season grazing requirements. The location of medical facilities were sources of contention, as was the alignment of the Bor–Kongor road. The outstanding example was the lack of consultation concerning the alignment of the canal itself (see Chapter 2).

People claimed, too, that the government frequently failed to give adequate support to rural 'self-help' schemes that it had itself initiated. For example, people sold cattle in order to raise cash for the building of schools and dispensaries. In many cases, perhaps the majority of them, the government then failed to provide adequate staff or equipment. There was a strong feeling in Nuer districts that the development schemes associated with the Jonglei canal were being over-concentrated in the Bor area and that their own interests were being neglected. Dissatisfaction reached such a level in 1981 that the Nuer members of the Regional Government resigned just before the dissolution of that government by Presidential decree.

Further resentment in the Jonglei area was caused by the actions of both the central and regional governments which drastically reduced opportunities for migrant labour and trade. In 1981–2 the Central Government, anxious to relieve Khartoum of the pressure of migrants, expelled thousands of Nuer labourers, along with many more thousands of other southerners as well as western Sudanese from the capital. Similarly, the Regional Government in 1982–3 took action which initially limited Dinka involvement in the Juba meat market, and ultimately led to the enforced repatriation of these Dinka and their herds to Bor District. These decisions severely curtailed the further involvement of the people of the Jonglei area in the development of the country, and this became one of the grievances underlying support for the SPLA in the Jonglei area. This was an indication of the extent to which the Dinka and Nuer peoples had become a part of the national economy, and their wish to continue to develop with it.

Development in perspective

It is necessary to qualify this picture. Socio-economic research in the Jonglei area as a whole has been uneven in both location and scope. Most of the work has been concentrated on the Dinka of Bor and Kongor Districts, and there are few reliable statistical data to quantify the trends discussed in this chapter. In most cases where data are available there is no long term database to substantiate observable trends. We have described in outline socio-economic change and its causes, though these generalisations may not have been universally applicable throughout the area. Although population estimates for the Jonglei area are of very doubtful accuracy, and it is not really possible to say whether the population has risen, fallen or remained static (see Appendix 6), it is important to stress again that during the 1970s and early 1980s the bulk of the population remained rurally based and primarily dependent upon subsistence agriculture. For most people, movement out of the area in search of work, or even into centres like Bor or Malakal, was only a temporary measure. Nevertheless, it is suggested that in the early 1980s Bor town made up approximately 23% of the estimated population of Bor District. Although this figure must be tempered by the fact that Bor gained migrants from other districts too, particularly from Kongor District, it remains a high proportion. In the northern part of the area Malakal also expanded dramatically.

Much of the infrastructure of the area was rebuilt following the destructive effects of war and flooding, but it also remained very rudimentary and inadequate (see Chapter 19). Some roads and tracks neglected during the civil war were repaired, the main road from Juba to Bor and northwards was reconstructed to a higher standard, and, of course, work began on the canal itself. The bulk of southern road building activity was concentrated in the immediate vicinity of the canal, and in the wet season most roads were impassable to motor vehicles. Health and veterinary facilities remained poor, struggling to provide effective services (see JEO, 1983; Redhead, 1984). Whooping cough, measles and endemic diseases such as tuberculosis and brucellosis continued to be common. Among cattle there were still recurrent epidemics of diseases such as rinderpest and contagious bovine pleuropneumonia (see Chapter 12). Educational facilities, despite their expansion, remained poor, and school attendance rates, particularly among girls, were low (El Sammani, 1984; Ilaco, 1979a).[8] By the early 1980s, however, employment opportunities were not sufficient to provide for the numbers of young people educated even under this system, and substantial pools of educated unemployed had developed in both Bor and Malakal. The period 1972 to 1983 saw more financial input to the Jonglei area than ever before, yet it needs to be stressed that Bor and Malakal were centres of a service economy entirely dependent upon financial resources derived from governmental or aid sources. Urbanisation was not accompanied by any parallel productive industrial growth, a common experience in Africa.

There seems little doubt that these somewhat sterile economic processes, accompanied by labour migration on an extensive scale, had begun to have profound, and in some respects damaging, effects on the social dynamics and rural economy of the Jonglei area. In addition to this, since 1983 the people of the area have once again become the victims of war, the long-term consequences of which will depend as much on the nature of the peace that follows as on the duration and severity of the fighting and its consequences. The purpose of this chapter has been to show that the effects of the Jonglei Canal must be seen against a background of communities that have already experienced many upheavals and changes, though nothing in the past is likely to match in scale the impact of the present struggle or its aftermath.

Notes

1. A period of 17 years of civil war followed the independence of the Sudan in 1956. The main area of military activity was in Equatoria, although by 1966 the effects of the war were being experienced in the Jonglei area. The current conflict in the south is not a straightforward resurgence of this war. In the present situation the main guerrilla force is the Sudan Peoples Liberation Army, the military wing of the Sudan Peoples Liberation Movement. By controlling large areas of countryside in Upper Nile and Bahr el Ghazal the SPLA currently holds the key to Sudan's economic development, as the richest oilfields yet located are there (see Mawson, 1984).
2. Over one-third of the rural population of Bor Athoic, for example, received financial support from relatives living in towns. This averaged out at a cash value per capita of about £S21.400 per annum. In return country people were sending goods to urban relatives to the approximate value of £S4.100. It has been estimated that the total amount annually remitted to the approximately 7,000 rural families in Bor District was between £S110,000 and £S125,000 (Ilaco, 1981).
3. In 1980 there were 983 men and women employed by the government in Bor town (Ilaco, 1981*b*), scarcely a large proportion of the district population, though there were others in smaller administrative centres.
4. No definite figures are available for numbers of Dinka and Nuer working in the northern Sudan, but it is thought that tens of thousands of young men were involved. In interviews conducted among Twic Dinka in Kongor District, El Sammani found that 21.7% of his respondents had lived for a period in Khartoum and 15.4% had visited Juba (El Sammani, 1984).
5. These percentages were calculated by taking the Bor and Kongor livestock populations presented in Appendix 8 and dividing them by (i) the 1979 figure for cattle emanating from Bor District sold in Bor auctions (Ilaco, 1981), and (ii) the extrapolation for Kongor sales adopted in the Range Ecology Survey (RES, 1983). The percentages are thus very approximate.
6. Figures for grain imports through Bor are available from the SDIT report (1955) for the period 1930–53 and from Ilaco (1981) for 1979 and 1980 only:

Year	1949	1950	1951	1952	1953
Tonnage	503	548	632	900	1,198

Source: From SDIT, 1955, p. 137

Ilaco (1981*b*) give figures in sacks, without stipulating the size of a sack. In a different context El Sammani (1984) reports that a sack is 'about 100 kgs':

Year	1979	1980
Number of sacks	87,000	108,000
Tonnage	8,545	10,607

7. Dinka called the services and facilities provided by the government *ke de aciek*, 'things of the creator'. *Aciek* is the creator god, one of the most important manifestations of the Divine (*nhialic*). It would seem that the perception of the 'things' of the government as *ke de aciek* is not the notion that the power of the government and that of the creator-god are the same, but that the experience of the power of the creator-god is similar to the experience of the

power of the government. Both move in mysterious ways, having an existence apart from people, and yet on occasion intervening in their lives in an apparently arbitrary way. The experience of both involves being on the receiving end of power, a force which needs to be held in balance (by taxation with the government and by sacrifice with God), and which can be influenced for the benefit of people, but which is also potentially dangerous.

8. In 1978 there were 25 primary and 2 junior secondary schools operating in Bor and Kongor Districts. These represented facilities for an estimated 17.8% of children of school enrolment age (7–12). In a situation such as this where school facilities have always been under-provided one would expect that many individuals outside of this age bracket would also wish to enrol. At that time the area did not compare favourably with the figures for the Sudan as a whole, where in 1975/76 22% of children between the ages of 7 and 12 were enrolled in primary school (El Sammani, 1984). By 1984 there were 24 primary schools in Bor District alone and 2 senior schools had been opened in Bor town. Teacher training lagged behind the rapid expansion in provision of schools.

Part IV

Agriculture and the exploitation of natural resources

12

Livestock and animal husbandry

Introduction

Cattle, and to a lesser extent sheep and goats, are at the heart of the economic and social life of the Nilotic peoples. It is the purpose of this chapter to complement the picture of the pastoral aspects of the economy already drawn in Chapters 10 and 11, with the basic biological information about the animals, how they are related to their environment, and the way in which management strategies are adapted to this.

Much of the information on which this chapter is based comes from the work of the Range Ecology Survey (RES, 1983), whose team was stationed at Nyany, 40 km south of Kongor, in the area of the Twic Dinka (see Figure 10.1). Here, a study group of Dinka cattle was selected at the start of the study, and then monitored for two years (see below). Difficulties of travel precluded similar detailed observations elsewhere. However, four visits, each of three weeks (in the dry seasons of 1981 and 1982), were made to Nuer camps near Woi (see Figure 10.1), and two shorter visits were also made to Shilluk villages near Malakal.

It was not easy to study the local cattle. The people are sensitive about their herds; excessive requests to examine a man's cattle, or even unexplained interest in them, would be regarded with suspicion. A man who discusses the numbers of cattle he owns is regarded as boastful, and likely to bring misfortune to his stock.

In April 1980, with the help of an interpreter, contact was made with local cattle camp leaders, and volunteers called for who would allow their cattle to be studied. From the herds in the care of these volunteers 138 cows, accompanied by a total of 124 calves, were selected to form the initial study group. Ear tagging was not permitted by the owners, but animals were individually identifiable using a combination of markings, earcut patterns, name of owner, and name of cow. At first the animals in the study group were visited once every three months, but in May 1981 this was altered to a monthly visit. At the same time, a second study group of 54 bulls and castrates was set up. Animals were lost from the groups for various reasons during the study (see below). Any calves born to study group animals were taken into the group. In addition, 27 male calves were purchased at auction and used for growth-rate studies. Unless stated otherwise, numerical information given later in this chapter is derived from the regular observations and measurements made on these study groups.

Livestock population estimates

The numbers of cattle, sheep and goats in the Jonglei area were estimated three times (in 1979, 1980 and 1981) by low-level aerial survey. The overall population estimates derived from these exercises are presented in Table 12.1, while a more detailed presentation of the numbers in each District of Jonglei Province (not the whole study area) is presented in Appendix 8.

The markedly higher totals for the late dry season count were due mainly to

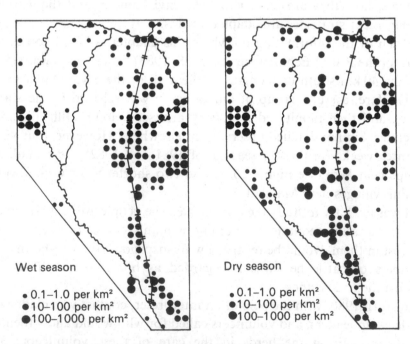

Figure 12.1. Cattle distribution in the study area in the wet season and in the dry season. Crossing points proposed for the Jonglei Canal are also shown (double lines: bridges; single lines: motorised ferries) (after RES, 1983).

considerable seasonal influx from east of the study area to the Khor Fullus, or farther south into Kongor District, from the west towards the west bank of the Bahr el Jebel and from the north to the north bank of the White Nile. This convergence of cattle is due to the extensive dry season grazing and water resources within the area. A further reason for the higher count in the late dry season (though a less important one) is that at this time most of the animals are out in the open *toic*, where there are few trees, and they are therefore more visible from the air.

The seasonal distribution maps taken from aerial surveys show cattle populations in the wet and dry seasons (Figure 12.1). This is important in demonstrating the movement of cattle during the annual cycle (see Chapter 10) and in connection with future crossing points for cattle described in Chapter 18 and also shown in Figure 12.1.

The types of stock in the area

The cattle in the area are of the Sanga type, probably derived from stabilised crossbreeds between *Bos indicus* and *Bos taurus* stock, but tending to resemble the former rather than the latter. Many authors (Mason & Maule, 1960; Payne, 1970; Payne, 1976) have regarded all Nilotic stock as forming a single breed. However, the JIT (1954, p. 301) commented 'It seems to us debatable whether Shilluk, Nuer and Dinka cattle constitute a single breed ...'. Local people definitely distinguish three different types, those of the Dinka, the Nuer and the Shilluk.

The typical Dinka animal is white, or pale coloured, with long legs, deep in the chest and long in the body, with a small udder. A massive pair of horns rising from a pronounced poll dominates the animal. The typical Nuer animal has shorter legs, is more barrel chested, and is more diverse in colour. The horns are similar in shape to those of the Dinka breed, but finer and shorter. The Shilluk cattle are smaller framed and finer boned than Dinka or Nuer

Table 12.1. *Livestock population estimates for the study area*

	Mid-wet season	Early dry season	Late dry season
Cattle	469,885 ± 39,602 (8.4%)	466,694 ± 58,173 (12.5%)	782,774 ± 103,787 (13.3%)
Sheep & Goats	101,636 ± 11,520 (11.3%)	96,696 ± 11,875 (12.3%)	176,442 ± 21,520 (12.2%)

[from RES, 1983]

stock, and are even more varied in colour. The horns are short and curved, like those of zebu cattle. The substantial differences in productivity between the types are considered later.

There is considerable mixing between these three types of cattle, and the influence of breeds from elsewhere may also be visible. What is interesting is that despite the great mixing of the peoples of the area over the preceding centuries (see Chapter 9), and the frequent redistribution of stock through raiding or marriage transfers, the types have still remained distinct, showing the power of human preference as an agent of biological selection. The great reduction in the herds caused by the floods of the 1960s led people, particularly after the first civil war ended in 1972, to acquire cattle for restocking from any possible source. In the southern part of the area, these were usually animals of the East African zebu type from the Murle and Taposa areas to the south-east; in the north, animals of the Northern Sudan zebu type, including the high-yielding Kenana breed.

The criteria used in selecting bulls for breeding differ between peoples (RES, 1983). The Dinka choose on the basis of coat colour, the shape and size of the horns, and body size. The Nuer appear to be more concerned with disease resistance and milk yield. The Shilluk are careful to obtain a bull from a different village to prevent in-breeding. They also prefer a small head, in the hope of avoiding problems at calving, and are apparently uninterested in horn or body size.

The sheep of the area are variously called Southern Sudan (Mason & Maule, 1960) or Nilotic sheep (JIT, 1954); the names appear synonymous. The goats are also referred to as Nilotic or Southern Sudan goats (Mason & Maule, 1960). Both sheep and goats are believed to be of very ancient breeds (Arkell, 1955). Sheep and goats have also been brought in from outside the area. Sheep showing varying degrees of fatness of tail, derived from the variously named Mongalla, Taposa or Murle fat-tailed sheep, and length of legs, the result of crossing with Northern Desert sheep, can both sometimes be seen. Goats show less evidence of cross-breeding but a few longer-legged, heavier-bodied animals, perhaps the result of crossing with Northern Desert goats, are to be seen.

The population structure of the herds

This varies from place to place and also during the year. In the dry season, when the main herds are in the *toic*, some cows with calves are kept at the homesteads to provide milk. Consequently, at this time (the only time when all animals are easily accessible to a researcher), neither a sample drawn from the camps nor one from the cattle-byres (*luaks*) is properly representative of the total population. Table 12.2 shows the cattle population structure in the Twic Dinka and Nuer areas. The most striking feature to emerge from this is that only 0.6% of the Dinka cattle population are adult bulls, giving a ratio of one bull to 65 cows (females that have bred). The Nuer keep far more bulls, giving a ratio of one bull to 16.5 cows. Further data are presented in Appendix 8.

Table 12.3 shows the population structure of the study groups of sheep and goats. There are very few males over 15 months old, because they are usually eaten earlier, and males of both species reach puberty at between seven and nine months of age, so that there is no need to keep older ones.

Table 12.2. *Cattle population structure*

Class		Number of animals	%
(a) Dinka			
Cows		517	40.5
Bulls	> 4 yrs	8	0.6
Castrates	> 4 yrs	47	3.7
Heifers	> 2 yrs	258	21.0
Bulls	2–4 yrs	29	2.3
Castrates	2–4 yrs	33	2.6
Males	7 m–2 yrs	99	7.7
Females	7 m–2 yrs	157	12.3
Males	< 7 m	54	4.2
Females	< 7 m	65	5.2
TOTAL		1,267	100.0
(b) Nuer			
Cows		330	43.3
Bulls	> 4 yrs	20	2.6
Castrates	> 4 yrs	63	8.3
Heifers	> 2 yrs	72	9.4
Bulls	2–4 yrs	13	1.7
Castrates	2–4 yrs	31	4.1
Males	7 m–2 yrs	70	9.2
Females	7 m–2 yrs	50	6.6
Males	< 7 m	61	8.0
Females	< 7 m	52	6.8
TOTAL		762	100.0

The annual cycle of management

In normal conditions, the Nuer and the Dinka move their herds according to a regular annual cycle. This has already been described, from the perspective of the people, in Chapter 10; the way in which it has been modified by changing conditions is discussed later.

At the end of the rains, in October–November, the grasslands of tall *Hyparrhenia* which surround the floodplain are burned. This, if the timing is right and the soil is still moist, leads to the production of a flush of young growth. Pools of water can still be found in *khor* beds and hollows. The cattle are taken out to these plains, their greatest distance from the river during the year, and remain there grazing the *Hyparrhenia* until surface water supplies run out in December–January.

By this time, the river flood waters have started to recede and the herds return towards the floodplain and the river. Traditional cattle camps, where as many as 5,000 animals are tethered each night, form the staging posts, in a series of moves, of this migration into the *toic*. The conditions of the pasture, the likelihood of disease transmission and the proximity of neighbours (and enemies) determine how long is spent in each camp. A typical stay lasts from

Table 12.3. *Population age and sex structure of small stock*

Age range (months)	<15	16–21	22–28	29–36	>36	Total
Teeth category	MT	1	2	3	4	
(a) Sheep (n=339)						
Males	30.7	1.4	0	0	0	32.1
(% of males)	(95.6)	(4.4)	–	–	–	(100.0)
Females	24.1	15.3	9.6	7.9	11.0	67.9
(% of females)	(35.5)	(22.5)	(14.2)	(11.6)	(16.2)	(100.0)
Age group totals	54.8	16.7	9.6	7.9	11.0	100.0
(b) Goats (n= 1,141)						
Males	26.4	0.6	0.1	0	0	27.1
(% of males)	(97.4)	(2.2)	(0.4)	–	–	(100.0)
Females	33.4	7.2	7.5	6.8	18.0	72.9
(% of females)	(45.8)	(9.9)	(10.3)	(9.3)	(24.7)	(100.0)
Age group totals	59.8	7.8	7.6	6.8	18.0	100.0

Note: The figures represent the percentage of the total population in each age/sex class. The figures in brackets are the percentages in each age class for males and females separately.

several days to a month. The Nuer use the same areas each year, but avoid the exact site of a previous camp, whereas the Dinka have evidently used precisely the same sites for centuries (see Chapter 9). In the *toic* the cattle are tethered at night around dung fires and released during the day to feed on the *Oryza* and *Echinochloa* grasslands of the floodplain. As the floods recede, the cattle and the people follow them. The Bor Dinka and the Shilluk may cross the river or move on to islands in the main channel of the Nile. This general pattern continues until the rains begin in April–May.

The people then move their cattle away from the river and out of the floodplain again to exploit the first flush of growth produced by the *Hyparrhenia* grasslands. This movement cannot, however, take place until enough rain has fallen to provide drinking pools, both for the livestock and the people. By July the *Hyparrhenia* grasslands have begun to become coarse and may also be flooded by accumulated rain water.

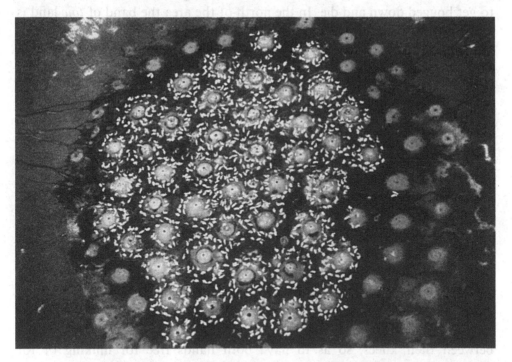

Plate 20. Vertical aerial view of a Dinka cattle camp in the early morning. The grey patches are the individual hearths. Surrounding each of these are the cattle, mostly white, belonging to each hearth. The two isolated pairs of cattle at the edge of the camp are almost certainly oestrous females tethered with a bull. Unoccupied hearths show that the camp can sometimes be larger. Tracks to grazing areas lead away to left and right. Photo: Alison Cobb.

The herds now move back to the better drained land around the settlements ('high land' of the JIT), and many of them are kept in cattle-byres (*luaks*) at night. They graze where they can in relatively unflooded areas, until the ending of the rains once again allows movement to the plains.

This is the typical picture of the annual cycle. From one part of the area to another, there is of course great variation in the distance covered each year, according both to the quantity and timing of the rains and the river flood, and the natural features of the topography, as described in Chapter 10.

In the south, raiding by the Murle from the east, described in Chapter 11, has meant that the Dinka of Bor and Kongor districts fear to move eastwards, and the *Hyparrhenia* grasslands there have remained virtually unused in recent years. The adjacent strip of *toic* land is relatively narrow, and abuts directly on permanent swamp. At the end of the dry season there can be a shortage of grazing, and it can also be difficult for cattle to reach water. Under these conditions cattle push into the swamp margins and weak animals tend to get bogged down and die. In the north of the area the band of *toic* land is much wider, and there are also shallow watercourses which carry useful grazing, so that dry season grazing shortfalls do not occur. In Kongor District, cattle may be restricted at the height of the rains by river flooding from the west and by rain flooding from the east which may lead to a shortage of grazing, though this is less serious than that in the dry season farther south.

Day-to-day management
Cattle

In the dry season, when cattle are in camps in the *toic*, and grazing is scarce, the Dinka herdsman's day begins at first light when the sheep and goats are released. Men and youths rub ash from the dung fires into the coats and on to the horns of their cattle. This is considered to deter biting flies and ticks; handling in this way increases familiarity between a man and his cattle, and thus his control over them. Other men go in search of cattle which may have strayed and joined neighbouring herds on the previous day.

Milking begins about 7am and takes about an hour. It is carried out by women, girls and uninitiated boys. They milk direct into a gourd gripped between their knees, so as to have both hands free for milking or for restraining the cow if necessary. The calf is allowed to suckle to induce milk let-down, and is then tethered by the cow's head while the first milk is taken for human consumption. The process is repeated, and then the calf is left with its dam until it is finally tethered to its own peg when the main herd moves off to graze. The time of release varies; in the dry season, when forage is limited, it may be 8.30 in the morning or even earlier, but in the wet season the cattle, whether in a camp or a *luak*, may not be released until 10 or 11am. The Dinka usually release their animals rather later than do Nuer or Shilluk at the corresponding time of year, a reason often cited for late release being that it diminishes contact with parasitic helminths, which move up and down the grass stems according to diurnal moisture fluctuations. Particularly in the dry season, however, when the risks of infection are reduced, both Dinka and Shilluk sometimes release their animals before milking, bringing them back to do so in mid-morning.

The herd is usually accompanied by two to six adult men and boys, the number depending on the perceived threat of raiders or predators. At least two herdsmen usually go to watch for cattle that have calving difficulties or get stuck in mud. There are times, however, when a herd hundreds strong may be untended for many hours at a time.

Calves that are still suckling are kept tethered in the camp while the young boys clear up the cattle dung produced during the night and spread it out to dry. Once this task is over, they take calves more than one month old to graze and drink. Calves younger than this are kept in the camp; if shade is available they are moved to it, but shade is rare in the *toic* and they will often remain tied in the full sun (or pouring rain) until evening. Rectal temperature

Plate 21. Early morning in a large dry-season Dinka cattle camp. The cattle, still tethered to their pegs, await release. Photo: Alison Cobb.

measurements on apparently healthy calves showed marked diurnal fluctuations. The older calves return around 4–5pm (although this, like the time of release, is variable). The adult cattle often return by themselves, but when grazing is short they are inclined to stay out late, sometimes not returning until after dark. Each animal in a family's herd is tethered to its own peg at night, the pegs arranged in an arc around the hearth.

The halters of the adult cattle are made from either rawhide or palm leaves and one of these is dropped over each peg. The dried dung is then gathered and poured over the smouldering hearths so that a dense pall of pungent smoke hangs over the camp until the next day. This smoke repels the often abundant flies and mosquitoes, and the cattle compete for prime spots in the smoke, downwind of the fires, when they return. After tethering to their pegs, they are left to chew the cud for about an hour before milking, after which they are left to settle down for the night. Cattle which stay in the *luaks* during the dry season are released earlier in the day because of the poorer grazing in the settled areas and because they usually have to go farther for water. They return at about the same time.

Plate 22. Dinka cattle drinking from a rain pool in the early wet season. The rest of the herd are moving through the grassland (largely *Sporobolus pyramidalis*) grazing as they go. The condition of the cattle is typical for the time of year. Photo: Michael Lock.

Sheep and goats

Such sheep and goats as are taken to the cattle camp are left to fend for themselves, but are usually tethered at night near their owner's hearth. Among the Twic Dinka, however, most goats remain throughout the year in the villages to feed the people who stay there throughout the dry season. Because they are vulnerable to predators, they are nearly always kept inside the *luak* at night and released at first light. If they are milked, it is usually once a day, in the morning. Herding of sheep and goats both in the *toic* and in the villages is somewhat casual, particularly in the dry season. In the wet season, however, more care is taken, because disputes can arise (and punishments be inflicted) if other peoples' crops are damaged.

Cattle recognition

Among all Nilotes, and particularly the Dinka, there is a rich descriptive vocabulary for cattle, and each animal is given a name which relates to its colour. The name depends not only on the colour combination, but also on

Plate 23. Goats tethered in a Dinka cattle camp. They lie on a large ash mound (hearth) which is partly surrounded on the right by logs set in the ground. Dung has been spread out to dry in the foreground, and behind there are temporary windbreaks woven from thin sticks. Photo: Alison Cobb.

the relative position of the colour patches. Thus a black-and-white male may be *Makuac, Mabil, Majok, Makuei,* or *Marial,* the *Ma* prefix denoting masculinity. Heifer names bear the prefix *Na,* but cows, on giving birth for the first time, lose the *Na* and acquire a further name, characterising a place or event particular to the birth, or a feature of her horns, for example.

Horn shape also varies greatly, and the names may denote this. This is particularly so in the case of song bulls (the castrate males) which play such an important role in the spiritual and social life of the people. Horns are trained by cutting the tip, when the animal is young, the desired shape being to curve the left horn forwards and horizontally, the right horn being left to grow naturally. Polled or hornless cattle occur occasionally; they receive the name prefix *Cod* or *Cot.*

The ears of cattle are often notched by cutting. This is usually done after weaning. The pattern of the ear cuts, often highly complex, is a mark of identity and is peculiar to each segment of the tribe (see Chapter 10).

Productivity of the herds
Fertility in cattle

Heifers of the Dinka breed calve for the first time at four to five years old. This is late by comparison with more productive cattle breeds, and is probably caused by their poor nutrition and thus slow growth to a rather large size at maturity.

On the basis of 90 observations (made during the Range Ecology Survey) the calving interval was calculated to be 24.4 $\pm$ 5.6 months (range 15 to 39 months). The average number of calves born to a randomly selected group of 154 cows was 2.7. Relatively few cows have more than four calves. Causes of infertility are discussed below. Of 119 pregnancies monitored, 15% ended in abortion. Figure 12.2 shows a marked peak of calving activity in the study group between July and October; 53% of calves were born in this period. Assuming a gestation length in *Bos indicus* of 290 days, this points to a main conception time between October and January. This is at the end of the wet season, when body condition is good and the grazing still adequate (see below, p. 302).

Mating takes place either while the cattle are at pasture or, if oestrus is detected while the cow is in the cattle camp, she is tethered in the morning at the edge of the camp with the bull of choice. This is the practice among the Dinka, but the Nuer either leave their breeding bulls untethered all night or let them roam through the camp seeking out cows in oestrus. In cattle with much *Bos indicus* blood, the period during which a cow will accept service from a bull (standing heat) can be as short as 2–5 hours (Rollinson, 1963). Thus a

Dinka cow tethered for up to 12 hours during the night and showing signs of oestrus may well have passed the time of standing heat, and refuse the bull when tethered with him in the morning.

The characteristics of the breeds and the criteria used in bull selection have been discussed above. Among the Dinka these criteria are strictly applied and only the best bulls are preserved entire. Indeed, if an owner keeps an entire bull thought to be inferior, it is not unknown for it to be killed by his neighbours. Thus random matings while the cattle are grazing are likely to be with a generally acceptable bull.

Nearly all calvings are assisted, more often than not unnecessarily. The calf is prevented by the Dinka from suckling the colostrum (the first milk produced after the birth), as it was said by them to induce *anuithok*, an illness whose identity remains a mystery. Nuer and Shilluk permit calves to take the colostrum. The calves are kept in the camp, and for the first ten days receive all the milk from the cow. Thereafter, milk is also taken from the cow for human consumption, and the calf is sent with others (see above, p. 287) to drink and graze during the day. No artificial weaning is practised; milk is taken both by the calf and for human use until the cow dries off naturally.

Milk yields

The average daily milk yields recorded (for human consumption) varied during the year, the main feature being the markedly lower yield during

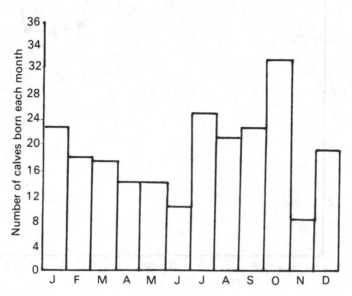

Figure 12.2. The distribution through the year of calf births to study group cattle (see text) (after RES, 1983).

the dry season (Figure 12.3). Values in Dinka cattle ranged from 1.73 kg per cow per day in June and July to 0.79 kg per cow per day in March. Comparative figures for Nuer and Shilluk cattle are given later. The average lactation, based on a sample of 319 lactations, was calculated as 352 days, during which an average of 463 kg of milk was taken for human consumption.

A feature of milk yield and milk use was that in the first six months of a lactation more milk was taken from cows with male calves than from those with female calves. In the first six months of life, a female calf received on average 12.4% more milk than a male, although the sexes attained the same weight at that age (RES, 1983). This weight is, of course, a greater proportion

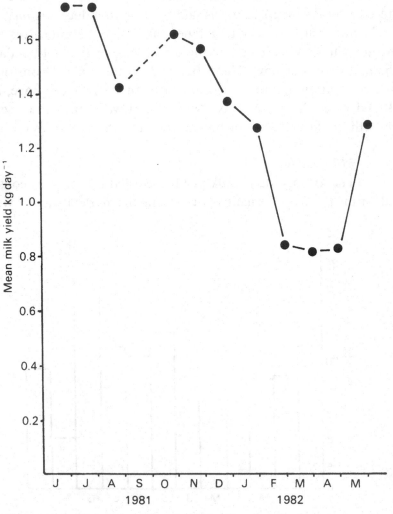

Figure 12.3. The mean daily yield of milk taken for human consumption from study group cattle (after RES, 1983).

of the final female weight than that of the male. This relatively poorer nutrition may have contributed to the higher post-weaning mortality found among male calves (see below).

Growth rates

The mean daily live weight gain of a group of 19 calves born in June and July was 0.14 kg per day during the first 300 days of life (Figure 12.4). The

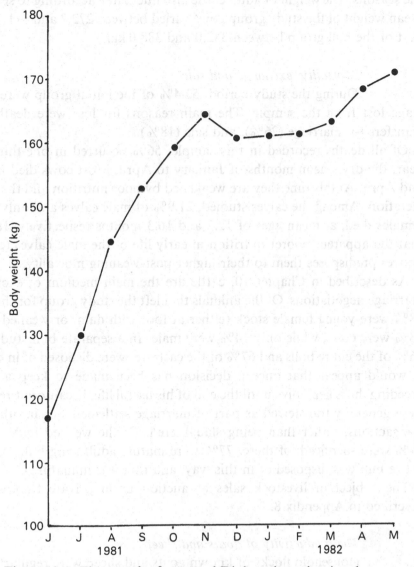

Figure 12.4. The growth of castrated Dinka calves, showing the contrast between wet (June–November) and dry (December–April) seasons. Each point is the mean of 18 weights (after RES, 1983).

average growth rate of calves in their second and third year was 0.16 kg per day. A group of weaned castrate male calves maintained a growth rate of 0.31 kg per day during the wet season between June and November. They lost weight in the next month, changed little during the dry season, and not until the following April was their November mean weight exceeded. It would appear that this pattern is repeated each year, so that the growth of Dinka cattle to mature body weight is not steady, but proceeds stepwise following the seasons. The weight of adult cattle also fluctuated according to season; the mean weight of the study group cows varied between 272.2 and 311.7 kg, and that of the bull group between 332.0 and 388.0 kg.

Mortality, exchange and sale

During the study period, 52.4% of the initial group were at some stage lost from the sample. The main reasons for loss were death (29%), transfers for marriage (26%), and sale (18%).

Of all deaths recorded in this sample, 56% occurred in one-third of the year, the dry season months of January to April. Most cows died in March and April. At this time they are weakened by poor nutrition and the stress of lactation. Among the calves studied, 21.9% of male calves and only 8.2% of females died, at mean ages of 12.2 and 16.3 months respectively. It may be that the apparent poorer nutrition in early life of the male calves (discussed above) predisposes them to their higher post-weaning mortality.

As described in Chapter 10, cattle are the main medium of exchange in marriage negotiations. Of the animals that left the study group for this reason 48% were young female stock (either at foot with dam, or weaned heifers), 38% were cows, while only 14% were male. In a separate bull study group, 36% of the entire bulls and 67% of the castrates were disposed of in one year. It would appear that once a decision has been made to keep a bull for breeding, he is kept alive until the end of his useful life, because entire animals were generally transferred as part of marriage settlements or in other social transactions, rather than being slaughtered. Of the weaned males studied, 24% were sacrificed; of these, 77% were mature adult song bulls. Only one entire bull was disposed of in this way, and that was immature.

The subject of livestock sales at auction or in private transactions is described in Appendix 8.

The productivity of goats and sheep

Household flocks of known goats and sheep were regularly visited in the Nyany area, and the main characteristics of their productivity were studied. The results are summarised in Table 12.4.

Two peaks of conception during the year can be clearly seen in Figure 12.5. These peaks coincide with the beginning and end of the wet season. At these times the grazing is both of good quality and accessible to small stock, as it may not be in the deeper water and mud at the height of the rains.

Goats are regularly milked; sheep very rarely. The yields are small, but goats give most when cows give least, an important factor in human nutrition noted in Chapter 10. The offtake from 10 goats at the height of the wet season in September 1981 had a mean of 144 g, but in the dry season the average yields were found to be twice as high. Goats' milk is drunk only by women and children; Dinka men refuse it, saying it is demeaning for grown men to take anything but the milk of a cow. This means that women and children have a very useful small supply of milk for themselves at the height of the dry season, and it also allows women and children to stay behind in the homesteads while most of the cattle are away in the *toic*.

Goats and sheep tend to die at different times of year. 76% of goat deaths from natural causes occurred during the wet season, whereas the dry season accounted for 62% of similar sheep mortality.

Owners are reluctant to part with female small stock, particularly sheep. During the two-year study, 35% of male goats and 32% of male sheep, but only 22% of female goats and 18% of female sheep were deliberately killed, sold, exchanged or given away by their owners.

Fewer sheep than goats were kept in the area, and this was reflected in the study group, which included 400 goats but only 121 sheep. Sheep were less robust than goats; during two years only 14.5% of the goats died of natural causes, while 26% of the sheep did so.

Despite the relative risks, sheep were clearly more highly valued. Traders investing in livestock bought sheep rather than goats. Sheep were rarely sold

Table 12.4. *Production characteristics of sheep and goats*

	Sheep	Goats
Age at first birth (months)	12.3	12.2
Birth interval (months)	7.2	8.2
Number of births per year per female	1.67	1.46
Number of offspring per birth	1.08	1.40
Twins (%)	8.5	36.3
Triplets (%)	0	1.8
Sex ratio at birth; females:males	1:1.67	1:1.05
Productivity per year	1.80	2.04
(offspring per female per year)		

at the District livestock auction centre at Kongor, but when they were, they made up to three times the price of an equivalent goat. Some of the enthusiasm for sheep may be attributable to the fact that the fat meat was preferred, and also to the remarkably rapid weight gain of young lambs (see Table 12.10). At 5 months old, for instance, male lambs weighed on average 14 kg (growth rate 93 g/day), which is nearly 40% more than male kids of the same age at 9 kg (growth rate 60 g/day). Even at 8 months, when the same lambs weighed 16 kg, they were still 24% heavier than the kids, at 12.4 kg.

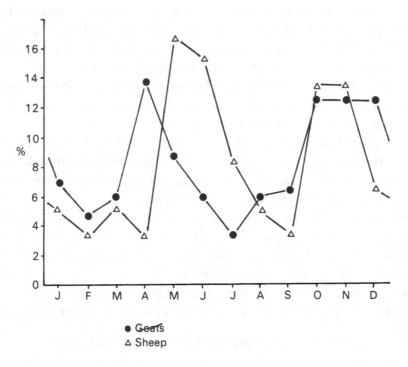

Figure 12.5. The seasonal distribution of conception in small stock (● – goats; △ – sheep). These data are derived from birth dates, assuming a gestation length of 5 months in both sheep and goats (Dahl & Hjort 1976) (after RES, 1983).

Variations in productivity within the study area

The main production characteristics of the cattle of the three peoples are shown in Table 12.5 below. As discussed at the beginning of this chapter, there are considerable differences in conformation between the different cattle types, and these differences are clearly reflected in the productivity of the three groups. Dinka cattle are the least productive, followed by the Nuer, with Shilluk cattle being the most productive. To some extent this may be due to better grazing in the Nuer and Shilluk areas, either through lower stocking rates or through easier access to better quality *toic* grazing, but there may also be genetic differences. This higher productivity is expressed particularly in the higher dry season milk yields. The longer lactation of Dinka cattle may be due to people persisting with milking even when the yield is very low, rather than to a physiological difference.

Nuer and Shilluk sheep and goats were not specifically studied, but they are similar in all areas, and are managed in much the same way. The Dinka leave more of their small stock behind in the villages during the dry season than do the Nuer. In the dry season, they often send their goats to friends and relatives who live near a well. In this way the goats can be safely guarded in *luaks*, sent to browse near woodland, milked by the women and children, and not put at risk from predators in the *toic*. All Shilluk small stock are kept inside the *luak* at night throughout the year.

Table 12.5. *Production characteristics of the three types of cattle in Jonglei. Growth rate for Shilluk calves very approximate.*

Characteristic	Dinka	Nuer	Shilluk
Age at first calving (yrs)	4.5	3.5	2–3
Calving interval (months)	24	18	16
Cow weight (kg)	285	231	230
Calf growth rate in 1st year (kg/day)	0.143	0.156	(0.16)
Lactation length (days)	352	190	240
Milk yield (kg/day)			
Dry season	0.83	1.28	1.37
Wet season	1.6	2.2	n.d.
Ratio – bulls : cows	1:6.5	1:16.5	n.d.
calves : cows	1:4.3	1:2.9	n.d.
Period when most calves are born	July – Oct	Nov – Jan	Oct – March

Limitations to productivity

Diseases and parasites

Diseases of cattle

Diseases which have the most significant effect on numbers or performance in the area are rinderpest (Nilotic: *nyapec*), contagious bovine pleuro-pneumonia (CBPP) and brucellosis. Haemorrhagic septicaemia and bovine ephemeral fever are also important, while foot-and-mouth disease may be so from time to time.

Rinderpest was the subject of a major attempted eradication programme in Africa in the 1960s and 1970s, under the aegis of the Organisation for African Unity. Although a considerable reduction in outbreak frequency was achieved, the disease was not eliminated, and it seems certain that the southern Sudan was one of the areas where it persisted. Numerous cases were observed in 1980 among both Dinka and Nuer cattle, many of them fatal. Rinderpest is potentially the easiest of the serious cattle diseases to control. An efficient vaccine providing life-long immunity from a single dose is available, and if all calves born each year are vaccinated, then complete control should be possible. The role of wildlife in the cross-transmission of this disease to cattle is discussed in some detail in Chapter 15. However, a regular and complete vaccination programme requires a settled political situation, transport, and fuel so that veterinary staff can circulate freely, conditions that cannot currently be met.

Plate 24. A Nuer dry season cattle camp. These occupy different sites each year, unlike those of the Dinka. Calves, too young to accompany the grazing herds, lie in the foreground among the tethering pegs of the adult cattle. Dung fires smoulder centre and left. The huts are temporary structures which are built each year. Photo: Alison Cobb.

CBPP causes coughing and other respiratory problems, symptoms the Dinka call *abut piou*, and which lead to quite a high mortality. Prolonged exposure is necessary for infection, but animals which recover can become carriers. The vaccines available against this disease have to be given regularly; a single dose does not confer lasting immunity.

Brucellosis causes abortion, and often some degree of infertility. In the area it seems to cause repeated abortion; elsewhere a cow may abort once and then produce normal calves although continuing to excrete the organism. Another local feature of the disease is swelling of the stifle, often associated with lameness. The Dinka refer to this as *'arem'*. Brucellosis also affects man, but it is unclear to what extent human infection occurs in the region; no statistics are available. Opportunities for infection are legion, for example, during assistance at calving and also during the inflation of the vagina (by blowing by mouth), a common Nilotic practice which stimulates milk let-down. Only a small number of tests were carried out on Nuer and Shilluk cattle, but the results suggested that this infection is less common than among Dinka cattle.

Parasites of cattle
In spite of the apparent absence of tsetse flies, over 65% of Dinka and Nuer cattle, and rather fewer Shilluk cattle, were serologically positive for trypanosomiasis, i.e. were showing evidence of past exposure to the disease. A much lower proportion were found to have the parasites in the bloodstream at the time of examination. It is difficult to assess the importance of trypanosomiasis in an area where there are other possible causes of debilitation in cattle. The Dinka recognise a wasting condition known as *'luac'* which is often translated as trypanosomiasis but which might equally be chronic fascioliasis. Many animals said to be suffering from *'luac'* were serologically and parasitologically negative for trypanosomiasis. Within the area, it is possible that the parasite (especially *Trypanosoma vivax*) is transmitted mechanically by the many biting flies such as stable flies (*Stomoxys*) and tabanids.

Tick-borne diseases such as babesiosis, theileriosis, anaplasmosis and heart-water appear to be unimportant in the indigenous stock. It is probable that most animals are exposed to these infections as calves, and acquire life-long immunity.

The liver fluke (*Fasciola gigantica*) is prevalent in cattle, causing anaemia and loss of condition. The alternation of wet and dry seasons, and frequent contact between cattle and suitable habitats for snails (the intermediate host of the parasite) together cause the high prevalence, and would make any control programme extremely difficult. Nematode worms are also often found to be abundant in animals, but the main species causing losses and debility is

Haemonchus contortus, a large nematode which feeds on blood from the wall of the abomasum. Bovine schistosomiasis, caused by *Schistosoma bovis*, is also highly prevalent but in most cases is probably not a major cause of debility. Details of the snails of the Jonglei area, and the trematode parasites they transmit, are to be found in Brown *et al.* (1984).

Diseases and parasites of sheep and goats
Much less information is available about the diseases of sheep and goats. The survey showed that there was serological evidence of bluetongue and *peste des petits ruminants*, but no clinical cases were seen. These two diseases could be a restraint on the success of any attempt to introduce exotic stock.

The parasites which have significant effects on cattle (*Fasciola, Haemonchus*, and *Schistosoma bovis*) are also important to goats and sheep. Fleas, lice and coccidia can also infest young kids and lambs and cause debility if not actual death.

General
Insect pests, notably tabanids and stable-flies (*Stomoxys*) by day and mosquitoes by night, constitute a severe constraint on animal production by the sheer misery they cause. They may reduce time spent grazing either by the continual worrying and distraction they cause, or by encouraging an early return to the shelter of the dung fires or the *luak*. There must also be an energy cost in terms of scratching, tail-swishing, head tossing and stamping. They add to the environmental stress on animals, as they do to humans. They may also spread disease, particularly trypanosomiasis and anthrax.

It is extremely difficult to quantify the effects of diseases and parasites on livestock production, especially in a mobile pastoral situation. When several conditions are contributing simultaneously to, for example, loss of weight at the end of the dry season, it becomes almost impossible. As a corollary of this, it would be equally difficult and misleading to attempt to translate causes of ill-health into monetary terms. It must be said, however, that the presence of rinderpest, CBPP and foot-and-mouth disease are a major constraint on the potential export of cattle from the area.

Tribal conflict
Conflict often arises over grazing rights in times of shortage, and fear of reprisals as well as actual pasture deficiency is a limiting factor to both animal numbers and productivity. In the Bor area, tensions recorded by the JIT (1954) and correlated with varying ecological conditions over the years, have become even more frequent since the rise in Nile discharges beginning in

1961. Friction has arisen between the Bor Dinka and the Mandari and Bari, who in times of greater plenty allowed the Dinka to share their resources. Some Bor Dinka take their cattle northwards into Twic Dinka territory, and use their dry season pasture by agreement, a movement that occurs in reverse in the wet season. Sections of the Lou Nuer, living immediately north-east of the Dinka, have almost always been short of water and grazing, and have been allowed to share *toic* pastures with the Gaawar Nuer south-west of Ayod, as well as with Nyareweng and Ghol Dinka. Disputes and hostilities may break out for many reasons other than grazing shortages, and the consequent state of insecurity may restrict cattle movement with the result that there is less than optimal use of available pasture.

A feature of recent times affecting the use of *Hyparrhenia* grasslands in the so-called Eastern Plain, has been conflict between Dinka and the Murle, and, to a lesser extent, the Toposa, two tribes of different ethnic origin from the Nilotes. The result of these raids and skirmishes has been described in the previous chapter. The raids have put additional pressure on the already overcrowded *toic*, to which the inhabitants are forced to drive their cattle much earlier in the season.

Animal nutrition

We have seen earlier that the weight and growth rates of livestock, particularly cattle, fluctuate markedly with the seasons. The most obvious similarly correlated factor is the amount of forage available. How are the two related?

The Range Ecology Survey included an investigation of the seasonal changes in standing crop and nutrient content of the grasslands in the Nyany area. A brief account of the methods employed, with the formulae used to derive additional information, is given in Chapter 7. The most important parameters obtained from the study were the total standing crop, the standing crop of green material, and the percentages of digestible crude protein and metabolisable energy in the standing crop. The fluctuations in three of these during the study period in the most important grassland types are shown in Figures 7.2, 7.3, 7.5, 7.7, 7.8 and 7.9. It must be remembered that these figures apply to the whole of the standing crop; no animal grazes unselectively, so that if there is sufficient bulk it should usually be possible for an animal to achieve a reasonable protein intake by exercising some choice. These four grassland types (*Echinochloa pyramidalis* (S), *Oryza longistaminata*, *Echinochloa haploclada*, and *Hyparrhenia rufa*) are described in Chapter 7, while the way in which they are used during the year is described earlier in this chapter.

The grass resource and annual grazing patterns

Examination of Figures 7.2, 7.3, 7.7 and 7.9 in relation to the seasonal migration pattern of the cattle herds already described makes it clear that their movements represent the only way of exploiting the area. During the dry season (January–April) only the *toic* grasslands (*Echinochloa pyramidalis* (S)) (Figure 7.3) have enough bulk to satisfy a cow. Their protein content is low, but by selecting material of relatively high protein content (mainly young leaves) a reasonable overall protein intake can be maintained. In May all grasslands respond to the rains with growth which, being young and leafy, has a high protein content. At this time grazing can be found anywhere, but it is the time when the *Hyparrhenia* grasslands (Figure 7.9) are traditionally exploited. At the height of the rains most cattle are grazing around the permanent homesteads where, although the standing crop is not high, protein content is good and flooding is least. At the end of the rains, the first move is traditionally back to the *Hyparrhenia* grasslands to use the flush produced after burning. By November–December the cattle are on the drier margins of the *toic*, where *Oryza longistaminata* (Figure 7.2) provide both bulk and a reasonable amount of protein, particularly if selective grazing is taken into account.

Small stock are more adaptable, mainly because their small mouths allow more selective feeding. Goats come through the dry season particularly well because they are browsers, and some shrub and tree leaves remain green and high in protein during the dry season. Sheep also manage well during the dry season, browsing less but selecting grasses such as *Cynodon dactylon* which continue growing at this time.

Mineral nutrition

Although both the soil (see Appendix 2) and the herbage are low in phosphorus and copper (RES, 1983), there are no indications of any mineral deficiencies. The same source suggests that the practice of rubbing ash from the dung fires onto the cattle may offset possible deficiencies, as the cattle tend to lick it from themselves and their fellows. Payne & El Amin (1977) and the JIT (1954) showed that the provision of salt (sodium chloride) may lead to increased milk yields.

Critical periods during the year

This is a topic which has raised some controversy. In the Range Ecology Survey area, the end of the dry season was undoubtedly the time when cattle were thinnest, when most deaths occurred, and when least milk was produced. Most births are in the wet season. This was true of the Dinka area near Nyany, which included a fair representation of all vegetation types, but in which cattle densities were relatively high. Evidence from other parts of the area, including from Nuer and Shilluk cattle, supported the interpretation that it is the dry season that is the least productive and most critical time for cattle. In view of the small amount of grass growth during the dry season, and the low quality of the residual dead grass, this is not surprising. This was also one of the conclusions of the JIT (1954), although, as described in Chapters 10 & 11, the peak of the wet season may sometimes so restrict mobility as to affect productivity. On the other hand, Payne & el Amin (1977) concluded, on the basis of a brief survey and of wet season aerial census figures (Watson *et al.*, 1977), that this latter was the critical limiting factor.

Undoubtedly there are differences in severity from one year to the next, and within the area. Both the years (1981 and 1982) of the Range Ecology Survey were years of relatively low Nile flood; in both 1980 and 1983 the flood was much higher. In years of low flood the *toic* grazing dries out quickly and there is little production during the dry season, so a shortage of grazing develops. In such years, also, the cattle are less restricted at the height of the wet season. It is possible that if the Range Ecology Study had coincided with two years of high flood, their data, though probably not their conclusions, might have been slightly different. Furthermore, in the Nyany area the *toic* belt is relatively narrow, and ends abruptly at the edge of the deep lake and river system. North of Kongor the *toic* belt is very much wider and its junction with the swamps much more gradual, so, as long as the cattle and their owners are prepared to keep moving, following the receding waters, they can find grazing much later into the dry season. While it is thus generally true that the dry season is the critical period for cattle, their ability to withstand this time of stress varies from year to year and from place to place.

The wet season is a most stressful time of year for small stock. Both sheep and goats appear to dislike feeding in standing water, and both dislike getting wet so much that they will run for the nearest *tukl* or *luak* when it starts to rain. There was catastrophic mortality of small stock in the floods of the early 1960s. Neither species thrives in the wet; both thrive in the dry. This is the reverse of the annual cycle of productivity in cattle, and highlights the advantage to people in the area of maintaining a mix of the three species.

Comparisons with other areas

The productivity of Jonglei livestock is clearly poor when compared with that of intensively managed stock in farmed or ranch conditions, either in Africa or elsewhere. To make a realistic assessment of the productivity of livestock in the Jonglei area, however, one must compare their

Table 12.6 *Environmental characteristics of ten pastoral areas*

	Sudan					Mali		Kenya		Uganda
	Twic Dinka	Nuer	Shilluk	Ngok Dinka Abyei[1]	Baggara[2]	Fulani[3]	Bambara[3]	Samburu[4]	Turkana[5]	Karamojong[5]
Mean annual rainfall (mm)	850	850	790	858	470	200–500	500	2–300	150–250	400–600
Floodplain use	**	**	*	**	*	**	***	–	–	–
Transhumance	++	+++	+	++	+++	+++	–	++	++	++
Vegetation type	Open grassland and floodplain	Savanna woodland and floodplain	Open grassland and floodplain	Savanna woodland and floodplain	Acacia grassland mixed annual and perennial and river valley	Annual grasslands with acacia and floodplain	Floodplain	Mixed annual grasslands with shrubs	Dry semi-desert shrubland. Narrow alluvial plain	Mixed annual and perennial grassland
Soil type	Clay	Clay	Clay	Cracking clay	Qoz sands	Sands and clay	Clays	Sandy	Sandy	Sandy loam

Sources: 1 Niamir, 1982
2 Wilson & Clarke, 1976a
3 ILCA, 1981a
4 Dahl & Hjort, 1976
5 Dyson-Hudson, N & R, 1982

Table 12.7. *Comparative characteristics of the productivity of cattle in African pastoral systems*

	Sudan				Mali			Kenya		Uganda			UK
	Dinka (Twic)	Nuer (Gaaweir)	Shilluk	Ngok Dinka[1]	Baggara[2]	Bambara[3]	Fulani[3]	Samburu (Boran cattle)[4]	Nandi Cattle (ranched)[4]	Turkana[5]	Karamojong[5]	Bahima[4]	Friesian[6]
Calving percentage	50	67	77	64	59	57.4	59	±24	42.7	38	34	70	
Age at first calving (months)	54	40	36		48	51	45	18	12		42–48		24
Calving interval (months)	24	17.9	15.6		18.4	16.8	19.0				14–15		12.5
Ratio Males : Females	1:3.7	1:1.9		1:1.6	1:2.2	1:1.2	1:1.6			1:1.5	1:3.4		
% cows in herd	40.5	43		30.0	42.8	35.7	36.1			35	37		
No. calves in lifetime	4–5	8–10			3	4–5						20	
1st year calf mortality (%)	24.8			14.3	15.5	21	17.5					5–10	
Annual cow mortality (%)	4.9					3.1	2.6						
Lactation length (days)	352	190		300			343	139	240		230		305
Daily milk yield (kg)	1.3	(1.5)	(1.7)	0.8			0.67	2.7–3.2	3.48	0.83	1.02		18.0
Lactation yield (kg)	462	(300)		100–450 (235)	193		235		831				5500

Sources: 1 Niamir, 1982
2 Wilson & Clarke, 1976a, 1976b
3 ILCA 1981a
4 Dahl and Hjort, 1976
5 Dyson Hudson N & R 1982
6 Brooke, R.A.H. pers. conum.

performance with that of stock from similar pastoral systems elsewhere in Africa. No two African pastoral systems are identical; each relates to an area with either a different rainfall regime, different vegetation, different soils, or a combination of these, and is affected by different social and political structures and traditions of land tenure and management. Table 12.6 gives a brief summary of environmental features for each of the areas compared, and Table 12.7 presents productivity data for cattle in each.

This table shows that cattle of the Jonglei area compare very favourably with those of other African pastoral peoples. Those of the Shilluk and the Nuer in particular do as well as any on the continent, though the Dinka cattle that were studied are less productive. This is true for the age at first calving, the calving interval, the number of calves produced and the daily milk yield. In terms of growth rates to one year, they are comparable with the Bambara and Fulani cattle of Mali, but lower than those of the Baggara of the western Sudan. However, growth rates (see Table 12.8) for the first two years of life are very similar, though in females the growth rate of Dinka cattle is better than that of others.

Data comparing the performance of Dinka small stock with others in Africa are presented in Table 12.9. This shows that they are as productive as any to be found in savanna areas of Africa under traditional pastoral management systems. Dinka small stock give birth at an earlier age, and have shorter inter-birth periods than any other (apart from Baggara goats, where the data are based on only a small sample). Their litter sizes are on a par with those from any other system, apart from Mubende goats and the East African Blackhead sheep, which were ranch-managed. The productivity of sheep, in terms of offspring per year (the product of lambing interval and average litter size), is better than any other reported. With the exception of Baggara goats, the same is true for Dinka goats.

Because the mature weights of different sheep and goat breeds are not the same, growth rates at given ages cannot be compared. Dinka sheep and goats are consistently lighter than are those from other pastoral areas. If, however, one expresses weight-for-age as a percentage of mature body weight, a comparison is possible and results are given in Table 12.10. In both the Baggara and Twic Dinka systems, sheep appear to grow faster than goats, and up to eight months old females grow faster than males. Within species, the percentage weight-for-age is very similar for females, although it is more variable in males.

In conclusion, it is fair to say that the Dinka, Nuer and Shilluk pastoralists of the area to be affected by the Jonglei Canal extract surprisingly good

Table 12.8. *Growth performance (kg) of cattle in different production systems*

	Males					Females				
	Weight at 1 yr	% Mature weight	Weight at 2 yrs	% Mature weight	Mature weight	Weight at 1 yr	% Mature weight	Weight at 2 yrs	% Mature weight	Mature weight
Twic Dinka (Sudan)	88	25	170	46	351	78	27	145	50	289
Nuer (Sudan)						74	31			236
Shilluk (Sudan)						74	32	137	59	233
Baggara (Sudan)[1]	105	29	175	49	360	105	37	150	53	285
Bambara (Mali)[2]	85	30	152	54	280	77	34	130	57	230
Fulani (Mali)[2]	84	34	126	52	244	82	40	116	57	203

Sources: 1 Wilson & Clarke, 1976b
2 ILCA, 1976a

Table 12.9a. *Comparison of production characteristics of sheep in African pastoral areas*

	Twic Dinka Sudan	Ngok Dinka Sudan[1]	Baggara Sudan[2]	Bambara Mali[3]	Macina (Fulani) Mali[4]	Black head Somali Sheep Ethiopia[6]	Oudah Fulani Chad[6]	E. African Black head sheep[5]	Indigenous Rhodesian[5] ewes
Age at first lambing (months)	12.3			15.3	16.5	18	13	17.5	15.3
Lambing interval (days)	216		275	242	251	521	372	296	257
Average litter size	1.08	1.04	1.14	1.06	1.35	1.05	1.07	1.33	
Lambing percentage per year	182		151	161	150	75	105	164	

Sources: 1. Niamir, 1982
2. Wilson, 1976*b*
3. ILCA, 1981*a*
4. Wilson, 1983
5. Dahl & Hjort, 1976
6. ILCA, 1981*b*

Table 12.9b. *Comparison of production characteristics of goats in African pastoral areas*

	Twic Dinka Sudan	Ngok Dinka Sudan[1]	Baggara Sudan[2]	Bambara Mali[3]	Sahel Upper[4] Volta[2]	Masai Kenya[4+5]	Mubende Uganda[4]
Age at first kidding (months)	12.2		(9.6)	15.3	12.0	18.3	18.6
Kidding interval (days)	246		238	261	329	292	296
Average litter size	1.40	1.06	1.57	1.20	1.27	1.25	1.33
Kidding percentage per year	208		241	168	141	155	164

Source: 1. Niamir, 1982
2. Wilson, 1976a
3. ILCA, 1981a
4. ILCA, 1981b
5. Peacock, 1983

Table 12.10. *Growth characteristics of goats and sheep in pastoral areas of Africa*

	Males					Females				
	5 month	% adult weight	8 month	% adult weight	Adult	5 month	% adult weight	8 month	% adult weight	Adult
(a) Goats										
Twic Dinka	8.9	32	12.4	44	28	8.7	30	12.4	42	29.5
Baggara[1+2]	12.5	25	17.5	35	50	12.5	38	16.0	49	32.7
Bambara[3]	12.0	45	17.5	66	26.5	12.0	45	17.5	66	26.5
(b) Sheep										
Twic Dinka	14.0	45	16.2	52	31	10.4	43	14.7	61	24.0
Baggara[1+2]	18.0	32	32.0	57	56.5	18	43	27	64	42
Bambara[3]	17.5	51	22.5	66	34	17.5	51	22.5	66	34

Note: All weights in kg
Bambara data were presented for both sexes together

Source: 1. Wilson, 1976a
2. Wilson, 1976b
3. ILCA, 1981a

performances from their cattle, sheep and goats, under what are very adverse environmental conditions, owing to seasonal extremes.

The future of the livestock industry

Attempts, over the last twenty years or so, to transform traditional African pastoral economies have met with almost universal failure (Goldschmidt, 1981). There is no reason to assume that things would be otherwise in the Jonglei area; change and improved methods of animal husbandry must be approached with caution, and preceded by careful investigation, research and trial.

The numbers of animals

One of the first questions to ask is whether the area could support a higher density of livestock in present circumstances; and then whether this would be the case after the canal is in operation. As we have seen, movement of the southern Dinka and their livestock into the Eastern Plain has recently been severely restricted by the threat of Murle raids (see Chapter 11). These grasslands were probably never the most crucial in the annual grazing cycle, providing 2–3 months grazing at the most. Nevertheless, the result has been an earlier descent on the river-flooded grasslands than might otherwise have been necessary and – at least in the southern part of the area – there is already a critical shortage of grazing at the end of the dry season. Lack of access to the Eastern Plain at the beginning of the rains may cause earlier, and hence over, exploitation of grasslands required during the wet season. Together these two circumstances imply that there is very little scope for increase in the grazing area available, and consequently for increase in numbers.

The recent constraint on the use of these 'intermediate' pastures to the east is a political issue. However, if the canal is in operation, while it may itself offer a measure of protection from hostile attacks, it will be another constraint on movement eastwards, necessitating in some cases four crossings each year (see Chapter 18). This, combined with the predicted reductions in *toic* pastures, will reduce the potential for increase in the herds.

In the lands occupied by the Lou and Gaawar Nuer, by contrast, there is, and, if the canal is operational, will probably still be, some scope for modest increase. On a drier Zeraf Island, post-canal, there is likely to be considerable opportunity for growth. For areas to the west and north of the Bahr el Jebel there is too little information to hazard any form of forecast.

In any event, any substantial increase in the numbers of cattle, sheep and goats should be matched by increase in the offtake, dependent upon the opening of suitable markets which the improved communications created by the canal should facilitate. Although overgrazing is not currently a constant or widespread problem, it could become so if reductions in dry-season pasture resources turn out to be as extensive as is suggested in Chapter 17, and increases in numbers or densities are not offset by parallel inducements for offtake through the market. Failure to link increased production with increased offtake has proved the downfall of many another livestock development endeavour in Africa, and there is no special cause for optimism for the Jonglei area.

Animal management and disease control

There are a number of aspects of Dinka cattle management which appear positively disadvantageous. The withholding of colostrum from the calf deprives it of a source of maternal antibodies which would confer some immunity to disease. The tethering of bulls in the cattle camp means that at least some cows are likely to pass through the period of standing heat without being served by a bull. The time of release of cattle is late; it might be better to allow cattle to graze in the cool of the early morning and to bring them into shade, if available, in the middle of the day. The Dinka believe, however, and with good reason, that releasing cattle early, while the dew is still on the grass, exposes them to worm infection. More milk could often be allowed to the calves, particularly males, which may suffer from poor nutrition. Crowded conditions in *luaks* increase the risk of the spread of respiratory diseases, and they are also a source of infestation with fleas and lice. But in the climatic conditions of the area there seems no alternative form of protection from exposure to rain and to biting insects, such as mosquitoes, stable-flies and tabanids.

Disease control will be a major objective when peace returns – nothing can be achieved until then – when the improvement of veterinary medical services and associated extension schemes should be given priority. The elimination of rinderpest – the most damaging of diseases – is probably feasible, given adequate supplies of vaccine, transport, and trained personnel, which are all currently non-existent. Control of Contagious Bovine Pleuro-Pneumonia (CBPP) is more difficult, but could be a long-term objective. Parasitic worms could be controlled by dosing at critical times of the year.

Conclusions

Considerable potential for improvement in the productivity of the livestock industry exists, but must be achieved by the cumulative effects of many small adjustments to the traditional system, which has evolved over the years as basically an extremely efficient way of exploiting a very difficult environment. The potential for improvement described in this chapter will exist irrespective of whether or not the canal is completed; the effects upon this potential and upon livestock numbers of the completion of the canal are discussed below in Chapter 17.

13

Crop production: traditional practice, constraints and opportunities

Introduction

There are many reasons why the Nilotic inhabitants of the Jonglei area regard crop production as a laborious and troublesome necessity, an attitude that is reinforced by a powerful attraction towards pastoral activities in which livestock, particularly cattle, have not only an economic role but also social and religious dimensions. There are fundamental reasons too, that relate to the physiological environment which have already been referred to briefly in Chapter 10. Limiting factors and constraints on production are such that additional labour inputs above a certain minimum do not produce commensurate extra returns. This is sometimes compounded by shortages of labour, perhaps more so now than in the past owing to the absence of a proportion of the adult male population on wage-earning expeditions. Too many risks and too many failures are disincentives to the expansion of areas of production. The strategy of mixed or sequential cropping, early-planted quick-maturing varieties balanced by second plantings of late-maturing sorghum, as well as the combination of varieties resistant to drought and food, reduce the risks. Yet total success is rare, sufficiency no more than average, a surplus unusual and total failure not uncommon, though some parts of the Jonglei area are less susceptible to the risks than others.

The environment is not one in which surpluses can be expected. The extreme variability of the rainfall from year to year, and the occurrence of rainless periods at unpredictable intervals during the wet season, often lead to crop losses and reduced yields (see Appendix 2 for further details of the climate of the region). The soils (see Appendix 1) are low in plant nutrients, and alternate between extreme hardness in the dry season and waterlogging during the wet. Diseases, birds and other animals attack the crops and take a substantial share. In the face of all these adverse factors, it is perhaps surprising that any crops at all are obtained.

In this chapter we attempt to describe the annual cycle of cultivation, the crops grown and their yields (where known), and also to give some idea of the variations found within the area and of the constraints upon production. We also deal briefly with some possibilities for improvement; the history of some recent attempts to achieve them are described in Chapter 19. The chapter is based on three main sources: the description of agriculture in the region produced by the JIT (1954, p. 323 *et seq.*); an account of farming systems in Bor District produced for this work by T. Struif Bontkes, formerly with ILACO at Bor, and a report on agriculture in Kongor District by McDermott (1984). It also draws on the account of agriculture among the Twic Dinka given by El Sammani (1984). Agriculture was not studied during the Range Ecology Survey but information was obtained from aerial surveys on the distribution of cultivations.

The annual calendar

A family cultivates, on the average, about one hectare of land, the size depending on the number of wives in the household, as each has her own plot. There are gardens surrounding the homesteads which are cultivated year after year while the main 'out-cultivations', where the bulk of the cereal crops are grown, will be as near as topographical circumstances allow in order to facilitate protection against birds and other pests. No rotation of crops is practised, and, although gardens round the houses receive a certain amount of incidental manuring from livestock pegged nearby at certain times of the year, shifting cultivation is practised and new land opened when the old is exhausted. Fields of this kind are therefore sometimes some distance from the homestead, though people will frequently move their houses to the site. The shape of plots is irregular, and takes account of microtopography to minimise flooding at the height of the rains. In some parts of the area low banks may be built around the plots to protect them from flooding.

The first clearance of land from the native vegetation is hard and time-consuming, requiring about 430 man–hours per hectare (McDermott, 1984), and is often carried out collectively by inviting neighbours and kin to assist, activities rewarded by meat and beer provided for the occasion. This may be done at any time of the year but is best done in the early dry season so that the uprooted vegetation dries out thoroughly. In subsequent years, only old crop residues and annual weeds need to be cleared, and the task is less hard. However, this is done in April and May, when the first rains have softened the soil, but have not filled the pools. Drinking water may be scarce, most of the previous season's grain has been eaten, while the cattle are still in the *toic*, so the hardest work of the year must be done when the people are not at their strongest.

Most of the cultivation is done with an iron-shod digging hoe or stick (Arabic: *maloda*), the chisel-like head being 10–20 cm wide. The heavier hoe, with the head set at right angles to the handle (Arabic: *turia*), is also found in the Nilotic area, but is not popular, being regarded as heavy to use without comparable returns. Spears and axes can also be pressed into service for land preparation and harvesting. Mechanisation is non-existent outside externally funded improvement schemes.

Seed is sown in the early rains, after the land has been prepared; a later crop is normally sown in September–October after the first crop has been harvested. Larger-seeded crops such as sorghum and maize are sown into holes prepared with a sowing-stick about 70 cm apart, 6–8 seeds being sown into each. Later, seedlings may be thinned and transplanted to give an even stand with about 80,000 plants per hectare (McDermott, 1984). Small-seeded crops such as sesame are broadcast, and the seed roughly covered by hoeing or by dragging a tree-branch over the soil. By hand-planting, the farmer can plant in very wet conditions when no machine could operate, and can plant or replant parts of a field at different times if necessary.

Once the main fields have been planted, attention turns to the small areas of higher ground close to the buildings. These are dug more deeply, and may be manured with cow-dung (McDermott, 1984). They are planted with maize, cowpea, pumpkins, okra and tobacco, among other crops.

Weeding is the main activity during June and July. One person can weed one hectare in 5–6 weeks. Delayed weeding increases the risk of water stress during periods of drought, because the crop must compete with the weeds for the available water.

By August, the sorghum must be protected from birds, one of the most serious hazards to crop production in the area (see Chapter 15). The usual method is to erect a high platform on which the bird-scarer sits, throwing or catapulting mud pellets, shouting, or rattling tins. This is often quite effective but is restricted to those fields not too far from the homesteads.

Harvesting of the main sorghum crop takes place in September–October. The heads are cut individually with a knife or spear head, and dried on racks in the sun. Prolonged rain at this time can cause losses through germination of the grain in the ear. A second crop of sorghum is then sown, usually after a gap of a month or so; this gap reduces damage by stalk-borers (McDermott, 1984). In dry years the first crop may be ratooned (i.e. cut and allowed to regrow from the base to give a second crop). All being well this second crop, which relies mainly on residual soil moisture for its growth, is harvested in January, and the land is then left fallow until the next planting season.

Plate 25. Dinka homestead with, from left, the *luak* (cattle byre), cooking house, and dwelling hut (*tukl*). The cattle are tethered in front of the *luak*, where a dark patch of dung, spread out to dry, is also visible. The water-filled depression at top left probably provided soil for the mound on which the buildings stand, and now acts as a water store. There is a dark patch of cassava by the *luak*, and to the left can be seen part of a large field of *dura* (*Sorghum*). Photo: Alison Cobb.

The crops

Sorghum (Sorghum bicolor)

This is by far the most important grain crop in the area. Its resistance to flooding, and its ability to stop growing for limited periods of drought, makes it particularly suited to the area. There is a wide range of varieties, several of which may be found in a single field. This mixing of varieties is deliberate, mainly in that different varieties differ widely in their resistance to flooding, so that a mixed planting is likely to give at least some yield whatever the conditions may be. No classification of varieties has as yet been attempted in the area, but it is possible to distinguish two broad classes. Varieties belonging to the first of these are of short stature, often with coloured grains, and mature quickly. The other class includes taller varieties which are slower to mature. These often yield more heavily in the south of the area; shorter, quick-maturing varieties yield best in the north, but both are planted throughout the area. In a trial of local varieties McDermott (1984) obtained yields of 500–900 kg ha^{-1}, but farm yields in the lower part of this range are regarded by the people as satisfactory (JIT, 1954, p. 367).

Maize (Zea mays)

This is planted by most cultivators in the area. Its importance lies in its quick maturation, so that it is ready to eat, albeit when unripe, before the first sorghum is ready. It thus plays an important role in reducing the length of the 'hungry gap'. It is limited by being less resistant than sorghum to both flooding and drought, and must be planted on flood-free sites, preferably well-fertilised by cattle herded previously in the area.

Cowpeas (Vigna unguiculata)

These are also planted by most families, usually close to the homestead. Although grown in small quantity, their high protein content must make them important in the diet. Infestation by weevils means that the seeds cannot be stored for long, and they are usually eaten soon after harvest. The leaves are used as a green vegetable during the growing season, as are those of *Corchorus olitorius* (Jew's mallow), which is sometimes grown in small irrigated plots close to the river during the dry season.

Groundnuts (Arachis hypogaea)

This crop is quite widely planted in well-drained sites, but harvesting of the underground fruits is difficult on clay soils. Ilaco (1980) obtained trial yields of 1,120–1,600 kg ha^{-1} on sandy well-drained soils.

Sesame (*Sesamum indicum*)

This oil-seed crop is grown in small amounts throughout the area, but is limited by its intolerance of flooding to well-drained raised areas, where yields of 530–590 kg ha^{-1} have been obtained, albeit from small trial plots (McDermott, 1984).

Pumpkins (*Cucurbita maxima*)

These are quite widely grown around homesteads, and the mature fruits can be stored for several weeks. Other cucurbits cultivated include watermelons (*Citrullus lanatus*), which are grown occasionally, mainly during the dry season, and bottle gourds (*Lagenaria siceraria*) and loofahs (*Luffa cylindrica*). The last two are grown to provide containers for milk and beer, and sponges, not as food crops.

Okra (*Hibiscus esculentus*)

This crop is another that is often grown around homesteads. The fruits may be eaten fresh but are also chopped and sun dried to give a powder which imparts a thick, glutinous texture to soups.

Tobacco (*Nicotiana tabacum*)

Most households grow a few plants of tobacco, mainly during the rains but also, if water is available, during the dry season. The leaves are dried and crudely cured in various ways (JIT, 1954, p. 349).

Geographical differences

Introduction

The only attempt at a comparison of the different agricultural practices within the area is that of the JIT (1954, pp. 366–72). This account is therefore based on theirs.

Southern Dinka

As explained in Appendix 2, this part of the area is particularly susceptible to unpredictable breaks in the rains during the wet season, often in July and August, while floods can be a problem late in the growing season. An early crop of sorghum is planted in April or May, with the first heavy rains. Losses from drought often necessitate resowing. This crop is harvested in August–September. A second crop is sown in late September or early October, but in dry years the early crop may be ratooned. In exceptional years both crops yield well, or both fail; usually a reasonable crop is obtained from either the early or late sowing. The parasitic weed *Striga hermonthica* (see below) is not a problem. Various vegetable crops are grown round the homesteads, and in the extreme south, on the sandy ridges beside the Nile, groundnuts can be successful. A few small farms can be seen on the river banks within the swamps; these produce crops during the dry season, probably for sale in Bor.

Nuer

The Nuer, in the centre of the area, have a particularly difficult environment to contend with. The ground is extremely flat, with few raised areas where cultivation is possible. Flooding by rain water is widespread, and often causes crop failures. Each household cultivates about one hectare of land.

Sorghum is planted at least three times in each wet season. A crop of quick-maturing varieties is planted as early as possible, usually close to homesteads or on small patches of higher ground. Once the soil has become wet enough for easy cultivation, a second crop of taller, slower-maturing varieties is planted in larger fields. This normally provides the bulk of the crop for the year. Finally, a late crop, or the same quick-maturing varieties sown early in the season, is planted either on specially cleared areas of land flooded earlier in the season, or among the residues of the early crop. Special attention is paid to this crop if the mid-season plantings fail. The small areas of high land available mean that a long rest period cannot be given between crops, and infestation by the parasitic weed *Striga hermonthica* (see below) can be a major problem.

Other crops, including maize, beans, cucurbits and tobacco, are generally planted close to the homesteads, with the early sorghum crop.

Shilluk

The Shilluk (see Chapter 10) are sedentary, have far fewer cattle, and are much more dependent on crop production than their southern Dinka and Nuer neighbours. Their villages are also more concentrated and permanent, and the density of population in relation to habitable land is very much higher. In consequence there is less scope for cultivation in the immediate vicinity of the houses. These cultivations (*pouthe thou'wot*) extend in clearly defined segments round the circular villages, and are subject to more intensively defined rules of tenure and are more closely managed. For example, at the beginning of the rains, large amounts of *toic* grass are cut and carried to these plots, spread thickly over the ground, and burnt. This practice, known as *path*, kills some of the weed seeds, destroys old crop residues, and doubtless adds nutrients in the ash, as well as improving the availability of nitrogen and phosphorus in the surface layers by heating them.

The most fertile areas, receiving run-off water from the house roofs and manure from the animals tethered there, are planted with tobacco, and with maize, cucurbits and beans. Farther away from the huts, an outer cultivated zone is planted with a mixture of quick-maturing sorghum varieties. Cowpeas and sesame may be interplanted with the sorghum.

The village plantings provide early crops to fill the 'hungry gap', but the main crops are grown in areas of intermediate land (*puothe wak*) where slow-maturing sorghum varieties are planted. These intermediate plantings, like those of the Dinka on the opposite bank, receive rather less, but certainly more evenly distributed, rainfall than areas farther south, so that crops are more successful. This applies to the more northern parts of Shilluk country; farther south intermediate land plantings suffer from excessive flooding at times, and there is a relative shortage of higher, better-drained land round the villages. Crop failure is more common than in the north. Throughout the area, *Striga* can be a major constraint and the distance between these 'out-cultivations' and the villages means that pests, such as birds and wild animals, are less easily controlled. Nevertheless yields are generally higher than among their Dinka and Nuer neighbours, 700 kg ha^{-1} being regarded as a successful crop.

The constraints on production
Introduction
In the sections above we have seen that crop production in the area is beset by many problems. These often lead to local or regional food shortages, at least at critical times of the year. In this section we discuss each of the main factors which cause, or combine to cause, crop failures.

Climate
The climate of the area is described in Appendix 2. The first difficulty encountered by the farmer is in judging when to sow. Often there are false starts to the wet season, with a period of heavy showers being followed by several weeks of dry weather. All crops are vulnerable at this stage, and a weather pattern of this kind will often make resowing necessary, sometimes more than once.

Once the crop is established, further problems can arise. Mention has been made in Appendix 2 of breaks in the rains. These decrease in frequency from south to north, but are otherwise unpredictable, and can do substantial damage to crops. Although sorghum is reasonably resistant to drought, and can stop growing for a week or two if water is short, it is impaired by longer periods of drought. Other crops are more easily damaged. The greater reliability of the rainy periods in the north of the area, and the lesser likelihood of dry periods during the wet season, especially in the areas occupied by the northern Dinka, explains the relative success of the crops in those areas, while more erratic climatic conditions farther south impose upon the inhabitants a more frequent need to import grain.

Among Nuer and southern Dinka, waterlogging and flooding succeed the earlier risks of periodic drought as the season proceeds. In addition to direct flooding from local rainfall, there can be flooding with water originating either from heavy storms some distance away, or spill from rivers and watercourses, or combinations of both, arriving in the form of creeping flow (see Appendix 3). The soils of the area, described in Appendix 1, are particularly susceptible to waterlogging. Sorghum is reasonably resistant to waterlogging, but other crops are much less so and some, such as sesame, are very intolerant. Some protection from flooding is sought by selecting higher ground for cultivation sites; where this is not available, low banks are often built to prevent the ingress of creeping flow. If accumulations of water from local rainfall become excessive, they are sometimes baled out over the banks. The Nuer also sometimes construct mounds on which crops are planted.

Finally, the end of the wet season is as unpredictable as its beginning, and once again there can be difficulties in judging the correct time for the sowing of late crops. Late rainstorms and flooding, or by contrast the early onset of the very dry winds from the north-east which quickly dry out the soil, can destroy late plantings. The rate of drying varies with the soil type, lighter soils drying out more quickly than clay-rich ones.

Soil fertility

Soils are discussed in Appendix 1. It is shown there that the soils of the region are generally low in most plant nutrients, with phosphorus and nitrogen being particularly low. It is likely that most of the long-established crops of the region are adapted to grow in soils of low nutrient status; recently introduced crops are likely to suffer unless fertilised. The possibilities for this are very limited; occasional use, more often than not coincidental, is made of animal dung on the homestead cultivations, but much of the dung is burned to produce insect-repellent smoke. Burning, while it leads to the loss of nitrogen and sulphur in the smoke, leaves other nutrients in the ash (see Chapter 7) so is not wholly destructive. The Shilluk practice of *path* (see above) is likely to have a beneficial effect.

Pests

The farmer in the Jonglei area protects himself to some extent from damage by pests by growing a range of crops, and by planting a mixture of varieties of some of them. In spite of this, crop losses are large.

Although there may be limited damage after sowing and during germination, birds do most damage to sorghum when the grain is ripening (see Chapter 15). As described earlier, a great deal of effort goes into bird-scaring, and as long as people can be spared to do this, it is reasonably effective. The size of the area that can be protected by a single family is, however, probably a limiting factor on the area that can be cultivated. Other crops, because they are usually planted close to the homestead, are more easily defended from birds, but the Range Ecology Team lost most of a maize crop, within their camp, to Cape Rooks (*Corvus capensis*)!

Some damage is done by wild mammals, but this is usually limited in extent. Elephant and hippo, because of their size, can destroy large areas of crops in a short time, but neither species is now common. Rats and mice can damage groundnuts, and also feed on stored crops.

Locusts were formerly a major problem, but were until recently kept under control by inter-governmental agencies throughout eastern Africa. Recent reports suggest that the Migratory Locust may have again become a major threat, but this is unconfirmed so far as the Jonglei area is concerned. Apart from locusts, insect pests are relatively insignificant, although stem-borers cause considerable damage to sorghum and maize, and the caterpillars of the moth *Spodoptera littoralis* (army worm) can destroy many plants.

Weeds

Weeds are abundant, and compete with the crop for water during dry periods, and for light when the crop plants are still small. The JIT (1954, p. 351) gives species lists. All are controllable by regular hoeing.

The parasitic weed *Striga hermonthica* has been mentioned earlier. It is almost absent from the southern part of the area but can be abundant and damaging in the north. It attaches itself to the roots of plants of sorghum and maize, and weakens them considerably; the JIT estimated losses of up to 30% of yield due to *Striga* infestation. Very large numbers of small seeds are produced which stay dormant in the soil until stimulated to germinate by the proximity of a host root. Some control can be achieved in sorghum by regulated flooding, as practised by Western Nuer Tribes, who subject *Striga* to levels of flooding which kill or weaken it, but damage the crop very little or not at all.

Diseases

The most serious disease in the area is sorghum smut (*Sphacelotheca sorghi*), which effectively destroys the heads of grain. Its prevalence varies considerably from year to year and from area to area.

In view of plans to increase rice cultivation, it is worth pointing out that wild rice (*Oryza longistaminata*) is extremely widespread, and that it might well act as a reservoir for diseases of cultivated rice.

In summary, the cultivator in the Jonglei Canal area has much to contend with. The vagaries of climate are probably the most important factors militating against regular successful crops; problems of soil fertility follow, with bird pests a good third. At present, other pests and diseases are relatively unimportant, but could become more prominent if other difficulties are overcome.

Possibilities for improvement
The crops and crop varieties used

Only a rather limited range of crops is grown in the region. Of these, only sorghum, cowpeas, okra and Jew's mallow are native to Africa. Other crops, such as maize, groundnuts and beans, have been introduced from other parts of the world in the last 3–400 years. As a general rule, it can be assumed that centuries of selection, natural and artificial, will have produced varieties of crops suited to the environment of the area. Local taste, also, is conservative, and newly introduced crops may not be readily acceptable, at least at first, even if they have higher yields. Thus varieties of sorghum tested by Ilaco yielded well but farmers did not like their taste. However, there may still be scope for the introduction of varieties from other parts of the world where climatic and soil conditions are similar. Trials by Dutch consultants (Ilaco) and by McDermott (1984) have shown that real possibilities exist. Paddy rice, tomatoes, and green gram (*Phaseolus aureus*) have all been successful to a greater or lesser degree, and are now cultivated by local people, but sweet peppers and eggplants did not win local acceptance. Ilaco also distributed fruit trees (citrus, mango, and guava) to schools for planting near boreholes, where they could be irrigated during the dry season.

Ilaco also tested the possibility of sowing mixed crops of sorghum and rice, the intention being that in wet years, or on flooded parts of the cultivated areas, the rice would thrive and provide the bulk of the combined yield, while in dry years and on elevated sites the converse would be true. The trials were only on a pilot scale but showed considerable promise, although a need emerged for a rice variety maturing at the same time as the sorghum.

Soil fertility

Appendix 1 describes the soils of the region and includes analyses which show that by normal agricultural standards they are deficient in major plant nutrients, particularly nitrogen and phosphorus. In limited trials, the JIT (1954) obtained little or no increase in growth through additions of single nutrients, but a doubling of yields when a complete fertiliser was applied.

In view of the high costs of fertiliser, and the very long distances that it would have to be transported, it is unrealistic to suggest any increased use of imported fertiliser unless economic circumstances improve in other respects. Rather, efforts should be devoted to demonstrating any improvement in yields that may follow the application of manure, household refuse (of which there is very little) and ash to cultivated plots. If and when the import of fertiliser becomes feasible and economic, then phosphatic fertiliser should have priority (not least because there are potential sources relatively close at Tororo in south-eastern Uganda), and should be applied together with the locally produced wastes mentioned above.

Cultivation and management

The JIT and later the Dutch agency both experimented with the construction of ridged or cambered beds in which crops are grown on ridges to protect them from flooding. Yields were increased, but the labour of construction of such beds probably places them beyond the reach of most people. However, if their advantages could be clearly demonstrated, they might become popular; in other parts of the world cultivators have, for example, been prepared to construct elaborate terrace systems when the advantages of so doing are sufficiently clear.

A major constraint identified by both the JIT and Ilaco is the difficulty of clearing enough land before the sowing season. The installation of boreholes or *hafirs*, by providing water at a critical time in the cultivation season, allows the earlier return of labour to the fields, and means that more time can be devoted to land clearance and less to walking long distances in search of water.

Mechanisation has been looked upon as a panacea which would allow the clearance and cultivation of large areas, but the experience of Ilaco is that this is not so. The soils are peculiarly intractable, and are only suitable for mechanised cultivation on the very few days in the year when they are neither too hard nor too soft. McDermott (1984) suggested the establishment of a government-operated tractor ploughing service which would charge enough to allow it to break even. Farmers would be encouraged to join into groups so that the tractor could plough a worthwhile area in one locality in a day. Past experience has been that such schemes rapidly founder for lack of fuel, spare parts, and the skills required for maintenance.

At a much simpler level, there must surely be scope for the testing of other patterns of hoe for surface cultivation and for weeding. The present tools are so inefficient that almost anything would represent an advance. Such a project would require only a small input of time, personnel and money.

Large and small-scale irrigation

Irrigation is another practice which is often looked upon as the answer to all the problems of crop production in the region. Such experience as there has been, however, has suggested that large-scale schemes are unlikely to be successful, because of the infertility and intractability of the soil. Experiences in this field are discussed in more detail in Chapter 19 below.

Small-scale irrigation, close to the river, is another matter. A number of very small-scale irrigated gardens exist, producing greens such as Jew's mallow, amaranthus, and cowpea, all harvested and sold when young and leafy. Some of the fishing settlements along the river also grow maize and other vegetables during the dry season, for their own consumption and for sale, and there is clearly scope for extending this. Ilaco were just starting one such scheme, with some signs of success, when renewed hostilities put an end to the work.

It should be noted in passing, however, that the JIT carried out fairly successful trials in the irrigation of many varieties of rice, sorghum, cotton and forage crops on a small experimental scheme at Malakal which was sufficiently promising for the Department of Agriculture to plan for its extension to 5000 feddans. The SDIT report made the point that water could be used – to supplement rainfall – more efficiently in the southern provinces than in the semi-arid conditions of the Gezira, though it needs to be stressed that this observation was made strictly on engineering considerations. Calculations of irrigation requirements to supplement rainfall showed a cropped area per 1,000 m^3 of irrigation water of 0.49 feddans at Bor, 0.56 in Malakal and 0.27 at Wad Medani (SDIT, 1955, p. 234).

The JIT, in searching for 'remedial measures' (see Chapter 1) recommended a rotation of rice with pasture grasses or, alternatively, cotton, sorghum and a forage legume (*Lablab purpureus*).

The JIT assumed the practicability of gravity irrigation and, indeed, stipulated that the functions of navigation and irrigation should be performed by separate canals, one of 5 million m^3/day capacity to be run parallel to the main canal or canals with cross-regulators sited at offtakes for irrigation canals for units of 100,000 feddans (42,000 hectares). The major constraint appeared to be difficulties of drainage, except where water could be carried to natural water courses, a matter of particular difficulty along the line of the canal. Mechanised production was recommended only in the transitional belt between the Semi-Arid and Flood Regions, e.g. in the Paloic and Dunjol Dinka areas, where rainfed crops have always been much more successful than farther south, and where the soils are less susceptible to waterlogging. The viability of irrigated crop production in economic terms was always open to question, depending on the types of crops and cropping patterns feasible, and costs of transport, quite apart from technical engineering and agronomic factors, but it needs to be said that the potential should not be summarily dismissed. Much further research, sustained experiment and trial in irrigated agriculture is needed, to take place concurrently with efforts to improve more traditional means of rain-fed cultivation.

Conclusions

The foregoing paragraphs might tend to suggest that there is no possibility of overall and consistent self-sufficiency in food in the area in the foreseeable future. However, while each of the possible improvements mentioned may seem small, taken together they could well have a substantial effect on the economy of the area. To date, there are no results from trials to suggest that a 'transformation' approach (Garang, 1981) to the area's problems is likely to succeed. There are, however, signs that the introduction of a number of small improvements might be of considerable cumulative benefit to the people of the area when peace returns.

14

Fish and fisheries

Introduction

Fish stocks in the *Sudd* wetland, to the west of the canal line, offer one of the few under-exploited resources which remain in the inland waters of Africa at the present time. Unlike some other African cattlemen, the inhabitants of the Jonglei area appreciate the value of fish and they have traditionally obtained a major seasonal food supplement from the floodplains and margins of the riverain swamps. Some have turned to fishing for subsistence and there is a growing awareness of the usefulness of fish as a cash-crop. Problems related to processing and marketing are, however, major constraints to fisheries expansion.

Boulenger (1907) catalogued the fishes of the Nile, and their zoogeographical status is reviewed by Greenwood (1976). If the highly endemic faunas of Lakes Victoria, Kioga, Edward and George are excluded, then the Nile basin, sharing a large number of species with rivers in western Africa, may be viewed as a northern extension of the Occidental Province. According to Sandon (1950) about 100 species have been recorded in the upper White Nile system which encompasses the *Sudd*, including 31 siluroids, 16 characoids, 14 cyprinoids, 11 mormyrids, 8 cichlids and 7 cyprinodontids. For these Girgis (1948) found local names from the 'swamp people' which covered 62 species. Representatives of the fish fauna are shown in Figure 14.1.

Until recently little was recorded about the distribution and ecology of fishes in the *Sudd*. Its size and remoteness have tended to restrict exploration to occasional forays along main channels and into large lakes. From studies associated with the earlier Jonglei investigation, Sandon (1951) discussed fragmentary reports on fish movements and reproduction and, with Tayib (1953) presented some data on the food of *Sudd* fishes. Further information on the diet of fishes in the area is supplied by Monakov (1969).

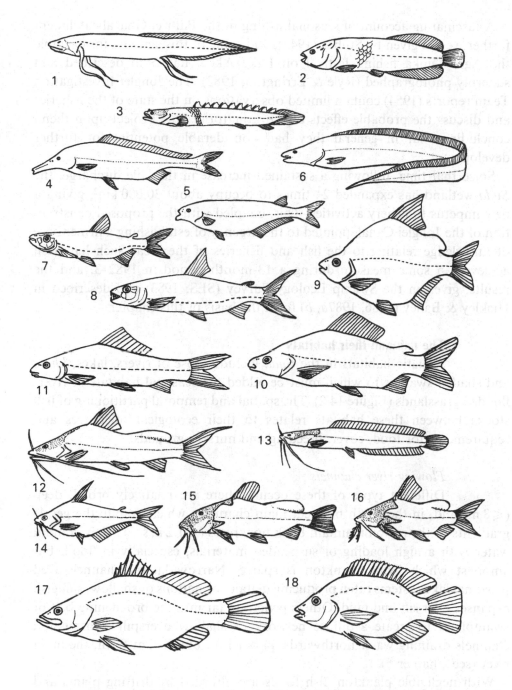

Figure 14.1 Representatives of the Sudd fish fauna: 1. *Protopterus*, 2. *Heterotis*, 3. *Polypterus*, 4. *Mormyrus*, 5. *Hyperopisus*, 6. *Gymnarchus*, 7. *Alestes*, 8. *Hydrocynus*, 9. *Citharinus*, 10. *Distichodus*, 11. *Labeo*, 12. *Bagrus*, 13. *Clarias*, 14. *Eutropius*, 15. *Auchenoglanis*, 16. *Synodontis*, 17. *Lates*, 18. *Oreochromis*.

A fascinating account of seasonal fishing in the Bahr el Ghazal catchment farther west is given by Stubbs (1949), and some of this activity, for example the Agar Dinka fishing festival on Lake Akeu, has been described and superbly photographed (Ryle & Errington, 1982). The Jonglei Investigation Team reports (1954) contain limited observations on the state of the fisheries and discuss the probable effects of the Equatorial Nile Project upon them, concluding that in general they had considerable potential for further development.

Since that time, following a sustained increase in the Nile discharge, the *Sudd* wetland has expanded $2\frac{1}{2}$ times to occupy about 30,000 km², giving a new impetus to fishery activities. This, coupled with the proposed construction of the Jonglei Canal, pointed to the urgency of establishing a firmer base of knowledge relating to the fish and fisheries of the area. Such has been achieved in some measure during a 13-month period in 1982–3, and the results, given in the Swamp Ecology Survey (SES, 1983) and described in Hickley & Bailey (1986, 1987*a, b*) form the basis of this chapter.

The fish and their habitats

Aquatic habitats in the *Sudd* include those of rivers, lakes, *khors* and shaded swamps to which must be added the seasonal habitats of river-flooded grasslands (Figure 14.2). The spatial and temporal partitioning of fish stocks between these habitats relates to their ecological tolerances and requirements for food, cover, breeding and nursery grounds.

Flowing river channels

Different types of these occur. There are relatively broad, deep ($\leqslant 8$ m) and, in the south braided, main channels, which despite the gentle gradients of the region maintain brisk flow rates ($\leqslant 1.2$ m s⁻¹). They contain water with a high loading of suspended materials, especially in flood, but amongst which living plankton is sparse. Narrow lateral channels feed proximal lake systems often producing deltaic formations of shallow, sluggish expanses over silt and mud. Others penetrate far into the broadening belt of swampland to create a diffuse network of shallow overspill transmission channels draining water northwards, parallel to, but distant from, the main river (see Chapter 5).

With negligible plankton, fish foods are afforded by drifting plants and animals, detritus, and aquatic invertebrates such as waterfleas, shrimps and insect larvae in bordering vegetation. Monakov (1969) also identified a very impoverished benthos of oligochaetes and midge larvae with sample bio-masses ranging from 0 to 0.2 g m⁻². Probably all fishes recorded in the *Sudd* may be found at some time in flowing water, but compared with lakes this is a relatively harsh habitat except in sheltered fringes and backwaters. Commonest in catches from main channels are omnivores or carnivores of one sort or another. They include the characids, *Hydrocynus* and *Alestes*, mochokid and bagrid catfishes, and the Nile perch, *Lates niloticus* (Table 14.1). The tilapia, *Oreochromis niloticus* and the pelagic catfish, *Eutropius niloticus*, occur in slack water and the tiny species, *Micralestes acutidens* and *Chelaethiops bibie*, are abundant in the water hyacinth fringe. Of all of these fishes only the

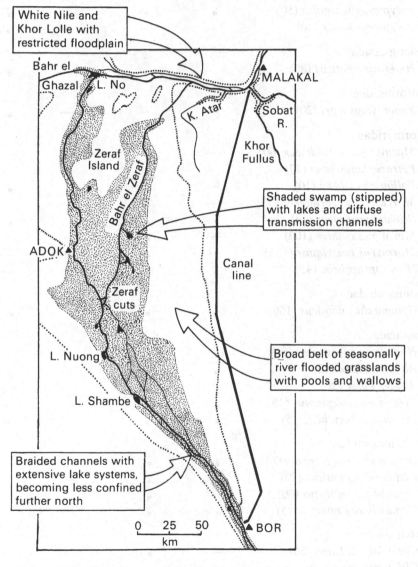

Figure 14.2. Sketch map of the *Sudd* wetlands showing the main ecological zones inhabited by fishes.

Table 14.1 *Fish species and their distribution recorded in the* Sudd *Wetland in 1982–3.*

Species list (approx. maximum length, cm)	River channels	Lakes and khors	Shaded swamp	River floodplain
Lepidosirenidae				
Protopterus aethiopicus (150)	—	•	•	•
Polypteridae				
Polypterus senegalus (50)	•	•	•	•
Polypterus bichir (70)	—	—	—	•
Osteoglossidae				
Heterotis niloticus (80)	—	•	•	•
Notopteridae				
Xenomystus nigri (20)	—	—	—	•
Mormyridae				
Mormyrops anguilloides (60)	—	•	—	—
Petrocephalus bane (20)	—	•	—	—
Pollimyrus isidori (10)	—	—	—	•
Marcusenius cyprinoides (30)	—	•	—	—
Brienomyrus niger (13)	—	•	•	•
Mormyrus cashive (100)	—	•	—	—
Mormyrus hasselquistii (35)	—	•	—	—
Hyperopisus bebe (45)	•	•	—	—
Gymnarchidae				
Gymnarchus niloticus (150)	—	•	•	•
Characidae				
Hydrocynus forskahlii (50)	•	•	—	—
Alestes dentex (40)	•	•	—	—
Alestes nurse (20)	•	•	—	—
Alestes macrolepidotus (45)	•	•	—	—
Micralestes acutidens (5)	•	•	•	—
Distichodontidae				
Distichodus engycephalus (35)	•	•	—	—
Distichodus niloticus (70)	—	•	—	—
Distichodus rostratus (60)	—	•	—	—
Nannocharax niloticus (5)	•	•	—	—
Citharinidae				
Citharinus citharus (50)	—	•	—	—
Citharinus latus (50)	•	•	—	—
Nannaethiops unitaeniatus (3)	—	•	—	•

Table 14.1 (*cont.*)

Species list (approx. maximum length, cm)	River channels	Lakes and khors	Shaded swamp	River floodplain
Ichthyboridae				
Ichthyborus besse (20)	—	•	—	—
Cyprinidae				
Labeo niloticus (50)	•	•	—	—
Barbus bynni (70)	—	•	—	—
Barbus perince (10)	—	•	—	—
Barbus stigmatopygus (2)	—	—	—	•
Barbus leonensis (3)	—	•	—	—
Barbus pumilus (3)	—	•	—	—
Barbus anema (3)	—	•	—	—
Barbus indet. (3)	—	—	—	•
Chelaethiops bibie (5)	•	•	—	—
Bagridae				
Bagrus bayad (70)	•	•	—	—
Chrysichthys auratus (22)	—	•	—	—
Clarotes laticeps (50)	•	•	—	—
Auchenoglanis biscutatus (40)	•	•	—	—
Auchenoglanis occidentalis (100)	—	•	—	—
Schilbeidae				
Eutropius niloticus (35)	•	•	—	—
Schilbe uranoscopus (30)	—	•	—	—
Siluranodon auritus (17)	—	•	—	—
Clariidae				
Clarius allaudi (30)	—	•	•	•
Clarius gariepinus (100)	—	•	•	•
Mochockidae				
Brachysynodontis batensoda (20)	—	•	—	—
Synodontis clarias (25)	—	•	—	—
Synodontis schall (43)	•	•	—	—
Synodontis frontosus (34)	•	•	—	—
Mochocus brevis (3)	•	—	—	—
Cyprinodontidae				
Epiplatys marnoi (3)	—	•	•	•
Epiplatys bifasciatus (4)	—	•	•	•
Aplocheilichthys loati (3)	—	•	•	•
Nothobranchius virgatus (5)	—	•	—	•

Table 14.1 (*cont.*)

Species list (approx. maximum length, cm)	River channels	Lakes and khors	Shaded swamp	River floodplain
Channidae				
Channa obscura (35)	—	•	•	•
Centropomidae				
Lates niloticus (150)	•	•	—	—
Cichlidae				
Hemichromis letourneauxi (9)	—	•	•	•
Hemichromis fasciatus (25)	—	•	•	—
Pseudocrenilabrus multicolor (8)	—	•	—	•
Thoracochromis loati (6)	—	•	—	—
Oreochromis niloticus (50)	•	•	—	•
Sarotherodon galilaeus (40)	—	•	—	—
Tilapia zillii (27)	—	•	—	•
Eleotridae				
Eleotris nanus (4)	—	•	—	—
Anabantidae				
Ctenopoma petherici (16)	—	•	•	•
Ctenopoma muriei (8)	—	•	•	•
Tetraodontidae				
Tetraodon fahaka (40)	•	•	—	—

surface drift-feeding *Chelaethiops* was caught in sufficient numbers to demonstrate a clear-cut preference for running water.

Lakes and khors

These provide the greatest diversity of fish habitats in the *Sudd* and, being easier to fish with nets, they offer the best scope for fishery developments. The large, established Lakes Shambe, Nuong and No, lying close to the Bahr el Jebel, have shown considerable persistence with time remaining little changed over many decades, whereas others, for example those recorded in early maps north of Adok, have largely disappeared. In general, however, there has been a marked increase since the rise in Nile discharges such that a range of 'lake' conditions now occurs in the *Sudd*. These vary from short-retention channel bulges and more extensive river-lakes, to swamp pools and sheets of apparently isolated water. All *Sudd* lakes are shallow with depths of between 1 and 3, maximally 4, metres. Their physico-chemical quality is affected by the relative contributions of young 'river-derived' and old 'swamp-influenced' waters in their make-up (Chapter 6). Lake beds have variable thicknesses of soft organic oozes with firmer substrates of sand and silt in areas exposed to continual bottom turbulence.

The term *khor* refers to a water channel, and is included here to mean a sort of elongated 'lake'. Typically towards the south and north of the *Sudd*, *khors* form lateral arms of the Nile which penetrate the restrictions of the river's trough and lead on to the adjacent floodplain. Filled with flood or ebb water during the wet season, they appear as static backwaters or dry creeks for the remainder of the year. An exception is Khor Lolle which persists as a perennially sluggish canal providing a northern parallel of the White Nile between Tonga and a point due west of Sobat mouth (Figure 14.2).

As is shown in Table 14.1, the greatest variety of fishes have been recorded in lakes and their vegetation fringes. A total of 62 species was identified in collections from gill-nets, electric-fishing and traps. Coefficients of similarity for gill-net catches from 20 lakes were calculated and subjected to average-linkage cluster analyses in order to force patterns of association. This showed that there was a considerable uniformity of population structures in terms of main species composition and their relative abundance (Hickley & Bailey, 1986). Overall the most numerous groups or species caught were, in descending order:

> characids, *Alestes dentex* and *Hydocynus forskahlii*,
> mochokids, notably *Synodontis frontosus* and *S. schall*,
> other catfishes, notably *Eutropius niloticus*, *Auchenoglanis biscutatus* and *Clarotes laticeps*,
> tilapias, chiefly *Oreochromis niloticus*,
> *Labeo niloticus*,
> *Distichodus* spp and *Citharinus* spp,
> mormyrids, notably *Mormyrus cashive*,
> *Heterotis niloticus*,
> *Lates niloticus*.

In terms of biomass of fish sampled, the last four groups assume greater significance.

Extensive feeding grounds for fish in *Sudd* lakes are afforded by the bottom, open-water and vegetation. Organic detritus and its associated bacteria, microflora and fauna provide a rich food supply. Larger benthic invertebrates requiring firmer substrates have generally proved elusive to bottom grabs so that only modest densities, with sample biomasses ranging from 1.0 to

4.7 g m^{-2}, have been recorded (Monakov, 1969; SES, 1983). By contrast bottom-feeding fish are frequently found with stomachs filled with chirono-mid-midge larvae or bivalve molluscs. Except in 'old' waters, recorded standing crops of plankton are typically higher in lakes than flowing channels. Those for net zooplankton are highly variable, 6×10^2 to 6×10^5 individuals m^{-3} or 0.001–0.246 g m^{-3}, and contain, in qualitative terms, 3–6 water-fleas including *Ceriodaphnia cornuta*, *Diaphanosoma excisum* and *Moina micrura*; 3–6 copepods including *Thermocyclops neglectus*, *Thermodiaptomus galebi* and *Tropodiaptomus* spp.; and up to 24 rotiferans of which *Brachionus*, *Lecane*, *Keratella* and *Filinia* appear to be the most speciose genera (Monakov, 1969; SES, 1983; Green, 1984).

The main grazing pathways in feeding inter-relationships, however, evolve around submerged plants and amongst the roots of the water hyacinth, *Eichhornia crassipes*. Beds of hornwort, *Ceratophyllum demersum*, and saw-weed, *Naias pectinata*, the latter often with a thick periphytic loading, are particularly widespread (Chapter 7) and contain diverse communities of crustaceans, insects and molluscs. Root mats of the floating hyacinth, in both fixed fringes and mobile rafts, trap suspended material, attract invertebrates and may be subject to heavy browsing.

There can be no doubt that this is a biologically rich and important habitat especially in view of the present distribution and quantities of water hyacinth in the system. Invertebrates found here include water-fleas, copepods, ostra-cods, the conchostracan *Cyclestheria hislopi*, and the larger crustaceans *Caridina nilotica* and *Macrobrachium niloticus*. In addition, there are oligo-chaetes, leeches, snails, water-mites and insects including odonatans, ephe-meropterans, trichopterans, dipterans and coleopterans. The latter are par-ticularly speciose – 22 species were identified from recent collections which, however, is only half the number recorded from the Nile cabbage, *Pistia stratiotes* (Rzóska, 1974), a former occupant of the hyacinth niche. In 1982, the major concentrations of invertebrates were found beneath the outer half of the hyacinth fringe giving fresh-weight biomasses of 8–19 g m^{-2}.

Some of the main feeding relationships identified in a typical river-lake are shown in the foodwebs of Figure 14.3 and discussed in detail by Hickley & Bailey (1987*b*). A broad partitioning of food resources is apparent but dietary specialisation amongst fishes was the exception rather than the rule. Herbi-vores are represented by *Distichodus*, *Citharinus* and the tilapias (*Oreochro-mis*, *Sarotherodon* and *Tilapia*). *Distichodus* chiefly consumes macrophytes, whilst the diet of *Citharinus* includes a high proportion of bottom debris. The latter formed the major component in the stomach contents of *Heterotis* and *Labeo* which may be classed as mud-feeders. The relative abundance of both

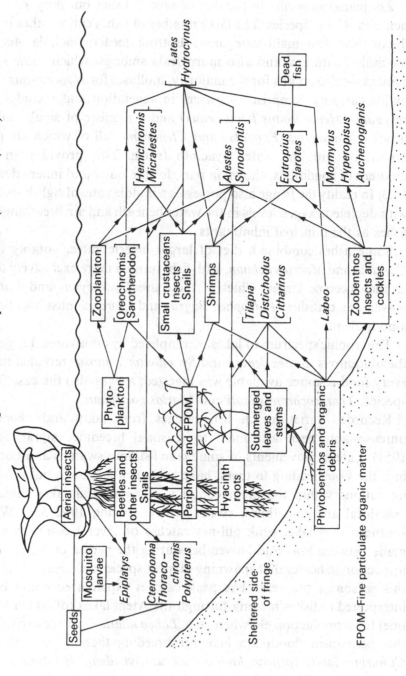

Figure 14.3. Some feeding inter-relationships in *Sudd* lakes (after SES, 1983).

groups increased significantly in samples from within and around submerged vegetation.

Zooplankton occurs in the diet of several fishes but only achieves prominence in *Alestes* species. The large number of fishes eating other invertebrates divide into two main categories. Bottom feeders include *Auchenoglanis*, mochokid catfishes, and also mormyrids amongst which some selectivity is evident – chironomids for the majority, molluscs for *Hyperopisus*. The second major category comprises browsers in vegetation and includes *Polypterus senegalus*, *Hemichromis letourneauxi* and a number of small and abundant fishes – *Micralestes*, *Epiplatys* and *Ctenopoma*, all of which are particularly associated with the water hyacinth fringe. This provides an interesting gradient of conditions, shown in part, for the outer and inner edges in Figure 14.3. In reality the fringe also possesses a middle zone of tightly-packed plants with depleted oxygen levels in the water beneath and air-breathing beetles and fishes as the principal inhabitants.

Some fishes combine a diet of larger invertebrates, notably the shrimps *Caridina* and *Macrobrachium*, with fish, turning more exclusively towards fish as they become larger. Chief of these are *Hydrocynus* and *Lates*, but the scavenging catfishes, *Eutropius*, *Bagrus* and *Clarotes* must also be numbered among them.

The trophic spectrum of lakes is completed by omnivores. Large samples of the ubiquitous *Alestes dentex* and *Synodontis frontosus* revealed that virtually every food resource available was ingested, as was also the case for the rarer species, *Thoracochromis loati* and *Clarias gariepinus*.

Recorded variations in fish catches from lakes and *khors* have not unreasonably been attributed to seasonal breeding migrations (Sandon, 1951). Lateral movements of some lake fish into swamps and floodplains, or into the *khors* leading to them, have been demonstrated, but information on migrations within the main complex of rivers and lakes remains largely anecdotal and sometimes, contradictory. During 1982–3 in Wutchung, a southern river-lake, peak gill-net catches of several common species were made between May and November during the period of high water. Gonad inspection indicates that spawning in most species coincides with, or includes, this season of the year. The peak catches experienced could be variously interpreted as fishes moving through the Atem lakes (of which Wutchung is one) for reproduction elsewhere, e.g. *Labeo niloticus*, or the arrival of fishes in this permanent floodplain lake for breeding therein, e.g. *Alestes dentex*, *Citharinus latus*, *Auchenoglanis biscutatus*, *Synodontis frontosus* and *Sarotherodon galilaeus*. The precise spawning locations remain to be discovered for a majority of species, but the capture of fry or small juveniles of *Polypterus*,

Heterotis, Hydrocynus, Alestes, Distichodus, Citharinus, Labeo, Synodontis, Auchenoglanis, Oreochromis, Sarotherodon, Tilapia and *Lates*, in the hyacinth fringes and marginal vegetation of Wutchung and elsewhere, indicates that these species breed successfully within the permanent aquatic system which also provides suitable rearing grounds (Hickley & Bailey, 1986).

Shaded swamp

This gives the *Sudd* its distinctive physiognomy. *Cyperus papyrus* forms a riparian belt, the breadth of which diminishes from south to north. Outside this in the centre, lie tracts of *Typha domingensis* which have spread enormously since the early 1960s (Chapter 7). Low levels of dissolved oxygen and a restricted invertebrate fauna of air-breathing bugs and beetles, and oxygen-thrifty midge larvae and snails, characterise interstitial water. Many fishes may move through transmission conduits in the swamp but 23 species are recorded as swamp-associated (Table 14.1). Of these several known (or suspected) air-breathers, for example *Protopterus, Polypterus, Heterotis, Gymnarchus, Clarias, Ctenopoma,* and *Channa* are potential penetrators of this seemingly inhospitable environment. There is good evidence that during the wet season the nest builders, *Heterotis* and *Gymnarchus*, find spawning grounds in shallow open-water within the *Typha* and that, with *Clarias, Polypterus* and *Ctenopoma*, in years of deep flooding, they may emerge on to river-inundated grasslands. In the event *Protopterus* has eluded capture in the *Sudd* swamps, but elsewhere its nests and young have been reported in the matted roots of papyrus.

Floodplains

These are wetted and sealed by rain which may fill ditches and pools prior to river flooding, which usually begins in July. It commences in the south via seasonal *khors* and continues on broader fronts as water spills through permanent swamp on to the flat landscape. Flooding is enhanced by rainfall and in the east by the creeping flow of surface runoff. Transient oxygen depletion from litter decomposition and swamp flushing characterises flooded habitats before algae develop and a succession of benthic, motile and periphytic associations of microscopic life gets underway. Rotifers and small crustaceans are abundant, aestivating molluscs, e.g. *Pila* spp., re-awaken and beetles come to dominate a diverse insect fauna which is also rich in mosquito larvae. Sixty-three mosquito species occur in the swamps and temporary wetlands of the *Sudd* (Lewis in Rzóska, 1974). Towards the end of their existence floodplain pools become increasingly turbid and overgrown by vegetation leading to a return of low dissolved oxygen concentrations.

Twenty-three species of fish were collected by electric-fishing and seine-netting in seasonal habitats, which in overall descending order of relative abundance included: cyprinodonts (notably *Epiplatys marnoi*); anabantids (notably *Ctenopoma muriei*); *Clarias gariepinus*; cichlids (notably juvenile *Oreochromis* and *Hemichromis letourneauxi*); *Nannaethiops* and small *Barbus* spp; *Polypterus senegalus; Channa obscura*; small mormyrids (for example, *Brienomyrus niger*).

Of the grassland invaders *Polypterus senegalus*, clariid catfishes and most of the accompanying small species are intent on spawning. In common with swamp species, some are air-breathers; the tiny cyprinodonts are credited with the utilisation of the well-oxygenated surface film and others, for example the small mormyrids and cichlids, may be physiologically tolerant to reduced levels of dissolved oxygen. The majority are insectivores but the wealth of microscopic organisms in pools provides important food resources and the larger fishes are, in part, piscivorous (Figure 14.4) (Hickley & Bailey, 1987*b*).

Whilst the activities of *Protopterus, Polypterus, Heterotis* and *Gymnarchus* may extend beyond the swamps on to flooded grasslands, it appears that only *Clarias gariepinus* amongst larger species, depends upon these habitats for spawning. In Lake Sibaya, South Africa, even this species has been shown to forego migration if the flood fails, and to breed successfully at lake and river edges (Bruton, 1979).

The restricted species penetration of grassland habitats differs from the situation in some other tropical floodplain rivers. In these, the general pattern of events associated with the flood-cycle, as reviewed by Lowe-McConnell (1975), Welcomme (1979) and Bruton & Jackson (1983), encompasses a lateral migration by many fishes, including species important for the fishing industry, to spawn in access channels or on the floodplain, and to grow-on in conditions of plentiful food supplies and cover from predators. High water on the floodplain in these systems affords the main feeding, growing and fattening season for nearly all species (Lowe-McConnell, 1975). By contrast it seems that in the *Sudd* the broadening band of papyrus and *Typha* swamps, enclosing channels, lakes and extensive shallows, provides a permanent floodplain, albeit swollen in the wet season, offering a variety of conditions to suit the needs of the majority of species. At the same time the peripheral tracts of shaded swamp vegetation in the centre and north must form an effective barrier to the lateral movement of most river and lake fishes except, perhaps, where passages have been forced by large animals and fishermen's canoes (Hickley & Bailey, 1987*a*).

A further important finding to emerge from studies in the south-eastern grasslands is that except along the interface with the swamps, there is little

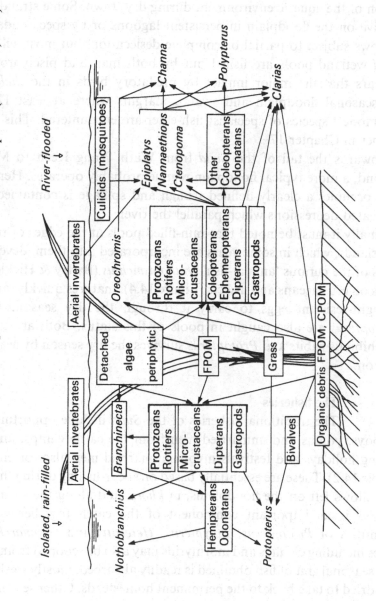

Figure 14.4. Some feeding inter-relationships in river-flooded and rain-filled pools (after SES, 1983).

return of water, materials or fish to the permanent system. Those fishes which do enter the seasonally river-flooded grasslands become concentrated in isolated pools, a phenomenon common elsewhere, but usually preceded by a substantial back-migration led by riverine fishes most sensitive to the deterioration of the aquatic environment during dry-down. Some stranded fish may survive on the floodplain in persistent lagoons or by special adaptations in wallows subject to partial or complete desiccation, but most will die. In the *Sudd* wetland pools are fished out by both man and piscivorous birds. It appears that the major impact by predatory birds in the *Sudd* occurs in the seasonal floodplain and swamp margins where at least 12 of the 16 commonest species of specialist fish-eater are encountered. This is discussed further in Chapter 17.

Towards the tail of the *Sudd* from Wath Wang Kech to Malakal and beyond, a more typical floodplain regime probably operates. Here the White Nile occupies a clearly defined trough and spillage is contained in narrow elongated depressions which parallel the river.

Finally it must be noted that rain-filled pools at the edges of river-flooded grasslands, which in some years are incorporated into them, develop populations of the curious 'annual-fish' *Nothobranchius* (Bailey & Hickley, 1985). It feeds on crustaceans and insects (Figure 14.4), matures quickly, and produces drought-resistant eggs to carry it through the dry season. *Clarias* and *Protopterus* are also caught in pools of this status. Both are credited with amphibious habits and *Protopterus* survives the dry season by aestivating in a cocoon.

The fisheries

The traditional fisheries of the *Sudd* include opportunistic fishing by boys, youths and unmarried men from the cattle camps, and organised fishing holidays and festivals involving married men also, or entire families (Table 14.2). These are essentially based on spear-fishing during the dry season in wallows left on the floodplains, in *khors*, and along the swamp margins. *Clarias* is an important component of the catch together with variable quantities of *Protopterus, Polypterus, Heterotis* and *Gymnarchus*. Smaller fishes including cichlids and mormyrids may also be scooped from relict pools. The seasonal glut of fish obtained is readily absorbed, mostly fresh, but in part sun-dried to take back to the permanent homesteads. Other seasonal activities range from trapping in rising floodwaters to prising *Protopterus* from cocoons in floodplains, long after the water has disappeared.

Larger areas of permanent open water were rarely fished until the arrival of nets in the late 1940s and early 1950s. At that time, migrant fishermen from

Table 14.2 *The main types of fishing activity in the Sudd wetland*

Activity	Participants (+ estimated numbers)	Season and fishing area	Product
1. Opportunist fishing	Dinka and Nuer pastoralists: males from cattle camps or in large organised parties: some entire families (100,000)	Dry season, chiefly in pools and wallows on the floodplains and swamp margins, with some penetration into the 'permanent' system	Chiefly fresh-fish
2. Subsistence fishing	Shilluk combining fishing with other activities (9,000)	Mostly dry season in river and *khors* between Tonga and Renk	Chiefly fresh-fish
	'Zeraf' Nuer (?)	Perennial, in centre and north of *Sudd* including Zeraf Island	Fresh and sun-dried fish
	Monythany Dinka (600)	Perennial in the southern part of the *Sudd*	Fresh and sun-dried fish
3. Commercial fishing	Shilluk & Monythany Dinka near towns (200)	Perennial in channels and lakes with close proximity to towns in north and south of *Sudd* region	Fresh and sun-dried fish
	'Zeraf' Nuer (?)	Perennial in centre and north of *Sudd*	Sun-dried fish
	Dinka & Nuer pastoralists (1,000)	Perennial in centre and south of *Sudd*	Sun-dried fish
	Fisheries Directorate & Fishermen's Co-operative at Malakal (70)	Dry season chiefly in the north	Sun-dried, salted fish
	Private entrepreneurs (?)	Mostly dry season in north and southern areas	Fresh-fish; sun-dried, salted fish; wet-salted fish

the north, and also from West Africa (Fellata), finding fishing a virtually untouched and uncontested resource, set up fishing camps on the larger accessible lakes and perennial *khors*. Often financed by northern merchants, they concentrated on the production of sun-dried and salted fish for export to the northern Sudan or southwards to present-day Zaïre. In these camps, as hired labour, local people learned to make and use nets. Shilluk fishermen along both sides of the White Nile from Tonga to Malakal now deploy gill, drift, cast and seine nets to make fishing an integral part of their subsistence economy.

Of the other pastoralists prior to the 1960s flooding, virtually none, except the *Monythany* Dinka in the southern *Sudd*, were full-time fishermen working from permanent waterside settlements. However, since the sustained rise in discharge and the massive enlargement of the permanent floodplain of the Bahr el Jebel system, other groups of Dinka and Nuer, especially those of the central area and on the Zeraf island, have turned to fishing (see Chapter 11). Many were forced into this way of life at least temporarily, as their grazing lands turned into perennial swamps in the 1960s. They operate from small camps on channels, lakes, and in the *toic* of the tribal sections of their origin,

Plate 26. A Nuer communal fishing party. Such gatherings occur regularly during the dry season as fish become concentrated in lakes and pools left isolated by the retreating river flood. Photo: Paul Howell.

wherever a solid base afforded by a mound or firm mat of floating vegetation is suitably located. All follow a mainly subsistence economy, in which fish are caught for home consumption and for sale or exchange as fresh or sun-dried products, in order to obtain other foods, cloth and essentials (Table 14.2).

The principal fishing craft are dug-out canoes. Trunks of local doleib and doum palms can be used to make cheap but short-lived canoes. Imported hard-wood canoes of *Khaya senegalensis* or *Cordia abyssinica* are more expensive but preferable because they have better payloads and a working life of up to 20 years. In three aerial surveys the highest estimate of the total number of canoes in the *Sudd* was $7,533 \pm 1,085$ in March, 1980 (SES, 1983). Square-transomed planked boats, able to take an outboard engine, are restricted to the Malakal area and narrow rafts of ambatch bundles are used along the lower Bahr el Zeraf and White Nile.

Gill nets, made up from durable twines to give stretched mesh sizes in the range of 13–16 cm, are the commonest gear used in the permanent system. They are staked out in lagoons and lakes or deployed as surface drift-nets in flowing water. Baited-hook and line fishing is also practised but it is

Plate 27. Fishing camp in the heart of the *Sudd*, occupying a small mound surrounded by water and *Typha* swamp. Fishing canoes can be seen in the water, and behind the camp are racks hung with strips of drying fish. Photo: Alison Cobb.

generally considered to be inferior to net fishing. Although there is some variation with season and locality amongst the species caught, they are, ranked in order of overall importance in terms of abundance: *Distichodus; Citharinus; Heterotis*; tilapias *(Oreochromis* and *Sarotherodon)*; *Lates; Gymnarchus*: large mormyrids, large catfishes, large characids; *Labeo*.

Throughout the year fish may be preserved by sun-drying, but losses from spoilage are high in the wet season. Prior to drying, large fish are typically filleted. The fillets remain attached to the tail and are cut into long strips. In the case of tilapias these are then braided to form plaits. Alternatively fish are split from top to bottom and opened out into a sheet; drying takes longer by this method, but there is less wastage.

Fisheries development projects in the northern zone have included the creation of a Fisheries Training Institute at Malakal and the setting up of boat-building and fishing co-operatives among the Shilluk. The fishing co-operative and, until recently, the Government's Directorate of Fisheries, have been concerned with the production of sun-dried and salted fish. Other commercial camps for salting, wet or dry, have declined but throughout the

Plate 28. A large Nile perch split and hung for drying. The strips of flesh remain attached to the tail (*top*). The head is being dried separately on the left at the end of the pole. Photo: John Goldsworthy.

Sudd prior to 1984 some entrepreneurial activity continued, usually financed by entrepreneurs from the northern Sudan (Table 14.2). At the same time the emphasis of government and aid agencies came to centre on encouraging the perennial 'indigenous' fishermen to produce a cash-crop of sun-dried fish.

For the most part the resource appears to have been in a healthy state. Based on large meshed gill-nets, the fishery selects for large individuals of widely acceptable species. The abundant and very spiny catfishes, *Synodontis* spp., are not retained in numbers, to the advantage of the net handler and the longevity of the net. Current levels of year-round fishing do not threaten the fish stocks. Those in the more established fisheries of Lakes Shambe and No have reportedly dwindled, but this is more than offset by the stocks available in the many new lakes, between Bor and the Zeraf Cuts and along the Bahr el Zeraf, since the water levels rose in the 1960s.

There is little satisfactory information on quantities of fish caught in the *Sudd*. One speculative approach using the averages of three estimates of the total numbers of canoes and some recorded daily landings of fish, gave an annual figure of 27,000 tonnes from the permanent system – that is excluding

Plate 29. Bundles of dried fish awaiting export from the *Sudd.* The plaited strips are tilapia; the bigger bundles below, with tails attached, are *Distichodus.* Photo: Michael Lock.

the seasonal floodplain. Some outlets for *Sudd* fish identified during the Swamp Ecology Survey are shown in Figure 14.5.

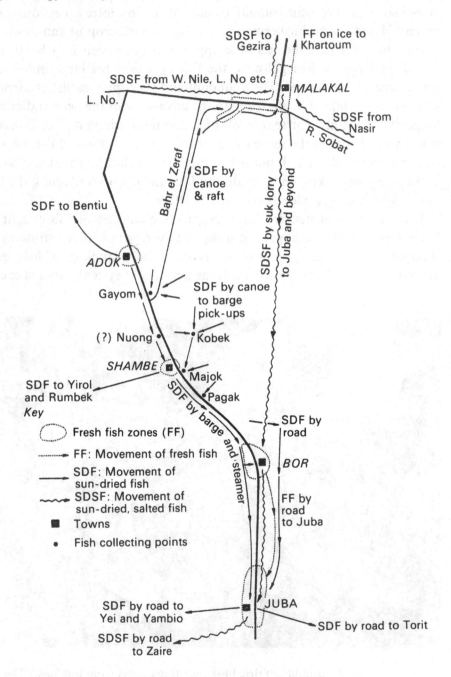

Figure 14.5. Outlets for products of the commercial fisheries of the *Sudd* (after SES, 1983).

Of the natural living resources in the wetland, fish are undoubtedly under-utilised. Some estimates of the potential annual yield, to the order of 100,000–200,000 tonnes appear unduly high, being based upon inflated evaluations of the area of the *Sudd* and assumptions about its floodplains extrapolated from elsewhere. However, much larger harvests could be sustained providing that the fishing effort is well distributed.

Marketing presents a major constraint to the growth of *Sudd* fisheries. Because of the distances involved and restricted payloads, only small quanti-ties of fish can be canoed, rafted, or head carried out of the swamp. Bales of processed fish require collection and bulk transportation to major wholesale and retail outlets. Some use is made of commercial river traffic and privately hired barges for this purpose. In recognition of the problem however, the United Nations Development Programme (UNDP) prior to 1984 had as its main objectives for the southern and central regions, the provision of a reliable fish transportation service by barge and lorry, together with the construction of simple shelters at collection centres in the *Sudd* and capacious storage facilities at Juba and Bor. In the future, when peace is restored, markets for *Sudd* fish products will need to be expanded both within the Sudan and in neighbouring countries.

In addition, as the result of consultation with fishing communities, the project managers and Government Fisheries Departments appreciated the need for an active extension service to promote the value of co-operative action and improve camp hygiene, health care, fish processing in the wet season, and the construction, hanging and repair of gill-nets. Beyond these important areas for development and without a fully-trained and equipped fisheries inspectorate to gather the information needed to supply a factual basis for management measures or to enforce them, a *laissez-faire* approach is the only one feasible. Although seldom officially admitted, such policies operate in many extensive fisheries to reasonable effect.

15

Wildlife

Introduction

Before the late 1970s, the extent of the wildlife resources of the southern Sudan was relatively unknown. Since gaining their independence in the early 1960s, most other Eastern and Central African countries (Kenya, Tanzania, Uganda, Zaïre and Zambia, for example) had investigated their wildlife resources, gazetted numerous National Parks and other protected areas (IUCN, 1986) and had developed sophisticated and profitable tourist industries, largely based on the wildlife. In contrast, after two decades of war, neglect and economic stagnation, which had diverted attention away from such matters as wildlife management and research, very little was known at that time, either by the southern Sudanese authorities or by the outside world, about the wildlife resources of any part of the southern Sudan. Jonglei Province was no exception.

It is extremely difficult to determine accurately the numbers and distribution of wild animal populations, particularly when censuses, for reasons of cost and practicability, must treat many species at once. It is the purpose of this chapter to describe some of the basic facts about the animal populations that were established (mainly in the course of the performance of other tasks) during the Range Ecology Survey, between mid-1979 and mid-1982. Much of the information described was gathered during the three aerial surveys, which were not primarily designed to make a census of wildlife, either mammalian or avian.

The information is only summarised here: a more complete account, including distribution maps for each of the bird and mammal species mentioned, was produced as the result of the Range Ecology Survey (RES, 1983).

Most African grasslands undergo substantial seasonal variation, and the *Sudd* floodplain is one such, as described in previous chapters. The effects of this on the wildlife are twofold: firstly, many large mammal species are obliged to be seasonal migrants, either to escape the deep flooding in the wet season, or to seek the only remaining water supplies in the dry, or a combination of both. Some of these species, while being migrants in similarly extreme environments elsewhere, also have sedentary populations in more stable environments. These species include the elephant, giraffe, tiang, reed-buck and Mongalla gazelle. The second consequence of the fluctuating environment, which exposes, between November and April, such huge expanses of *toic*, is that this area becomes extremely favourable habitat for birds, particularly aquatic ones. Furthermore, the position of the *Sudd* on the major eastern migratory route from Europe and Asia to Africa makes the area one of the most important passage and wintering grounds in Africa for palaearctic migrants, as well as being essential to some millions of intra-African migrants. Ripe grass seeds, crustacea, molluscs, aquatic insects, and enormous numbers of fish, all contribute to the diversity and high productivity of this part of the system at this time of year.

The most abundant and interesting species are discussed below, as a prelude to a description of the role of wildlife in the present-day economy of the area, both in terms of disease transmission to domestic animals, and as a source of food and other benefits to people.

The mammals
The Tiang

Tiang are the most abundant of the wild herbivores in the Jonglei area. Table 15.1 shows that their numbers varied greatly between counts. Figure 15.1 shows the distribution of the species at the three counts. From this two points are clear. First, there is a scattered, fairly low population of tiang in the area at the height of the wet season. Secondly, there appears to be a large influx of animals from the south-east of the area at the beginning of the dry season; by the end of the dry season the main part of the population has moved north and west into the river-flooded grasslands. This pattern of distribution is best explained by postulating two tiang populations in the area – a more-or-less sedentary, sparse population found mainly in the woodlands of the north-central part of the area, and a very much larger migratory population which leaves the area during the wet season. This is similar to the behaviour of two tiang populations in the Serengeti in Tanzania (Duncan 1975). A programme of reconnaissance flights during the year established the route of this migration, which is shown in Figure 15.2. It is intriguing that both tiang and white-eared kob spend the wet season in more-or-less the same area, but migrate in different directions in the dry season, the white-eared kob moving north-east to the floodplains of the Pibor (Fryxell, 1985), while the tiang move north-west to the Nile floodplain.

The tiang are animals of the rain-flooded grasslands and, in the dry season, the *toic*. They also use the woodlands, but to a lesser extent. They seem to need dry ground during the wet season, and this need may partly explain their migrations. If the area of suitable dry season grazing were the critical limiting resource controlling tiang numbers, a relationship already demonstrated in other East African herbivores (Sinclair, 1974), then the increase in the area of *toic* since the early 1960s, already described in Chapter 8, might satisfactorily explain their apparent increase. The evidence for this, admittedly circumstantial, is that members of the JIT did not record huge herds of tiang in the *toic* to the west of Kongor and Ayod, despite five years' fieldwork in the area. Herds of several thousand at a time, such as are commonplace now (and were regularly seen by the Range Ecology team in 1979–83), could scarcely have escaped the JIT's attention, had they existed.

Table 15.1. *Population estimates of large wild herbivores*

	Mid wet season	Early dry season	Late dry season
Bushbuck	164 ± 72	461 ± 277	268 ± 153
Buffalo	10,182 ± 5,579	8,518 ± 2,590	4,501 ± 1,734
Duiker (grey)	98 ± 55	99 ± 56	234 ± 98
Elephant	3,938 ± 3,619	1,725 ± 796	2,964 ± 1,629
Mongalla gazelle	1,130 ± 870	55,032 ± 30,617	65,937 ± 14,961
Giraffe	3,202 ± 996	6,025 ± 1,289	4,527 ± 996
Lelwel hartebeest	0	65 ± 64	234 ± 165
Hippopotamus	2,823 ± 501	3,524 ± 663	2,252 ± 612
White-eared kob	10,584 ± 5,776	2,035 ± 916	11,672 ± 5,052
Uganda kob	2,489 ± 1,370	0	0
Nile lechwe	12,711 ± 5,596	11,924 ± 3,351	32,279 ± 15,368
Oribi	2,507 ± 388	1,553 ± 306	6,006 ± 1,328
Reedbuck	2,547 ± 516	15,207 ± 7,107	33,380 ± 15,610
Roan antelope	1,056 ± 375	2,087 ± 575	4,124 ± 1,401
Sitatunga	1,094 ± 211	200 ± 79	1,108 ± 242
Tiang	34,633 ± 10,338	117,531 ± 31,401	359,496 ± 94,686
Waterbuck	1,782 ± 443	2,284 ± 890	8,851 ± 2,939
Common zebra	0	4,533 ± 2,526	3,889 ± 1,767

Figure 15.1. The distribution of tiang (*Damaliscus korrigum tiang*) in the mid-wet, early dry and late dry seasons (after RES, 1983).

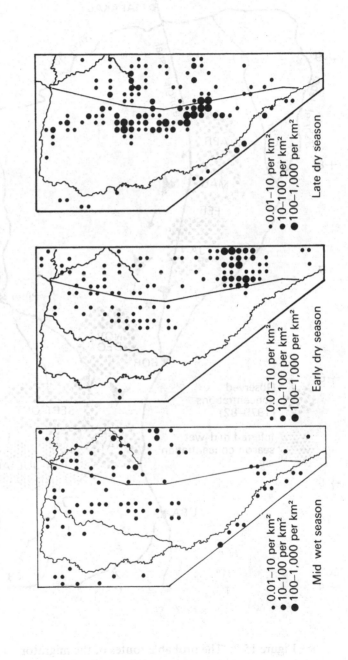

Mid wet season

· 0.01–10 per km²
● 10–100 per km²
⬤ 100–1,000 per km²

Early dry season

· 0.01–10 per km²
● 10–100 per km²
⬤ 100–1,000 per km²

Late dry season

· 0.01–10 per km²
● 10–100 per km²
⬤ 100–1,000 per km²

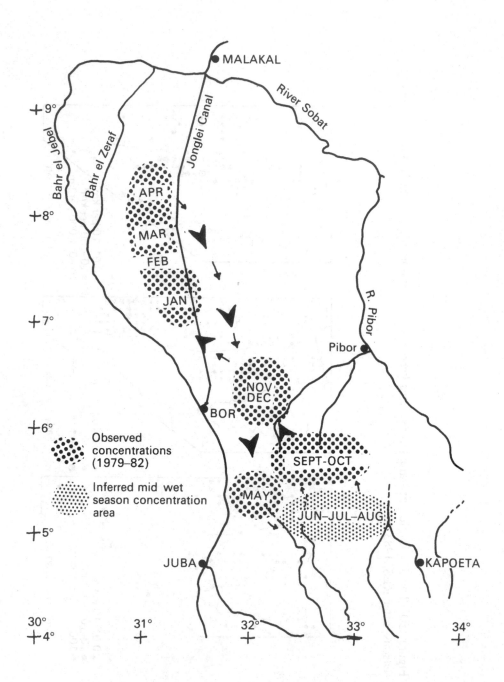

Figure 15.2. The probable routes of the migratory tiang population, their known dry season concentration and calving area, and their presumed wet season concentration area (after RES, 1983).

Other large herbivores
Mongalla gazelle

Like the tiang, Mongalla gazelle show very large fluctuations in numbers between counts, and this, too, can be explained by their migration (Figure 15.3). However, only 15–20% of the population appears to cross the canal line, the rest remaining to the east of it (though quite where they spend the wet season remains a mystery); only a few reach the *toic* for the dry season. Why should this be? Like many African herbivores Mongalla gazelle do not have to drink, though sometimes they may do so. They survive on the water obtained from the grass they eat, probably supplemented by dew when it is available. In the heat of the day they shelter under trees, thus reducing evaporative water loss. They are therefore not obliged to seek water in the *toic*.

But if Mongalla gazelle can survive in the *Hyparrhenia* grasslands throughout the dry season, why do they migrate at all? It has been suggested (RES, 1983) that they depend on the tiang to improve the physical structure of the grass sward for them. Such a relationship has been documented in the Serengeti (Bell, 1969; Sinclair & Norton-Griffiths, 1980) for the conspecific Thomson's gazelle, which relies on the topi (the same species as the tiang) and, in particular, the wildebeeste to improve the accessibility of the long grass so that the gazelles have access to small nutritious grass shoots near the ground. The small mouth of the gazelles allows them to be very selective in what they eat, and to take advantage of scattered nutritious shoots in generally poor-quality grassland. If they depend on the tiang in this way, their migrations are partly explained.

Nile lechwe

The Nile lechwe is found in the floodplains of the Bahr el Jebel, the White Nile above its confluence with the Sobat, and the Bahr el Ghazal. It also occurs in the Machar Marshes and along the Pibor, Akobo and Baro rivers as far as Gambela in Ethiopia. The study area thus contains a high proportion of the world's population of this species.

Lechwe live in the shallow waters of the *toic*, moving into the edge of the swamps at the height of the dry season if river floods are low (Figure 15.4). They seem to prefer areas with water 10–40 cm deep. The meat of lechwe is much prized. However, its habitat makes it largely immune to hunting with dogs, but it is easily stalked and killed with firearms, which must pose some threat. Nevertheless, a herd of lechwe some 200–300 strong on Fanyikang Island, to the west of the confluence of the White Nile and the Sobat, has traditionally received royal protection from the Shilluk *reth*; they are killed

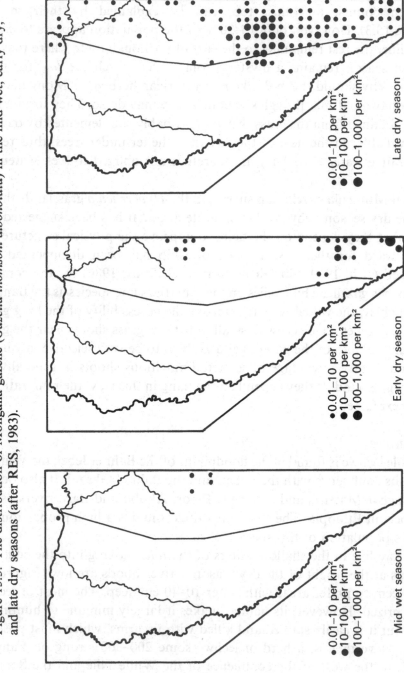

Figure 15.3. The distribution of Mongalla gazelle (*Gazella thomsoni albonotata*) in the mid-wet, early dry, and late dry seasons (after RES, 1983).

Mid wet season
•0.01–10 per km²
•10–100 per km²
●100–1,000 per km²

Early dry season
•0.01–10 per km²
•10–100 per km²
●100–1,000 per km²

Late dry season
•0.01–10 per km²
•10–100 per km²
●100–1,000 per km²

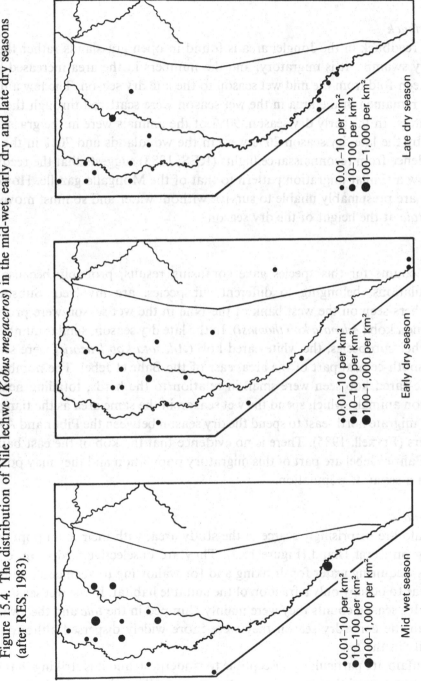

Figure 15.4. The distribution of Nile lechwe (*Kobus megaceros*) in the mid-wet, early dry and late dry seasons (after RES, 1983).

only when a new king (*reth*) is installed, to provide skins to be worn by members of the royal house during the attendant ceremonies (Howell & Thomson, 1946).

Reedbuck

The reedbuck in the Jonglei area is found in open grasslands rather than in reedy swamps. It is migratory, and the numbers in the area increased some thirteen-fold from the mid wet season to the late dry season. The few animals that remained in the area in the wet season were scattered through the drier habitats. In the early dry season, 90% of the animals were in the grasslands, but by the late dry season 20% were in the woodlands and 70% in the *toic*. Evidence from reconnaissance flights (RES, 1983) suggests that the reedbuck follow a similar migration pattern to that of the Mongalla gazelle. However, they are presumably unable to survive without water and so must move into the *toic* at the height of the dry season.

Kob

The counts for this species gave confusing results, probably because two populations, belonging to different subspecies, are involved. Substantial numbers seen on the west bank of the Nile in the wet season were probably Uganda kob (*Adenota kob thomasi*). In the late dry season, similar numbers of another subspecies, the white-eared kob (*Adenota kob leucotis*) were seen in the south-central part of the area, east of the Bahr el Jebel. The numbers of white-eared kob seen were small in relation to the herds, totalling nearly a million animals, which spend the wet season in the same area as the tiang and then migrate north-east to spend the dry season between the Pibor and Akobo Rivers (Fryxell, 1985). There is no evidence that the kob of the east bank of the Bahr el Jebel are part of this migratory population and they may possibly form a separate population.

Buffalo

Buffalo are surprisingly scarce in the study area, with their main population being on Zeraf Island (Figure 15.5). They are unselective feeders on coarse grass, requiring water for drinking and for wallowing to keep cool, and they appear to occupy only a fraction of the suitable habitat. In the wet season and late dry season counts they were mainly clumped in the *toic* and the swamps, but in the early dry season they were more widely dispersed although still mostly in the *toic*.

Buffalo are particularly susceptible to rinderpest, and it is striking that their distribution and that of cattle are almost mutually exclusive. In the Serengeti

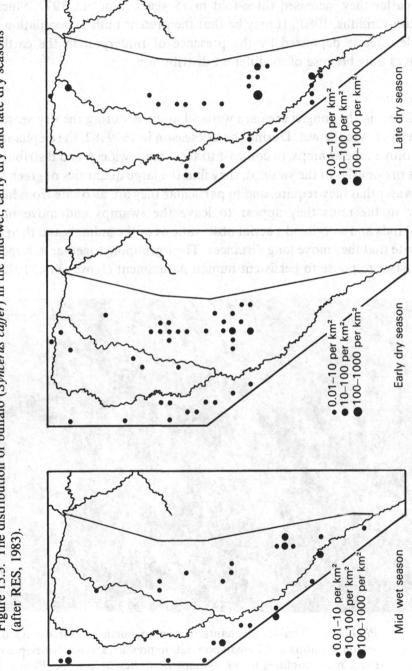

Figure 15.5. The distribution of buffalo (*Syncerus caffer*) in the mid-wet, early dry and late dry seasons (after RES, 1983).

Mid wet season

Early dry season

•0.01–10 per km²
•10–100 per km²
●100–1000 per km²

Late dry season

in northern Tanzania, the buffalo population was suppressed until rinderpest was controlled in the early 1960s, though the relationship between the distribution of the two species, outside the National Park, is not known. Thereafter they increased three-fold in 15 years (Sinclair, 1977; Sinclair & Norton-Griffiths, 1980). It may be that the present buffalo population in the Jonglei area is depressed by the presence of rinderpest in the cattle, and survives only because of its different distribution.

Elephant

Elephants in the Jonglei area are wetland animals during the dry season and dryland ones in the wet. During the dry season in 1979–82, the elephants were seen only in the swamps, in contrast to their more widespread distribution 30 years previously. In the swamps, they find the large quantities of green fodder and water that they require, and in particular they are also safe from hunting. Early in the rains they appear to leave the swamps and move into the grasslands and woodlands; aerial observations of the animals and their tracks indicate that they move long distances. Their grouping together in large herds suggests a response to persistent human harassment (Laws *et al.*, 1975).

Plate 30. A herd of elephants in swamp north of Bor during the dry season. Hunting and human population pressures mean that elephants are much more confined to the swamps than they used to be. Photo: Alison Cobb.

Hippopotamus

This animal is found predominantly in the swamps and *toic* throughout the year. Hippo have been much hunted in the area in the past, and are still shot or speared when the opportunity arises. They are thus scarce or absent along the main river channels, and are found mainly along small channels well away from the main river. Elsewhere in Africa, hippo need short grass for feeding and water to rest in. During the Swamp Ecology Survey, hippo were seen far from any dry land, and were heard feeding at night on grass mats along river channels. Many of these floating grass mats (mainly *Vossia cuspidata*) showed signs of feeding by hippo. It is possible that at least some of the hippo in the region feed entirely within the swamps and rarely reach dry land. The low hippo numbers in the *Sudd* remain, however, surprising.

Other herbivores seen during the aerial counts were bushbuck, giraffe, grey duiker, lelwel hartebeest, oribi, roan antelope, sitatunga, waterbuck and common zebra. Of these, the sitatunga is an animal of the swamp and *toic*, extremely difficult to see. Bushbuck, duiker, oribi, roan and waterbuck are probably more-or-less sedentary. Zebra move into the south of the area in the dry season from the east. Lelwel hartebeest were seen in very small numbers in the south of the area; their main habitat, broad-leaved woodland with a tall grass understorey, is absent from the survey area. Giraffe, rather surprisingly, are often found in the *toic* (in quite deep water in the wet season), and in open grassland, sometimes many kilometres from the nearest shrub or tree.

Carnivores

Lions were seen on several occasions from the air, and there were also several reports of them being either shot or speared, always during the dry season. Their density is much lower than would be expected from the numbers of prey available, perhaps because they appear to be killed whenever opportunity arises. The larger number of records in the dry season suggests that they move into the area from the east, following the migrating antelope herds.

The leopard is one of the most secretive and unobtrusive animals in Africa. The Range Ecology team saw only one in four years and received reports of several others. It is probably widespread.

The spotted hyaena is not uncommon, and is not infrequently seen and much more often heard. It is feared and disliked by the Dinka as it undoubtedly sometimes attacks cattle, often damaging the udder, and is often also blamed for unexplained losses of calves.

Hunting dogs occur but were not seen during the Range Ecology Survey. The Dinka name of the species, *jong nyapec*, means 'rinderpest dog', and it was said that they arrived in the area at the time of the first rinderpest outbreaks (a time when there are many sick animals to prey upon). Jackals of two species (common & side-striped) occur, but are not very common.

The small carnivores of the area are numerous and diverse, but mostly nocturnal. Several species of mongoose occur, as do wild cats, servals, ratels, civets and genets. Otters (*Lutra maculicollis*) occur occasionally in the swamps.

The birds
Introduction
The Range Ecology Team (RES, 1983), working unsystematically and making no collections, recorded 270 species of bird between 1979 and 1982. Literature records suggest that another 200 species may occur within the study area. By African standards this is not a rich avifauna. Equatorial regions, particularly those with a greater variety of habitats, often support at least twice as many species, often in smaller areas. The particular feature of the water birds of the area is their enormous numbers.

Populations and seasonal variations
During the aerial surveys any conspicuous and identifiable birds were recorded, allowing estimates, albeit highly tentative, to be made of their populations and their distributions to be mapped. The population estimates are shown in Table 15.2. Some species, such as the white stork and pallid harrier, are migrants from Eurasia and are therefore absent from some counts. Others, such as the Abdim's stork and wood ibis, are also sometimes absent because they are intra-African migrants. Other species, not normally thought of as migratory, leave the area almost completely during the wet season. Such are the goliath heron, black kite (Figure 15.6), pied crow and marabou stork. On the other hand, some larger, solitary and largely sedentary species such as the black-headed heron, saddle-billed stork, shoebill, and fish eagle gave remarkably consistent population estimates at all three counts. Even these, however, showed seasonal movements correlated with the advance and retreat of floodwaters (Figure 15.7). It must be remembered that many smaller species such as the garganey, and the very numerous smaller waders (ruff, black-tailed godwits, sandpipers, etc.) could not be reliably identified from the air and so are not included in the estimates.

The herons, storks, ibises and similar water birds are the most conspicuously abundant component of the avifauna. Most of them use a particular habitat – the swamp margin. During the dry season they are nomadic,

Table 15.2. *Estimates of the population sizes of some larger birds in Jonglei*

	Mid wet season	Early dry season	Late dry season
Ostrich	1,486	4,961	6,240
Long-tailed cormorant	232	8,883	6,006
White pelican	0	0	5,643
Pink-backed pelican	3,649	6,110	11,187
Grey heron	994	0	0
Black-headed heron	1,652	1,460	1,716
Goliath heron	0	3,819	3,234
Purple heron	2,587	2,091	5,049
Great white heron	7,506	9,530	19,074
(Cattle) egret	172,359	65,253	86,724
Squacco heron	3,845	9,402	18,414
Shoebill stork	6,407	5,143	4,938
White stork	0	16,500	0
Wooly-necked stork	1,350	2,475	1,485
Abdim's stork	0	0	858
Openbill stork	13,469	288,536	344,487
Saddlebill stork	3,640	4,017	4,158
Marabou stork	196	359,719	194,007
Wood ibis	0	3,775	11,154
Sacred ibis	16,201	4,419	17,688
Hadada ibis	697	429	231
Glossy ibis	787	1,695,240	8,778
White-faced tree duck	7,150	0	51,810
Fulvous tree duck	0	0	8,775
Knob-billed duck	394	9,611	9,075
Spurwinged goose	1,153	88,220	150,216
Lappet-faced vulture	298	33	462
Hooded vulture	164	149	396
Black kite	66	4,655	5,841
Black-shouldered kite	0	231	66
Long-crested hawk eagle	0	100	165
Grasshopper buzzard	66	982	1,287
Dark chanting-goshawk	197	330	100
Pallid harrier	0	0	132
Harrier hawk	0	429	396
Bateleur eagle	758	594	1,089
Fish eagle	1,320	1,263	1,650
Crowned crane	36,823	22,715	14,685
Arabian bustard	299	945	728
Black-bellied bustard	994	396	297
Pied crow	32	1,599	2,574
Cape rook	595	867	3,465

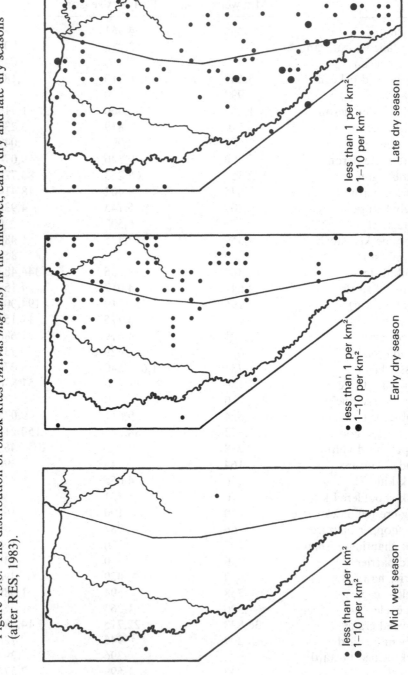

Figure 15.6. The distribution of black kites (*Milvus migrans*) in the mid-wet, early dry and late dry seasons (after RES, 1983).

• less than 1 per km²
● 1–10 per km²

Late dry season

• less than 1 per km²
● 1–10 per km²

Early dry season

• less than 1 per km²
● 1–10 per km²

Mid wet season

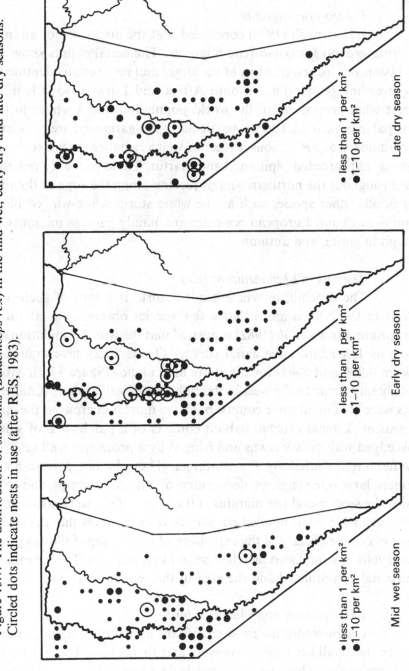

Figure 15.7. The distribution of shoebills (*Balaeniceps rex*) in the mid-wet, early dry and late dry seasons. Circled dots indicate nests in use (after RES, 1983).

moving from place to place to exploit the fish that become available as pools dry up. Some (Table 15.2) are extremely abundant.

Palaearctic migrants

Euroconsult (1978) concluded that the area was not an important wintering ground for palaearctic migrants. The aerial counts show that this conclusion was incorrect. Most of the larger and more easily identified species have breeding populations in both Africa and Eurasia, so it is difficult to suggest what proportion of the world population might winter in the area. Some palaearctic duck species, particularly the garganey, are present in very large numbers, as are various waders as well as smaller species such as yellow wagtails, red-throated pipit and sand martin. These probably remain in the area throughout the northern winter, feeding along the edge of the retreating river flood. Other species such as the white stork, white-winged black tern, common swift and European bee-eater are mainly passage migrants, passing through in spring and autumn.

The shoebill (Balaeniceps rex)

The shoebill, or whale-headed stork, is a bird of such particular interest in the Nile swamps that a few special observations about it seem appropriate. Brown *et al.* (1982) suggested that the world population might be as low as 1,500 birds; the aerial surveys (Table 15.2) gave figures varying between 4,900 and 6,400 for the study area alone (Figure 15.7). More birds undoubtedly occur to the west in the Bahr el Ghazal swamps (Guillet, 1978). Nests were seen in all three counts, but were most numerous at the end of the wet season. Typical shoebill habitat consists of a patchwork of small open pools edged with grassy lawns and fringed by a protective wall of *Typha*, the very habitat favoured by the sitatunga. They do not favour main river channels, large open lagoons, deep water, or papyrus swamps, and only move outside the swamp and *toic* margins at the height of the wet season when there are fish available in the flooded grasslands. It seems likely that the shoebill has been a major beneficiary of the expansion of the swamps following the floods of the 1960s. It would also therefore seem likely that it will suffer some decline if the canal is completed and the area of the swamps decreases.

The economic importance of birds

Undoubtedly the most significant agricultural pests in the Jonglei area are the small seed-eating weavers and their allies. The most numerous species are quelea (*Quelea quelea*), masked weaver (*Ploceus taeniopterus*), red bishop (*Euplectes orix*), black-winged red bishop (*Euplectes hordeacea*), yellow-crowned bishop (*Euplectes afer*), and the fan-tailed widow-bird (*Coliuspasser axillaris*). These species breed during the wet season, feeding mainly on the natural grasses, but at the end of the rains the ripening crops become an attraction, and local people expend a great deal of energy in keeping the birds from their crops. Doves (*Streptopelia* spp.) are also important pests of sorghum fields during the wet season.

These species, particularly *Quelea*, are already a menace to crop production and would be an important factor to be considered in any development of cereal agriculture. They are highly migratory, occur in vast numbers, and usually roost in inaccessible parts of the swamps, so that control would be extremely difficult.

On the credit side, the huge flocks of water birds that form during the dry season at the edge of the retreating flood line and around drying pools, could be a major attraction to tourists, should the political and economic conditions of the area make tourist development a viable proposition. These water birds probably also have an important role in controlling the population of disease vectors, such as snails.

Wildlife and disease
Diseases of wildlife

Very few diseased wild animals were encountered during the Range Ecology Survey. On several occasions, generally in the dry season, carcasses of tiang and giraffe were seen in some numbers. These deaths were probably due to a combination of water and nutritional stresses at this time. No evidence of a major disease outbreak was detected.

An alternative approach to the study of disease in wild animals is to examine blood serum for antibodies to diseases from which the animal has long since recovered or never overtly suffered. The Range Ecology Survey included the post-mortem examination of a sample of animals which had been shot for the purpose. The tiang was the main species sampled in this way. Table 15.3 shows the result of tests on blood samples taken from tiang. These

Table 15.3. *Results of serological testing of Tiang*

Disease	Number tested	Number positive	% positive
Bluetongue	20	7	35
Brucella abortus	19	0	0
Herpes virus	21	8	38
Rinderpest	14	2	14

suggest that tiang are susceptible to infection by the viruses of rinderpest, bluetongue, and the alcelaphine herpes group (related to malignant catarrhal fever in cattle), although they do not necessarily indicate that the animals ever had observable clinical signs of the disease.

Wildlife as a reservoir of disease

There is little, if any, evidence for the exchange of infection between wild and domestic animals in the area. In the outbreaks of rinderpest observed by those concerned with the Range Ecology Survey (RES, 1983), in which a large number of cattle died, no tiang or other wild species was seen to be involved. The question of whether wild animal populations can act as a long-term reservoir of rinderpest virus is currently under review. Following intensive vaccination of cattle during the 1950s and 1960s in northern Tanzania, the disease also seemed to disappear from the wild animal population (Sinclair, 1977). It was concluded that for the virus to persist in game, continual transfer from cattle was necessary (Plowright, 1982). How-ever, the recent reappearance of rinderpest in buffalo and cattle in northern Tanzania, and the finding of antibody titres against rinderpest in various species of wildlife in Kenya, has renewed speculation about the role of wildlife in the epidemiology of the disease (Woodford, 1984). Many species of wildlife are known to be susceptible to rinderpest (Scott, G. R., pers. comm.). However, the range of susceptibility is wide (Plowright, 1982). Of the wild species in the Jonglei area, buffalo are the most susceptible and the possible significance of this in relation to their distribution has been discussed above. The tiang and reedbuck are moderately susceptible, while the Mongalla gazelle and waterbuck are much less so. The white-eared kob and the Nile lechwe have not been studied in this context.

Vaccination cover of cattle in recent years has been so poor in the southern Sudan that there will have been sufficient numbers of susceptible animals without there being the need to postulate a wildlife reservoir.

CBPP and haemorrhagic septicaemia are essentially diseases of cattle and water buffalo. Records of CBPP in wild species are extremely rare (Leach, 1957; Shifrine & Gourlay, 1965). Pasteurellosis has been found in a wide range of wild animals. However, the reports often do not make it clear whether the precise strains which cause haemorrhagic septicaemia in cattle are involved (Davis, Karstad & Trainer, 1981). There is no suggestion that wild hosts are an important reservoir of the disease.

A wide range of wild animals can suffer from brucellosis (Wither, 1981). However, the high prevalence of this disease in the cattle of the area is most unlikely to be affected by continual transmission from wild species. The infection is spread mainly by discharge from a recently calved or aborted cow, and the crowded conditions of cattle camps provide plenty of opportunities for this to happen.

Wild animals are also susceptible to the viruses of bluetongue and foot-and-mouth disease. In the case of the former, cattle are the principal maintenance host, and serology showed that most of the cattle in the Jonglei area had experienced infection. Tiang also showed evidence of infection but are not likely to be influencing the domestic animal situation. In any case, infection with bluetongue virus in the indigenous stock is not causing disease. Only if exotic breeds were introduced would it become important. Wild animals may be involved in the epidemiology of foot-and-mouth disease and indeed buffalo have been shown to be symptomless carriers of the virus elsewhere in Africa. However, in the Jonglei area the distribution of buffalo scarcely overlaps that of cattle at the present time. In indigenous cattle foot-and-mouth disease usually takes a mild course. However, the disease would become very important if the export of stock from Jonglei were to be contemplated; the role of wildlife would then have to be investigated in more detail.

Bovine ephemeral fever and lumpy skin disease were observed in Dinka cattle by the Range Ecology team. There is no evidence for the involvement of wild hosts in the epidemiology of these diseases.

As mentioned above, there is serological evidence for the occurrence of herpes virus in the tiang. The transmission of malignant catarrhal fever from wildebeeste to cattle in Kenya and Tanzania is well known. However, although the tiang calve at a time and place when frequent contact with cattle is possible, there is no indication that such a transfer of disease occurs in the Jonglei area. Moreover, alcelaphine herpes viruses are known to occur in topi and hartebeeste elsewhere, but transmission from these species to cattle has not taken place (Plowright, 1981).

Parasites

In areas where tsetse fly occur, trypanosomiasis is the major disease of cattle for which wild hosts act as reservoirs of infection. Tsetse, if present at all in the Jonglei area, are at a very low density. The assumption has been, and until good evidence to the contrary is found must remain, that the main method of transmission in the area is mechanical, by biting flies other than tsetse (RES, 1983). For mechanical transmission to occur, such a fly interrupted in its blood meal on one host must rapidly attach to another. This is more likely to occur from cow to cow within a herd than from, for example, a tiang grazing some distance away. The few blood smears of tiang which were examined during the Range Ecology Survey were negative for trypanosomes. Serological testing was not carried out.

Helminth parasites common to both domestic and wild animals (RES, 1983) included the liver fluke *Fasciola gigantica*, the schistosome *Schistosoma bovis*, and the gastro-intestinal nematode *Haemonchus contortus*. None of these was found abundantly in the wild species, whereas heavy infestations occurred in cattle, sheep and goats. In the domestic species they were a cause of debility and, in the case of *Fasciola gigantica*, of condemnation of livers at slaughter places. It seems unlikely that the burdens in wildlife pose any significant additional threat to the health of the domestic stock.

Wildlife and men
Hunting
Hunting plays an important part in the life of many people in Jonglei, particularly during the dry season, when one of the major attractions of moving to the cattle camps in the *toic* is the opportunity to hunt. At this time wildlife biomass can exceed 10,000 kg km^{-2} in the *toic* and there are plenty of opportunities for hunting. All participate; often the young boys are the most avid hunters, and women may also join in. Dogs play a crucial role in hunting, and are kept and bred specially for this purpose. This enthusiasm for hunting seems to be something new; the JIT (1954), Evans-Pritchard (1940) and Lienhardt (1961) all describe a general disdain for hunting in both Nuer and Dinka, although according to Howell (1945a) the Nuer were formerly enthusiastic elephant hunters, mainly to obtain ivory for trade. It is possible that the floods of the 1960s, and consequent lean years of famine and reduced stock numbers, and the difficulties encountered during the years of civil war, encouraged a start to hunting, and that an enthusiasm for hunting and its benefits has persisted. It is also possible, as described above, that there are more animals to hunt.

The Bor Athoic organise large hunting camps in the dry season in the Eastern Plain. The quarry is mainly reedbuck and gazelles, and the meat is hung up in strips to dry and later carried as head loads to the settlements.

Hunting also takes place in the swamps. Hippo are speared from canoes, using spears with detachable heads joined to a rope with a float made of ambatch wood (*Aeschynomene elaphroxylon*). The meat is eaten at once, or dried, the skin is sold for making whips, and the bones are often of necessity used as weights for fishing nets, as there are no stones in the swamps. Another hunting method in the swamps is to burn an area of vegetation and wait in a canoe downwind of it to spear anything driven out by the flames. Sitatunga are probably the quarry here.

The Range Ecology team (1983) studied hunting in the south of the area in the dry season of 1982. There is little hunting in the heat of the day, but, in the early morning, when animals which have fed and drunk during the night may still be close to habitation, boys look for animals and actively hunt them. The dogs chase the animal, and either seize it or bay it until the hunters arrive with spears to dispatch it. The carcase is quickly cut up, shared out, and carried back to the village or cattle camp.

Most of the hunting takes place in the *toic*, but opportunities also occur as animals migrate through the settlement belt or attempt to use water-holes there, and are not neglected. During the study, an attempt was made to watch all hunts in a limited area, and to record the quarry, the success or failure of the hunt, and the numbers of people and dogs involved. The study lasted 43 days. During this time, just over three animals were killed each day.

Several points emerged from the study. First, single animals of any species were much more vulnerable than herds. Overall, a lone animal had only a 12% chance of escape when hunted, while an animal in a group had a 50% chance of escape. Secondly, reedbuck were found to be the most vulnerable; 37 out of 43 hunted were killed. Mongalla gazelle were also vulnerable; 39 out of 62 were killed, though an additional uncounted large group departed unscathed. Young animals were particularly vulnerable. Tiang were least vulnerable; 26 were killed, and 340 and two other uncounted large groups escaped unharmed.

As a rule, people alone or dogs alone were noticeably unsuccessful in hunting. The combination was, however, much more effective, with four to six people with three dogs appearing to be the most effective grouping. Most hunts occurred early, but the greatest percentage of successes was later, in the heat of the day. The reasons for this are clear; herds, which we have seen to be difficult to catch, tend to pass through early in the day, but the later single stragglers are more likely to be caught.

Hunting and nutrition

This information permitted a calculation of the role in human nutrition played by game meat. Using the data from the study summarised above, and liveweight information derived from the Range Ecology Survey and Coe *et al.* (1976), a mean weight of 50.2 kg for each animal killed is arrived at (RES, 1983). The rate of killing was 2.6 animals per day in an area of 32.3 km^2. Thus wild animals were being killed at a rate of 4.05 kg km^{-2} per day at the time of the study, in the latter part of the dry season. At the peak of the tiang migration the figure could well have been higher.

Assuming that hunting occurred only in the *toic* and settled areas, which together make up 18.2% of Kongor District, and only during the dry season (November–March; 185 days), an annual harvest of wildlife meat in Kongor District amounting to just over 1,000 tonnes could be calculated. On the assumption that 70% of each carcase is eaten, then that figure is reduced to 700 tonnes. On the basis of this figure, the 91,000 inhabitants of Kongor District (JEO, 1983) each have an annual consumption of wild mammal meat of 7.7 kg. Ilaco did not consider wildlife meat at all in their surveys of nutrition in Bor District (Ilaco, 1981). This is clearly an oversight, since hunting appears to be as widespread in Bor District as it is in Kongor. Using the market price of beef as a guide, the monetary value of the annual harvest of wildlife meat in Kongor District alone was calculated to be £S1.57 million (RES, 1983).

Plate 31. A group of Nuer youths about to cut up a tiang that they, with the help of their dogs, have run down and speared. Game is an important source of protein for some sections of the community, particularly during the dry season. Harvesting in this way, for local consumption only, represents a sustainable way of exploiting the wildlife stocks. Photo: Alison Cobb.

Hunting, the law and firearms

Hunting in the southern Sudan was in 1982 regulated by the Wildlife Conservation and National Parks Act of the Southern Regional Government. Under this, species are divided into groups. Those in Schedule III include tiang, reedbuck, bushbuck and mongalla gazelle; these may be hunted at any time, as long as firearms are not used. Schedule II animals (including white-eared kob, Nile lechwe, hippopotamus, giraffe, zebra and lion) may only be hunted with a licence. Section 28 of the Act prohibits the use of dogs for hunting at all times, evidently an unenforceable (and pointless) piece of legislation.

Hunting permits are issued to anyone on payment of the scheduled fees. Firearms permits are issued by the Police; the likelihood of their being issued varies according to prevailing security conditions.

Despite worsening conditions, firearms control is lax, and prior to 1983, many ordinary people possessed guns. The impact of the current civil war and the wider circulation of firearms will now undoubtedly mean that their ownership and attendant use for hunting will increase. Hunting with spears and dogs means only a reasonable offtake; hunting with firearms, on a large scale could be devastating. A number of incidents were recorded during the Range Ecology Survey in which members of the uniformed forces used their firearms to kill large numbers of wild animals; on one occasion more than 400 tiang were shot. A combination of education and improved discipline is the only means to control such misuse of firearms.

Hunting with firearms is a particular threat to two species, the Nile lechwe and the elephant. The lechwe is normally protected by its habitat from hunts with dogs and spears, but it is easily stalked and shot. Elephants have the misfortune to carry tusks which are a valuable export commodity, yielding much-sought-after foreign exchange. In 1982, the Sudan was Africa's largest exporter of ivory. The extensive hunting of elephants is only feasible if firearms are available.

Wildlife and the future

Tourism has never developed in Jonglei Province because of its remoteness. In the future, given a new international airport at Juba, an improvement in fuel supplies and a return of international confidence, it has some potential.

Sport hunting by foreigners was reasonably well established in the Southern Region, and was the most important single earner of foreign exchange. However, the difficult terrain of the Jonglei area, and the scarcity of water in the dry season, has meant that few hunting groups use it. The discovery of oil in the area will one day not only push hunting out of first place as a foreign exchange earner, but will also have adverse implications for wildlife; oil-drilling crews around the world are notoriously lawless poachers.

Commercial cropping of wildlife has been recommended for the area (Watson *et al.*, 1977) but not attempted. The experiences of government-run wildlife cropping schemes elsewhere in East Africa have suggested that large ungulate populations cannot necessarily sustain large cropping operations. Problems include the preservation, storage and transport of carcasses, distance from markets, and the mobility of the wildlife herds. It seems most unlikely that such an operation in the Jonglei area could be both financially viable and biologically sustainable. Traditional hunting, as described above, harvests much meat close to the place where it is needed, contributing substantially to the subsistence economy. It seems a sensible way of exploiting the wildlife of the area in the present state of economic development. It is uncontrolled and undocumented, so there can be no question of enforcing quotas or similar controls. Regular monitoring (see Chapter 20) must be an important part of future management.

The report of the Range Ecology Survey team makes various recommendations for the future (RES, 1983). They suggested an eastward extension of the Badingilu Reserve (outside the Jonglei area, to the south) in order to safeguard the wet season habitat of the tiang and white-eared kob; development-free zones in the Mayom area (north of Kongor) to maintain a corridor for migrant wildlife; and the setting up of a reserve to include the major habitats at the heart of the Jonglei wetlands. They also proposed a no-hunting zone extended for three kilometres on either side of the Jonglei Canal, if and when it is completed, to prevent casual shooting of animals from passing vehicles and boats. They emphasised the great international importance of the Jonglei wetlands as a habitat for migratory and sedentary wild animals, and migratory birds.

Conclusion

This chapter has tried to give a brief outline of the abundance and diversity of bird and mammal species in the area, and of the way in which they are economically important to the people. In Chapter 17 we shall consider how this situation may be affected by the advent of the Jonglei Canal. Future developments manifestly depend upon a return to peace in the area; given this, wildlife could make a very positive contribution to the non-subsistence economy of the area, and the area itself could make a valuable contribution to global conservation efforts.

Part V

The local effects of the Canal

16

The effects on climate, water and vegetation

An assessment of effects on climate

Some anxiety has been expressed that one of the environmental effects of operation of the canal in reducing the area of swamp and seasonally inundated grassland might be to reduce water losses by evaporation and transpiration, and so to have an effect on rainfall downwind.

The JIT (1954) suggested that extra convective activity from the heating of more dry land surface would counterbalance the decline in the amount of water entering the atmosphere by evaporation and transpiration, so that the net effects would be negligible. Since then there have been suggestions that a reduction in the area of seasonal and permanent swamp could lead to a reduction in rainfall both locally, over the *Sudd*, and also farther away, downwind (Mann, 1977). This view is at least partly based on the area of higher rainfall marked over the *Sudd* on many rainfall maps; this exists on the strength of a single gauge which has been shown (Sutcliffe & Parks, 1982) to be unreliable.

Since 1961 the amount of water flowing into the *Sudd* from the Bahr el Jebel has increased greatly (see Chapters 4, 5 & 8). This has led to an expansion of the area of permanent and seasonal swamp by about two and a half times. This massive expansion has not led to any detectable increase in local rainfall at stations close to the swamps (see Figure 5.8), nor is there any evidence of increased rainfall farther north in the Sudan.

While there still remains much scope for studies of the interactions of the atmosphere with large inland lakes and swamps, and their effects on local rainfall, there seems to be no evidence at present that the very large changes in swamp area in the last 30 years have affected rainfall at local or remote sites. It therefore seems unlikely that a reduction in swamp area caused by the operation of the canal would have any effect either.

An assessment of the effects on areas of flooding

Adaptation of the hydrological model

The natural conditions of flooding were described in Chapter 5, and a hydrological model was used to deduce areas of flooding from historical measured inflows and outflows. It is possible to use a similar model to simulate the effect which the Jonglei Canal would have had on flooding during the period for which there are records. Because the canal will reduce river inflows to the swamp and thus reduce evaporation, the direct outflows from the swamp will be altered, and the previously measured outflows are therefore irrelevant. The model can be adapted for use without outflow measurements by deriving a relationship between inflow and outflow from the records, and then assuming that the same relationship will hold when part of the flows are diverted. This approach has been used in previous studies to assess the canal benefits in terms of increased flows in the White Nile, and the results of statistical analysis of inflows and outflows are consistent with these studies (e.g. El Amin & Ezeat, 1978).

In addition, in order to estimate the effect of the canal, the model has to be adapted from measured inflows at Mongalla to deduced flows at Bor, where the canal offtake is sited. Although much of the analysis is described in tables and equations, these have as far as possible been confined to Appendix 9.

Relationship between inflow Q and outflow q

Monthly inflows measured at Mongalla were compared with outflows measured at Malakal (from which the Sobat flows are deducted). The flow records at Mongalla were less reliable before 1922, as few discharge measurements were made, but the comparison was based on records from 1916 to 1972 in order to include the high flood of 1916–18. The four years 1963–6 were omitted as a gap in the measurements at Mongalla coincided with an abrupt change in the rating curve and gave rise to a number of outliers in the relationship.

Monthly outflows were compared as a dependent variable with monthly inflows using various time lags, and swamp storage estimated from a preliminary model. The regression was carried out in arithmetic form and in logarithmic form, and the results are summarised in Appendix 9. Although there is a lot of scatter or unexplained variance, much appears to result from deriving outflows from the differences between Malakal and Sobat flows.

The relationship improves as the lag between inflow and outflow increases to three months but improves little for longer lags. Introducing storage as an independent variable provides little improvement, and, as it is calculated rather than measured, it was omitted. The plot of the inflow–outflow relationship is trumpet shaped, which suggests that the logarithmic form of relationship is preferable.

Because comparisons of dates of maximum and minimum flows have supported a three-month lag, this interval was chosen and an improved relationship was obtained by grouping the flows into quarterly means, giving a complex empirical equation linking outflow to inflow in the previous quarter.

As shown in the Appendix, however, this equation implies that outflow exceeds inflow at low flows and this is an unreasonable extrapolation below the range of measured flows. A more realistic curve was derived for flows below 1.73 km^3 per month. The joint curve is compared with annual total inflows and outflows from 1905 to 1980 in Figure 16.1, which suggests that a reasonable relationship has been derived.

These equations were used to obtain from Mongalla inflows a sequence of predicted outflows with a lag of three months for the period 1905 to 1980. These were substituted for measured outflows in a second trial of the model to provide another sequence of estimated areas of flooding. Comparison with previous results showed that the timing of the seasonal fluctuations, the magnitudes of the maxima and minima, and the comparison of estimated with

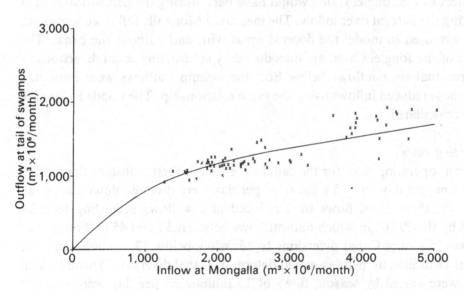

Figure 16.1. The relationship between swamp inflows at Mongalla, and swamp outflows (Malakal gauge, minus Sobat River flows at Hillet Doleib). The curve was derived by logarithmic regression of quarterly flows, and extended below 1730 by argument as described in the text. Each point represents the annual inflows and outflows, expressed in monthly terms (from Sutcliffe & Parks, 1987).

measured areas on specific dates, were little changed. Thus the model can be adapted to predict flooded areas using measured inflows but deduced outflows and then, by subtracting Canal diversions from the inflows, to estimate the effects of the Jonglei Canal.

However, as stated above, the river inflows have been measured at Mongalla, and the canal offtake will be at Bor; the canal diversions therefore have to be subtracted from river flows at the latitude of Bor. To convert Mongalla flows to those at this latitude, the average net evaporation was multiplied by the area flooded between Mongalla and Bor to give the relatively small loss. This loss provides a relationship between Mongalla and Bor inflows which can be used to relate Bor inflows to swamp outflows.

The hydrological model can now be repeated for the swamps below Bor, using the adjusted Mongalla inflow records and the relationship between Bor flows and swamp outflows. This gives estimates of areas flooded below Bor under natural conditions and again corresponds with measured areas.

Use of the hydrological model for prediction
The model has been tested by its ability to reproduce historical areas of flooding over the range experienced. It can be used to estimate what the effect of the Jonglei Canal would have been during the period 1905–80 in reducing the natural river inflow. The measured Mongalla inflows and rainfall series are used to model the flooded areas with and without the canal. The effects of the Jonglei Canal are introduced by subtracting canal diversions to give residual river inflows below Bor; the swamp outflows were estimated from these reduced inflows using the same relationship. The flooded areas are then recalculated.

Operating rules
Different operating rules for the canal were tested. First, constant flows of 20 million m³ per day and 25 million m³ per day were diverted down the canal. Secondly, these canal flows were reduced at low flows according to rules tested by the PJTC, in which natural flows between 33 and 45 million m³ per day would reduce Canal diversions to 15, while below 33, a minimum river flow of 18 million m³ per day would determine Canal diversion. Thirdly, canal flows were varied by season; flows of 15 million m³ per day were diverted during the dry season (November–April) and 25 million m³ per day during the wet season (May–October), and the reverse. As these rules resulted rarely in very low river flows, in order to avoid unrealistic results the outflow was not allowed to exceed the simultaneous Bor inflow when the flooded area was less than 500 km². The operating rules tested are summarised in Table 16.1.

Table 16.1. *Summary of estimated effects of canal on average areas of flooding (km²), Bor to Malakal, and percentage reductions*

Canal flow m³ × 10⁶/day	1905–61 Permanent swamp	1905–61 Seasonal swamp	1905–61 Total	1961–80 Permanent swamp	1961–80 Seasonal swamp	1961–80 Total	1905–80 Permanent swamp	1905–80 Seasonal swamp	1905–80 Total	1905–80 Mean area	1905–80 Predicted outflow[2]	1905–80 Predicted benefit
–	6,700	6,200	12,900	17,900	11,000	28,900	9,500	7,400	16,900	12,800	15,400	
20	3,500	4,600	8,100	14,000	9,100	23,100	6,200	5,800	12,000	8,800	20,600	4,500
	47	26	37	21	17	20	35	22	29			
25	2,900	4,300	7,200	13,100	8,700	21,800	5,500	5,400	10,900	7,900	21,800	5,700
	57	31	44	27	21	24	43	27	36			
20[1]	3,600	4,600	8,200	14,000	9,100	23,100	6,200	5,700	11,900	8,800	20,600	4,500
	46	26	37	21	17	20	35	23	29			
25[1]	3,000	4,200	7,200	13,100	8,700	21,800	5,500	5,300	10,800	7,900	21,700	5,600
	56	32	44	27	21	24	42	28	36			
25 Nov–Apr 15 May–Oct	3,300	5,400	8,700	13,700	9,900	23,600	5,900	6,600	12,500	8,900	20,500	4,400
	51	12	32	23	10	18	38	11	26			
25 May–Oct 15 Nov–Apr	3,800	3,900	7,700	14,400	8,400	22,800	6,500	5,000	11,500	8,700	20,700	4,600
	43	38	41	19	24	21	32	32	32			

Notes: [1] Flow down canal reduced at low flows according to PJTC rules: if river flow at Bor is Q and canal flow q, in m³ × 10⁶/day, then if 33<Q<45, q = 15; if Q<33, q = Q – 18
[2] Mean measured inflow = 33,000 m³ × 10⁶; mean measured outflow = 16,100 m³ × 10⁶

Canal losses and benefits

The swamp outflows used in the model were added to the canal flows, with canal losses estimated as 0.2 million m^3 per day (El Amin & Ezeat, 1978), to give the combined outflows. Comparison with the natural outflows provides an estimate of the canal benefit over 1905–80, which is consistent with other estimates. The backwater effect of raised levels at the canal outfall cannot be taken into account in this model, but analysis has suggested that losses and flooding will not result (Ibrahim & El Amin, 1976).

Summary of trials

Each trial provides monthly flooded areas for 1905–80, which may be compared with natural conditions in graphs similar to Figure 16.2, which shows that the timing of seasonal fluctuations remains, but with the amplitude reduced. Quantitative presentation of the trials requires selection of the important features of the graphs. One method is to present flooding statistics – maximum, minimum and range – in histogram form, as these correspond to total area, permanent swamp or *sudd*, and seasonal swamp or *toic*.

Discussion of results

Figure 16.3 presents the natural regime in histogram form, showing that the areas of flooding have varied greatly because of changes of river regime over the 75-year period. The permanent swamp has varied more than the seasonal swamp. Similar diagrams for different canal rules show that these natural variations remain important.

The average effects of the canal are given in Table 16.1. The reduction in average permanent swamp varies from 32 to 43%, while the seasonal swamp decreases by 11 to 32%. The reductions are naturally greater with the canal operated at 25 rather than 20 million m^3 per day.

Varying the canal flow seasonally could affect the relative change in permanent and seasonal swamp. With high diversions in the wet season and low in the dry season, estimated reductions in the areas of permanent and seasonal swamp were 32% in each; the reverse regime reduced the areas of permanent swamp by 38% and seasonal swamp by only 11%.

Figure 16.2 also shows that reductions are relatively greater in the early years of low inflows than in the recent years of high flows. Thus in Table 16.1 the estimated effects of the canal are summarised for the low years 1905–61 and the high years 1961–80 as well as the whole period. For example in Table 16.1, the permanent swamp is reduced by 47% in the early years, and 21% later; the corresponding reductions in seasonal swamp are 26% and 17%.

Although the reductions are greater in terms of area in the later years, they are smaller in percentage.

One cannot deduce which of the two flow patterns is more typical of the long-term regime or forecast future flows. Analyses of lake levels and other long-term records suggest that wet and dry periods alternate and that changes are random in their timing.

The simplicity of the hydrological model implies that undue precision should not be attached to the results, but physical measurements have been used wherever possible and the predicted benefits are in general agreement

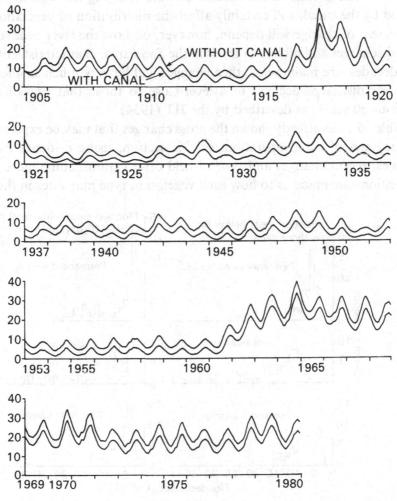

Figure 16.2. Estimated areas of flooding, Bor to Malakal (km^2 × 10^3), without canal flow, and with canal flow (20 m^3 × 10^6 per day) according to PJTC rules (see text) (from Sutcliffe & Parks, 1987).

with previous studies. It is suggested that the results provide a reasonable indication of the degree to which the permanent and seasonal swamps will be affected by the operation of the canal.

In terms of location it is possible to predict that flooding will regress from the present extended area towards the area flooded during the earlier period of low flows, but the effects of the canal must be placed in the context of the changes which occurred after the rise in East African lake levels.

The effects of the canal on vegetation

The decline in the volume of water passing through the swamps, caused by the canal, will certainly affect the distribution of vegetation types. The extent of change will depend, however, on how the river behaves. If the high discharges and high river levels in the *Sudd* area, characteristic of the last two decades, are maintained, then changes will be very much smaller than if the river discharge declines to a level close to those that prevailed in the previous 50 years, as described by the JIT (1954).

Table 16.1 has already shown the gross changes that may be expected under two regimes of canal operation. In this section, using evidence from past changes, recent changes and general field observations, some more detailed suggestions are made as to how each vegetation type may alter in the future.

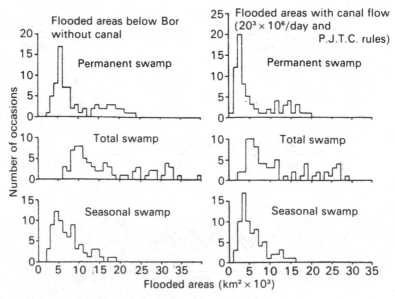

Figure 16.3. Flooded areas below Bor without canal flow, and with canal flow (20 m^3 × 10^6 per day) according to PJTC rules (see text). The histogram bars represent the number of years in which flooded areas fall within the given limits (from Sutcliffe & Parks, 1987).

Table 16.2 lists the main vegetation types recognised, the area covered today by each, and the description given to each by the JIT (1954); they have already been described in Chapter 7. An indication of the likely appearance of the distribution of vegetation types, post-canal, is shown in Figures 16.4 and 16.5.

Flowing waters

At present the river channels are almost free of submerged vegetation, probably because the river is too turbid to allow enough light to reach the river bed. Garstin (1904) describes the water of the main river near Lake No as clear but coloured; water of this kind in the present river system supports submerged plants, but the water in the main river channel is now much more turbid. This may be the result of more turbid water leaving the highlands than previously, or of a higher rate of flow preventing particles from settling, or of the removal of obstacles, which might have filtered the water directly, or have diverted it from channels into swamps. Flow rates measured in 1982–3 averaged about 4 km per hour; in 1874 they averaged about 3.5 km per hour (Watson, 1876) – not a large difference, but, since the particle size that can be moved by water varies as the fourth power of its speed, it could be significant. It is much more likely that filtration of the water through the many *sudd* blockages which existed at the time of Garstin's visit

Table 16.2. *The major vegetation types of the region*

Name used in this account and by RES (1983)	Area (km²)	Name used by JIT (1954)
Swamps		
Cyperus papyrus swamp	3,920	Permanent swamp
Typha domingensis swamp	13,570	
Open water with aquatics	1,510	
Vossia cuspidata	250	
River-flooded grasslands		
Oryza longistaminata	13,080	Riverain and *Khor*-bed grasslands;
Echinochloa pyramidalis	3,070	Shallow-flooded swamp grasslands
Rain-flooded grasslands		
Hyparrhenia rufa	15,820	*Hyparrhenia rufa* grassland of intermediate lands in the flood region
Woodlands and wooded grasslands		
Balanites aegyptica	6,140	*Acacia seyal–Balanites aegyptiaca* forests of high lands
Acacia seyal		
Mixed woodlands		Poorly developed mixed deciduous
Other woodlands		Broad-leaved forest type

Notes: These areas are derived from the final version of the Range Ecology Survey map (1983), by planimetry.

removed much of the particulate matter. It is thus unlikely that a reduction in river flow will allow any increase in the very few plants which manage to

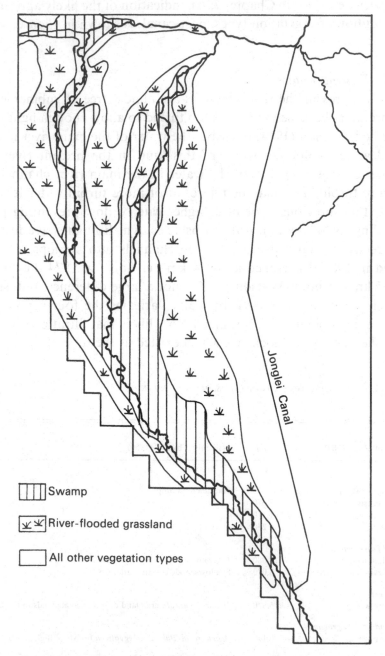

Figure 16.4. The expected extent of permanent and seasonal swamp if the Nile continues to discharge at 1963–80 levels. Compare Figure 8.3 (after RES, 1983).

survive in the main channel, although it may allow more to develop in side channels remote from the silt-laden main stream. The fringing vegetation is

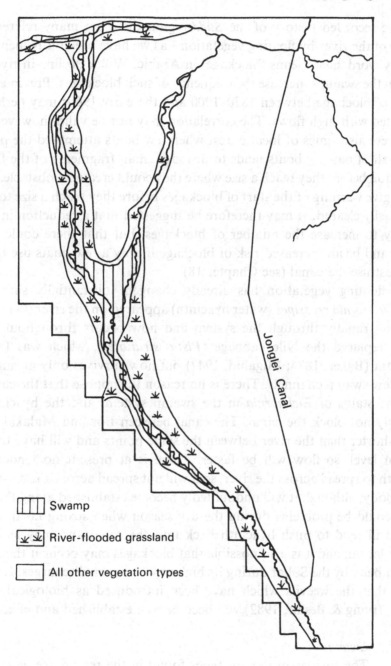

Jonglei Canal

▥ Swamp

⚶ ⚶ River-flooded grassland

☐ All other vegetation types

Figure 16.5. The expected extent of permanent and seasonal swamp if Nile discharge returns to 1900–60 discharges. Compare Figures 8.2 and 8.3 (after RES, 1983).

also unlikely to remain unchanged although, insofar as it reflects the adjacent swamp types, some changes may be expected in the south of the area if river levels fall.

In the recorded history of the *Sudd* region there are many references to closure of the river by floating vegetation – as we have previously mentioned, the very word *sudd* means 'blockage' in Arabic. Will a decline in river flow through the swamps increase the frequency of such blockages? Previous major phases of blockage between 1870–1900 and the early 1960s may perhaps be correlated with high flows. The correlation may not be valid, however, since these were also times of local unrest when few boats attempted the passage. The wash of passing boats tends to dislodge small fragments of the fringing vegetation before they reach a size where they could create an obstacle, and of course give warning of the start of blockages before they reach a size too large to be easily cleared. It may therefore be suggested that a reduction in flow is unlikely to increase the number of blockages, but that there could on the other hand be an increased risk of blockage if only a few boats use the Nile and most use the canal (see Chapter 18).

Free-floating vegetation has already changed substantially since 1957, when *Eichhornia crassipes* (water hyacinth) appeared on the river (Gay, 1958). It spread rapidly through the system and now occurs throughout. It has largely replaced the Nile cabbage (*Pistia stratiotes*), which was formerly abundant (Baker, 1874; Migahid, 1947) but now survives only in rain pools and a few swamp channels. There is no reason to suppose that the canal will alter the status of *Eichhornia* in the swamp system, and the hyacinth will probably not block the canal. The canal between Bor and Malakal will be much shorter than the river between the same points and will have the same drop in level, so flow will be faster. There is at present no tendency for hyacinth to spread across the river, so it will not spread across a faster-flowing water body, although it will undoubtedly become established along the edges. There could be problems during the dry season when strong north-easterly winds will tend to push hyacinth back into the canal and may lead to its accumulation, and it is also possible that blockages may occur if the canal is ponded back by the Sobat during its highest wet season discharges. It is to be hoped that the weevils which have been introduced as biological control agents (Irving & Beshir, 1982) will become well established and effective.

Swamps

The four main swamp types found in the region are described in Chapter 7 and their characteristics are summarised in Table 7.1. Changes in the total swamp area, and in the proportions occupied by different vegetation types, are discussed in Chapter 8. Since the great rise in river discharge of the early 1960s, the total area of swamp has increased from about 7,000 km² to nearly 18,000 km². Much of the increase is believed to be in the *Typha* swamps, although there is some doubt about this because the JIT (1954), lacking aerial facilities, were unable to look at swamp areas remote from the rivers, and it is in just such areas that *Typha* is now abundant. It now covers about 13,500 km², while papyrus covers about 3,900 km² and *Vossia* about 250 km². How will the canal affect this?

First, because of the reduction in discharge through the natural river channels, there will be a reduction in river spill and therefore a reduction in the area flooded by this each year, as well as in the area permanently flooded. For reasons which are explained in Chapter 5, a fall in river level means that spilling will start farther downstream than before, both seasonally and permanently. The consequences for the vegetation will be, first, that *Vossia*, which occurs in sites which are subject to substantial seasonal fluctuations in level, will extend downstream to an extent which will depend on the river flow. As has been shown in Chapter 8, *Vossia* formerly extended much farther downstream than it does now. Similarly, *Cyperus papyrus* will retreat downstream, and the width of the belt that it forms on each side of the main river will decline, again to an extent dependent on the change in the river flow. A substantial reduction may also be expected in the zone occupied by *Typha*. Much of this appears to be occupied by rather shallow water, and it may be surmised that a substantial proportion of this is contributed by rain running off from the east, as creeping flow. A fall in the river level will cause a marked decline in the river-derived water level in these swamps remote from the river; at the same time the creeping flow from the east will be able to move closer to the river, though the creeping flow may itself be reduced because of interception by the canal. All in all a substantial reduction may be expected in the zone occupied by *Typha*.

All the swamp vegetation types are alike in having a much greater standing crop than have the seasonally river- and rain-flooded grasslands. This is because the bulk of the above-ground biomass in the latter types is destroyed each year by burning; by no means all the swamp vegetation is burned each year although large parts are. However, in the swamps, naturally, only those parts above the water are burned. The rhizomes of *Typha* are generally buried in the mud, and would not be burned even if the swamp dried out completely, but those of papyrus often form a floating mat which, if completely dry, can burn. The complete burning of papyrus and its rhizomes is accompanied by the release of large quantities of nutrients, particularly phosphorus (SES, 1983). It is possible to envisage a situation in which large areas of papyrus swamp dry out as a result of the opening of the canal, combined perhaps with a low flood. If all the dry material burned, it would release large quantities of nutrients into the river system. While it is likely that much of this would quickly be absorbed by the remaining swamp plants it is possible that it could provoke a dramatic flush of growth of *Eichhornia*, and possibly of phytoplankton in those lakes receiving water from the river.

Seasonally flooded grasslands

The zones of seasonally river-flooded grasslands dominated by *Oryza longistaminata* and by *Echinochloa pyramidalis* lie outside the *Typha* zone, farther from the river. The calculations earlier in this chapter suggest that these grasslands will decrease in area by between 10 and 32%, depending on the canal operating regime and the river discharge. It has been shown in Chapter 7 that these grasslands occur in sites with fairly precisely definable flooding regimes, and this confirms the conclusions of the JIT (1954). These zones will thus not only decrease in total area, but will also move closer to the main river as the flooding regime farther away becomes unsuitable for them, and as the area now occupied by *Typha* dries out each year and becomes unsuitable for its growth. Both species have the ability to spread rapidly by rhizomes and by seed, and there can be little doubt that they will spread into new areas easily.

What will replace them? The maps, and the observed zonation, suggest that they should be replaced by rain-flooded *Hyparrhenia rufa* grassland, or sometimes by woodland. While this may be true in the long term, the short term changes may well be rather different, a conclusion based on changes that have taken place as a result of the exclusion of river flood water from areas of *Oryza* grassland by dykes.

Figure 16.6 shows the differences between the two sides of three dykes of different ages. Dyke A (Maar-Panpiol) is of uncertain age; it formed a barrier of varying effectiveness from the mid-1950s, and was raised in 1980. Sampling in July 1981 showed striking differences between the two sides. Dyke B (Jalle–Yomcir) was rebuilt in 1981 on the line of an older hand-built dyke of varying efficiency; there are similarly striking differences between the two sides. Dyke C (Maar-Kongor) was sampled shortly after its completion in 1981 and no difference was apparent between the two sides, in July 1981. A second sampling, after only one growing season, already showed marked changes. In all cases *Sporobolus pyramidalis*, rather than *Hyparrhenia rufa*, is the species which first replaces *Oryza*. Why should this be? In many parts of Africa there are reports of this species increasing with heavy use by cattle (Lock, 1972). This is mainly due to its resistance to trampling and grazing, through its dense tussock form and strong shoots and roots, but the very large number of small easily dispersed seeds must also be important in ensuring wide and effective dissemination of the species. *Hyparrhenia rufa* produces fewer seeds than *S. pyramidalis*, so might be expected to spread more slowly; however, it is already present in the *Oryza* grasslands in very small quantity on raised ground such as old termite mounds. Elsewhere in Africa, members of the Andropogoneae, including species of *Hyparrhenia*, have been shown to appear late in the succession from abandoned farmland to mature grassland (Bews, 1917), and it is possible that a similar succession will occur on previously annually flooded sites, with *Sporobolus* being the pioneer species and *Hyparrhenia rufa* forming the climax. The provisional prognosis, therefore, is for an initial colonisation by *Sporobolus pyramidalis*, with a gradual change to *Hyparrhenia* grassland if grazing pressure is not too high.

Within the main area of *Hyparrhenia* grassland, there should be no changes due to alteration of river levels by the operation of the canal because this

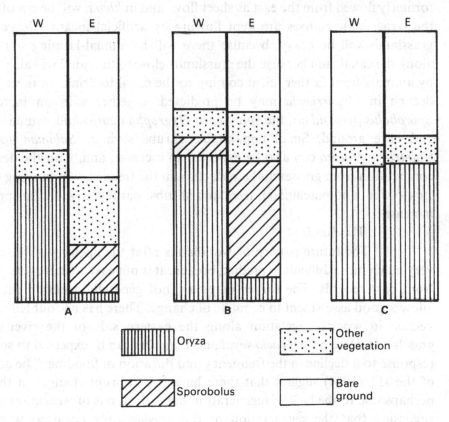

Figure 16.6. Species composition of the vegetation on the river-flooded (west) and unflooded or rain-flooded (east) sides of dykes. For additional information, see text.

grassland type is rain-flooded, not river-flooded. However, in the immediate area of the canal, as we shall see in Chapter 18, there will be ponding of creeping flow on the eastern side, so that flooding will probably be too deep and prolonged for *Hyparrhenia* to survive. An increase in *Oryza* is likely; the species is already present at low density in *Hyparrhenia* grassland. Species of *Echinochloa*, such as *E. stagnina*, and forms of *E. pyramidalis*, which already form floating mats on the seasonal watercourses in the eastern grasslands, may well appear in this newly flooded area. *Vossia* is not a common constituent of rainfed water bodies and is unlikely to appear, and, for the same reason, *Eichhornia crassipes* will probably not be common. It may be predicted, in fact, that the flooded strip on the eastern side of the canal will be very similar in its vegetation to that of the broad seasonal watercourses that occur in this area at the present time.

On the western side of the canal, settlement and usage will inevitably be heavy, and some deterioration of the grasslands seems likely. Water which formerly flowed from the east as sheet flow, and in *khors*, will be cut off unless the larger watercourses are kept flowing by artificial means. Usage of the grasslands will be heavy, because there will be animal-keeping settlements along the canal, and because the grasslands close to the canal will also be used by animals from farther afield coming to the canal to drink. In these areas a decline in *Hyparrhenia* may be predicted, together with an increase in *Sporobolus pyramidalis*, the spiny herb *Hygrophila auriculata*, annual grasses and bare ground. Small unpalatable shrubs such as *Solanum* spp. and *Calotropis procera* can also be expected to increase, and, with the decline in competition from grasses and a reduction in the frequency of burning (due to lack of fuel and patchiness), trees and shrubs, particularly *Acacia* spp., may increase.

Woodlands

The future pattern of woodlands after the opening of the canal is very much more difficult to predict, because it is not easy to detect clear trends from past records. The earlier maps are not generally accurate enough to allow a good assessment to be made of change. There has undoubtedly been a decline in woody vegetation along the eastern side of the river-flooded grasslands. Here both *Acacia seyal* and *Balanites* can be expected to spread in response to a decline in the frequency and duration of flooding. The accounts of the JIT (1954) suggest that there have been abrupt changes in the past, perhaps due to the lack of regeneration beneath stands of *Acacia seyal*. They suggested that the regeneration of this species only occurred when fires declined in frequency as a result of a dry year leading to lack of grass fuel. Regeneration and spread of woodland may therefore be independent of any changes produced by the canal, although it is certain that some new areas will become available for colonisation, and that pressure on existing woodlands for fuel and timber will remain high.

In summary, past fluctuations in the flood regime have so influenced the patterns of vegetation that it is possible to predict the future, as influenced by the Canal, with more confidence than would have been the case in an area where vegetation change had not been documented. Isolated years of exceptionally high or low flood will undoubtedly have local effects, particularly on the species composition of the grassland, but it will be the long-term mean flood levels, controlled both by the Canal and the natural discharges of the river, that determine the major vegetation boundaries.

Other hydrological effects of the Canal

We have so far been concerned with effects likely to reduce areas of natural vegetation that are of direct value to the inhabitants of the Jonglei area. In the next chapter we shall attempt to evaluate these reductions in terms of economic resources: livestock, fisheries and wildlife. Before doing so we must turn briefly to the more positive and possibly advantageous results of a canal that will on the average carry additional water past the *Sudd* and outside the natural channels of the river, resulting in a gain of 4–6 km^3 below.

The canal as a possible flood relief measure

In previous chapters we have referred to the damage caused by periodic, substantial rises in White Nile discharges, in this century notably the peak flow of 1917 and the high flows that began in 1961, reached a maximum in 1964, and have continued at high levels since. Damage was caused in the mid-1960s in almost all parts of the Jonglei area, but more particularly west of the Bahr el Jebel in the vicinity of Adok, on the Zeraf Island, and on areas immediately east of it. Huge numbers of livestock were lost from exposure and lack of access to grazing; flooded cultivatable land and village sites had to be abandoned, and in many cases people were forced to turn to fishing and hippo hunting as the sole source of livelihood. Others, by then virtually cattleless, were forced to move eastwards and resettle in areas away from the floods, with a consequent over-concentration on remaining resources. We have also related what happened farther south among Dinka peoples in Kongor and Bor Districts.

Claims for the canal as a fortuitous flood relief measure in these circumstances have often been made, but not substantiated in anything approaching quantitative terms, a curious omission in weighing in the balance future local benefits and disadvantages. The issue may, perhaps, be put into perspective by attempting to calculate the effect of the canal diversion on the natural peak flow of 1964. The highest 10-day mean flow measured at Mongalla, which occurred in October 1964, was estimated at 251 million m^3 per day and the monthly mean flow as 237 million m^3 per day. A canal diversion flow of 20 million m^3 per day is about 8% of these river flows.

The hydrological model, described in Chapter 5 and applied in Chapter 16 to estimate reductions in areas of flooding under varying flows when the canal is in operation, can provide an approximate estimate of the effect of the canal on the extent of flooding under the very high discharges recorded in 1964 by comparing areas inundated in that year with and without the canal. According to the model the estimate of the maximum area of flooding in 1964 was 39,500 km^2. With a canal diversion of 20 million m^3 per day, this would have been reduced to 33,500 km^2, a decrease of 15%. If the canal diversion was increased to 25 million m^3 per day, the flooded area would be reduced to 32,100 km^2, a decrease of about 19%. Though this may not be as great as the unsubstantiated claims made about the local benefits of the canal in the form of flood protection, it is substantial, and would likely be sufficient to prevent the inundation of some important areas of land, particularly cultivation areas and higher, better-drained land to which the people and their herds normally retreat at the height of the rains. This would be a considerable benefit, though it might be that the reduction in heavily flooded areas would vary from place to place. All factors need to be considered, including possible backwater effects up the Bahr el Jebel, Bahr el Zeraf and the Sobat resulting from concurrent high discharges from these rivers as well as the canal. These might diminish the benefits in reduced flooding in just those areas that suffered most in 1964, notably the Zeraf Island. These are priority matters for serious investigation.

Effects of peak discharges downstream

Finally, there is the question of high discharges below Malakal from Melut northwards, where the river has a limited hydraulic capacity. The effects of the Jonglei canal would be to increase the mean monthly flow in October, usually the peak month, to something approaching the recorded maximum of 128 million m^3 per day. It appears that studies have been suggested to determine the method and costs of increasing the capacity of the natural channel in this reach to carry the increased flow. This would, of course, be an even more crucial requirement if all the other plans for water conservation – Jonglei Phase II, and Machar and Bahr el Ghazal canals (see Chapter 20) – were ever put into effect. The discharge at Melut could then reach 175 million m^3 per day. A word of caution is here necessary; in no circumstances should the *Sudd* ever be used as a depository for excess water in order to avoid flooding downstream, a notion that was current at one stage of planning for a diversion scheme in the early years (see Appendix 4). This would cause the reflooding of parts of the Jonglei area reclaimed and resettled as the result of the drainage effect of the canal and could be very damaging to local interests.

17

Effects on livestock, fish and wildlife

In Chapter 16 we have estimated the reductions in areas of seasonally river-flooded grasslands, or *toic*, that can be attributed to the operation of the canal, rather than to periodic fluctuations in discharge originating in the East African catchment (see also Chapter 4). These riverain pastures are, as we have also seen, essential sources of livestock fodder in the dry season, and investigation has revealed that only meagre and certainly not matching substitutes can be found in the rainfed grasslands at that time of year. We must now consider how much these reductions will affect the existing pasture component of the local economy as well as how far they will curtail opportunities for expansion of the livestock industry.

Effects on livestock

The previous chapter dealt with the likely changes in the water regime of the Jonglei area which will take place if and when the canal starts to operate, and also examined the likely changes in the vegetation. This chapter turns to the ways in which the canal will affect the livestock of the area. First, there will be a reduction in the area of dry season grazing in the *toic*. The extent of this will depend upon the future discharge levels of the river, which in our present state of knowledge cannot be predicted, although it is likely that what has happened before will happen again. The extent of this reduction under two extreme regimes of river flow is discussed below, as is its likely effect upon livestock numbers. A second extremely important effect of the canal, its interference with seasonal migrations of the people and their cattle, is described in Chapter 18. Other possible effects, including those on animal disease, are discussed here.

Dry season grazing areas

It has been shown in Chapter 12 that in most years, and in most parts of the area, the dry season is the most stressful time of the year for cattle. Mortality is highest, milk yields fall, and conceptions are relatively low during the dry season. While it is true that the height of the wet season may also be a difficult time for cattle, especially in years of high local rainfall or high river flood, the latter at least will be mitigated by the operation of the canal. It is to the dry season pastures that we must look for the most critical of the local effects of its operation.

In Chapter 12 it has been shown that the most satisfactory pastures at the height of the dry season are those to be found in the *toic* – the seasonally river-flooded grasslands occupied by *Oryza* and *Echinochloa* grasslands. In Chapter 16 it has been shown that the model developed by Sutcliffe & Parks (1982) can not only demonstrate a clear relationship between river discharges at Mongalla and the extent of flooding, but can also predict areas of flooding with some degree of accuracy, starting from inflow figures at Bor. Table 17.1 shows the predicted changes in river-flooded grassland (*toic*) and permanent swamp areas under two regimes of river flow – continuing high levels, and a return to the low levels that pertained before 1961. The Table shows that the extent of the reductions depends on the discharge volume of the river. If it remains high, then the area of river-flooded grasslands will decline by between 17 and 21%; if it returns to pre-1960 levels, then the same vegetation type will decline in area by 26 to 31%. It must be remembered that some swamp will change into river-flooded grassland so that there will be gains as well as losses; the figures given take account of this.

The vegetation zones will take some time to adjust to changed flooding regimes – although, as has been shown in Chapter 16, quite large changes can occur within a year or two. One general point must, however, be made. If the river remains at its present high level, then the change resulting from the opening of the canal will be a rapid one (unless an operational regime allowing for a gradual opening over a period of years is adopted – and this is an unrealistic expectation). However, if the river falls to its pre-1961 levels, it will do so slowly over a period of years during which the people will have time to adjust their migrations to the changing vegetation patterns.

Taking the figures for change in floodplain area, and using the results of the aerial surveys, it is possible to calculate densities of cattle in the floodplain under the present regime of river flow, and under the two possible future regimes, with the canal in operation, mentioned above and in Chapter 16. The results of these calculations are shown in Table 17.2. It is clear that there are enormous variations in livestock density within the area.

Table 17.1. *Estimated effects of the Jonglei Canal on areas of flooding under three possible natural discharge regimes*

	River returns to pre-1961 discharges (1905–60 data)			River remains at high discharges (1961–80 data)			River returns to mean discharges (1905–80 data)		
	Swamp	*Toic*	Total	Swamp	*Toic*	Total	Swamp	*Toic*	Total
No Canal (Area, km²)	6,700	6,200	12,900	17,900	11,000	28,900	9,500	7,400	16,900
Canal (20 million m³/day) (Area, km²)	3,500	4,600	8,100	14,000	9,100	23,200	6,200	5,800	12,000
Decrease	47%	25%	37%	21%	17%	20%	35%	22%	29%
Canal (25 million m³/day) (Area, km²)	2,900	4,300	7,200	13,100	8,700	21,800	5,500	5,400	10,900
Decrease	57%	30%	44%	26%	20%	24%	42%	27%	35%

Source: Sutcliffe & Parks (1982)

The extent to which the present areas of river-flooded grassland could carry extra livestock is not easy to determine. The assessment of the carrying capacity of seasonal tropical grassland is always difficult. As we have seen in Chapter 7, the standing crop of grass varies enormously through the year, as does its nutrient content. This annual variation is overlain by irregular fluctuations from year to year caused by differing annual rainfall totals, and in the distribution of rain within the wet season. In the case of river-flooded grasslands, there is the additional complication of the independent fluctuations of river discharge caused by rainfall in the East African catchment far from the Jonglei area.

In the extreme south of the area, in Bor District, there is at present river levels essentially no *toic* grazing in the sense of river-flooded *Oryza–Echinochloa* grasslands on the east bank of the Nile. The swamp has a sharp boundary. Here a fall in river levels will probably allow some of the present swamp to become river-flooded grassland, increasing the area available to cattle in the dry season.

Cattle densities in this area are, however, already extremely high, and the consequent small reduction in livestock density will probably not make a great deal of difference. The area will continue to be overcrowded by the end of the dry season.

In Kongor District shortage of *toic* land for dry season grazing is probably a limiting factor in many years. At least south of Jonglei village, the swamp is delimited once again by a relatively abrupt line over which the river now spills to produce extensive *toic* grazings in years of high flood. A reduction in river levels would mean that this flooding would no longer take place and that cattle would have to look for dry season grazing in the swamps themselves. However, early records suggest (see Chapter 8) that this area was mostly free of papyrus before the floods of the 1960s and may well have provided

Table 17.2. *Estimated numbers of cattle per square kilometre of toic grazing*

	Estimated 1952	Present day (1980)	Canal with high discharges	Canal with pre-1961 discharges
Bor	209	456	728	523
Kongor	66	84	91	205
Ayod	24	31	33	54
Fangak	24	30	25	65
West of Bahr el Jebel	42	31	34	83

extensive dry season grazings, which it could do again given a fall in river level. The calculations (Table 17.2) suggest a large increase in dry season livestock densities and it is likely that the area will be able to support rather fewer cattle than at present.

Northwards from Jonglei village, the edge of the swamps becomes much less well defined, and the width of the *Oryza-Echinochloa toic* grazings becomes much greater. Here it seems unlikely that the dry season pastures are fully utilised at present, so that a reduction of say, 20%, would probably not be significant. Here the future behaviour of the Nile becomes much more important than in the south; the maps (Chapter 16) show that there is likely to be a very much larger decrease from the present areas of *toic* if the river falls back to its pre-1960 level than if it remains high. Nevertheless, by comparison with the areas farther south, even the highest predicted cattle densities look quite low, and there are unlikely to be problems here. There could even be the possibility of a modest increase.

To sum up, the changes in the area of river-flooded grassland and hence dry season grazing, although large, will vary considerably over the area and may not be as drastic as has sometimes been feared, particularly if the river stays at its present high level. People may have to move farther, and they may have to move to different areas (which could result in conflict) but they are unlikely to run out of dry season grazing completely. It is noteworthy that problems have arisen in the past at times of high flood, not at times when the river has been low (see Chapter 16).

Alternative dry-season grazing resources
The use of rain-flooded grasses
It has been demonstrated that the operation of the canal will lead to an important reduction in the area of critical dry season pastures. This will affect not only those who live close to the canal, but also those who move into the area from some distance away to use the river-flooded pastures at this time; this includes those living east of the Bahr el Jebel, those living west of it, and those who occupy the Zeraf Island. There is therefore a need to look for possible ways of alleviating pressures on the reduced area of riverain pastures at this time.

Most of the grasslands of the Jonglei area are rain-flooded (intermediate *sensu* JIT, 1954) but, as explained in earlier chapters, they have only limited seasonal use at present. They can provide good-quality pasture early in the rains, but later become long, coarse and unpalatable (see Figure 12.5). A second opportunity for grazing occurs at the beginning of the dry season, when burning can stimulate a flush of new growth if there is sufficient residual soil moisture. At both times, drinking water can be a problem; early in the wet season rain can stimulate grass growth before sufficient has fallen to fill surface pools, and at the beginning of the rains such pools evaporate quickly during intermittent dry periods.

Increasing the exploitation of the *Hyparrhenia* grasslands is therefore mainly a matter of providing adequate water supplies. It must, however, be stressed that there is no realistic possibility of using these grasslands year-long; in spite of occasional favourable reports, aerial surveys by both the JIT (1954) and during the Range Ecology Surveys showed that regrowth over most of the plains at the height of the dry season is at best sparse and often quite absent.

At the beginning of the wet season, the provision of water could either be from wells, or from pools provided with improved catchments so that they fill more quickly (see Chapter 19). At the beginning of the dry season, extra water could be supplied either from wells, or from deepened natural pools, or from wholly artificial pools. It is unlikely that shallow wells could reasonably supply the needs of large numbers of cattle, so improved *hafirs* are probably the most practical method.

The Eastern Plain is crossed by a large number of very shallow water-courses, obvious from the air by virtue of their different vegetation, but very much less obvious on the ground. It has been suggested (RES, 1983) that low bunds could be constructed across these to impound water, with the hole from which the dam material was excavated acting as a *hafir*. We have shown that such vegetation changes are rapid, and are often towards species of higher nutritive value which, because of the higher moisture content of the soil behind the bund, are likely to be of particular value in the dry season. The construction of such bunds, particularly on the eastern side of the canal, was one of the major recommendations of the Range Ecology Survey (RES, 1983). The FAO/UNDP project at Kongor (see Chapter 19) was just beginning work on experimental water capture systems of this kind when hostilities brought the project to an effective end in mid-1984. It is to be hoped that this will be an early feature of any new development programme when peace returns.

Little improvement in grass growth was seen in the brief period during which the Kongor experiment was monitored, but the *hafir* provided a valuable focus for more prolonged use of the *Hyparrhenia* grasslands. One season of trial was insufficient to produce any change in grassland composition but small artificial '*toics*' might eventually be produced, in the same way as on the eastern side of the canal. The value of such *hafirs* would be enhanced if burning of the surrounding grasslands was better controlled. At present there is a conflict between the need to preserve some areas of grassland unburnt to provide thatch, and to burn early enough for there to be sufficient soil moisture to support regrowth. There is a need for long-term experiments to determine the effects of frequent early burning on the productivity and species composition of *Hyparrhenia* grassland. A further possibility is to encourage heavy grazing of the *Hyparrhenia* at the end of the dry season, breaking down the tall stems and encouraging regrowth from the base without burning. Such a method of grazing management is successful in *Hyparrhenia* pastures in Venezuela (Lock, personal observation), although the form of *Hyparrhenia rufa* cultivated there is different from that in the Eastern Plain of the Jonglei area.

An increase in the time spent in the *Hyparrhenia* grasslands early in the wet season is more problematical, even if water supplies could be improved. Sustained grazing at this time of the year might prolong the season of use, but the soil is very susceptible to damage by trampling and quickly becomes flooded and muddy, and unsuitable for grazing livestock. It is also likely that prolonged grazing at this time would lead to undesirable changes in grassland composition. Experiments are needed.

Fodder preservation – hay-making or silage
The super-abundance of fodder in both the rain- and river-flooded grasslands at the end of the wet season makes forage harvesting and storage a tempting prospect. It does not, however, stand up to close examination. As Figures 7.2, 7.3, 7.5, 7.7, 7.8 and 7.9 show, the nutritive value of most of the local grasslands falls drastically towards the end of the wet season, and, by the time they are sufficiently accessible to be harvested on anything but a very small scale, their nutritive value is extremely limited. The harvesting and gathering-in of hay would be impossibly labour intensive, or would need special machinery, and termite-proof storage would be essential. It can probably be dismissed as a realistic possibility.

Silage production is more promising, and has been successfully attempted on a small scale at the Mafao farm near Juba. Even so, it may well be that the labour needed would be better expended on the production of food crops. There is a host of practical problems to be overcome, including the distance to sites where suitable species for silage production flourish, and the need for efficient and pest-free storage. An alternative is to make silage *in situ* in the grasslands most suitable for its production and to take the cattle to it, but this would tend to defeat the object of the exercise by taking stock away from the villages where their milk is needed. It could also raise all sorts of problems of preservation, controlled use and management, and the implications of ownership and tenure in a society in which grasslands are held as common property. Experiments would, however, not be out of place, using both wild grasses and also cultivated species specially grown for the purpose.

The growing of trees, legumes and other plants as fodder crops has also been recommended, and may be much needed if, as is to be expected, the population of smaller livestock, particularly goats, increases to match the expected growth of human populations near the canal.

Other effects on animal numbers

There are other ways in which the canal will affect the numbers of livestock. However, the magnitude of the effects, and the precise causes, are difficult to define and quantify because so many factors are involved. First, the canal may (or certainly should) improve access to the area by veterinary and other advisory staff. Vaccination should become more widespread and effective, and there is a very real possibility of eliminating rinderpest, the most important disease. This would remove a major constraint on cattle numbers, and also make conflicts less likely, as as much inter-group fighting has in the past occurred because of fear of disease rather than the need to defend grazing rights. Secondly, by improving communications both by water and by land, the canal should make trade easier and increase the possibilities of exporting cattle from the Jonglei area both to the north of the country and to Equatoria. This could be seen as a factor likely to reduce cattle numbers, but with good management it should lead, rather, to an increased rate of turnover without an overall increase in numbers (see Appendix 8 for further discussion).

Thirdly, putative changes in the distribution of settlements, with increasing numbers of people staying close to the permanent water provided by the canal throughout the year, may lead to a change in the proportion of species. Although there will be water available throughout the year on both banks of the canal, settlement will be much easier on the unflooded west bank. Here, however, there may not be sufficient grazing to support cattle in the dry months of the year (see Chapter 12), but browsing goats will be able to survive. There may thus be a change in the balance of species of livestock maintained, with a tendency for goats to increase at the expense of sheep and cattle.

So far this section has examined only the effects on numbers of livestock as they stood in 1983, and it has been suggested that the numbers that can be maintained are likely to remain more or less the same. It remains to examine the effects on the potential for increase in livestock numbers, and for the development of the livestock industry. The important regulating factor to animal numbers is the area of river-flooded grasslands. These grasslands are entirely dependent for their survival on the annual fluctuations in river discharge. We shall stress later (see Chapter 20) the importance of investigating the feasibility of seasonally regulating discharges between canal and natural channels in order to modify the reduction in dry season grazing that the operation of the canal will otherwise cause.

Effects on fisheries
Introduction

The *Sudd* is divisible into a number of broad ecological zones extending from the main river through permanent swamps to seasonally river-flooded grasslands. The distribution of open water and plant communities within them results, to a large extent, from an equilibrium between inputs of river water and dry and wet season conditions. As a consequence of increased Nile discharges since 1961, a shift in this equilibrium distribution has induced major movements of the ecological zones over a comparatively short period of time. The important effect so far as fish and fisheries are concerned has been an overall enlargement of the permanent floodplain. The papyrus belt with interconnecting channels and chain-lakes has broadened in the south and *Typha* swamp with diffuse channel networks and extensive shallows has shown massive expansion in the centre and north. Future shifts can be expected in response to reduced river flows into the *Sudd* as a result of an operational canal, diverting 20–25 million cubic metres each day, either acting alone or in combination with the effects of further natural climatic changes in the catchment of the upper Nile. Accordingly, recent investigations have attempted to discover how fishes are partitioned between habitats in the ecological zones and the extent of movements between them (Chapter 14; SES, 1983; Hickley & Bailey, 1986, 1987a). From this work and the discussion in the preceding chapter on possible changes in areas of flooding and vegetation types under different hydrological regimes, the implications for fish resources and fisheries activities can be extrapolated.

Rivers

The Bahr el Jebel and Bahr el Zeraf will persist as they have throughout this century. In the north where the river leaves the *Sudd*, return water from the canal combined with the River Sobat will increase perennial downstream flow and, when the Sobat floods, produce some back-up of water in the White Nile towards the Zeraf mouth. Both events should ensure the existence of *khors* in this area and may expand them. With discharges from the Great Lakes maintained at present levels the braided nature of channels in the south between Mongalla and Shambe will continue, but depths will be reduced and distant conduits in the *Typha* swamp will disappear.

Thus, avenues for longitudinal fish movements will remain but, in the south and centre of the *Sudd*, barge access to fish collecting centres, currently possible for example along the Atem and Awai rivers, will be precluded by shallowness. Processed fish will have to be canoed out to the main channel or small, shallow-draught vessels powered by outboard engines will be required by fishery developers. This situation will be further exacerbated by low river discharges. The southern Zeraf system and proximal channels in the *Typha* will be reduced to seasonal routes for overspill on to a readjusted floodplain and resort to overland outlets for fish-crops will be necessary.

When the canal is open, most commercial river boats will bypass the swamp. Reduced river traffic could lead to regular *sudd* blocks although it must be noted that, historically, blockages have generally been associated with high rather than low river levels. It is quite probable, however, that the use of fish barges on the main navigable channel will become doubly important to *Sudd* fishermen in the future.

Lakes

The established fisheries of some large lakes are said to have been adversely affected by increased water depth, but, overall, the flooding of the 1960s has multiplied the number of perennial lakes in the system and, thereby, the fishing potential. The open waters, variable substrates, submerged water plants and hyacinth fringes of lakes, offer a variety of niches for fish life, which is more abundant and more easily netted than in running waters. The popular species of the fishery, *Distichodus, Citharinus, Heterotis, Lates*, the tilapias and some larger catfishes and mormyrids, thrive well in lacustrine conditions. During the high-water season many of them find suitable spawning and rearing grounds in these shallow waters of the permanent floodplain.

A moderate change in the hydrological regime will bring shallower lakes liable to encroachment by surface vegetation, and an increased incidence of side-channel blocks by water hyacinth bringing problems of lake access for fishermen. A severe decrease in the discharge into the *Sudd* will bring about the total disappearance of many lakes in the papyrus zone and reduce others to the status of seasonal lagoons, with a serious loss of year-round fish and fishing potential. Any of the predicted changes in the hydrological regime will result in a shrinkage of lakes in *Typha* on Zeraf island and to the east of the Bahr el Zeraf and many will be lost as perennial bodies of water.

Shaded swamp

Two clear predictions to emerge in Chapter 16 were that any reduction of the input into the *Sudd* will affect the permanent system to a greater extent than the seasonal floodplain, and that losses of swampland will be proportionally larger in the north and centre than in the south. *Typha* swamp is the candidate for greatest contraction.

The extent to which *Sudd* fishes inhabit lakes and the channels connecting them deep in the *Typha* swamp, has still to be determined. It is clear that some fishes can penetrate shaded swamp vegetation and that some of these, for example *Heterotis* and *Gymnarchus*, make use of more open areas within it for spawning. Moreover, there are established fishing camps on swamp channels close to lakes east of the main river and the Bahr el Zeraf, and on Zeraf island. Thus, as the swamp contracts and reverts to grasslands, some fishing grounds will be lost and others will become seasonal in nature. In this context the Nuer on Zeraf island appear to present a special case. The fondest hope of many of them, who were forced into subsistence fishing by the floods, is to return to pastoralism. Yet an inadequate drawdown of the mean water level of the *Sudd* could leave them without the resources for survival – neither sufficient fishing grounds nor adequate grazing for cattle. Their best interests would be served by a return to the pre-1961 discharge levels of the Nile coupled with an operational canal.

Seasonally river-flooded grasslands

Under any of the predicted hydrological regimes annual overspill through a reduced permanent wetland onto a seasonal floodplain will continue. Excluding small fishes, *Clarias* species are currently most dependent upon the seasonal habitat and, with varying quantities of *Protopterus*, *Polypterus*, *Heterotis* and *Gymnarchus*, many continue to form the bulk of spear-fishing catches by men and boys in the cattle camps. However, the seasonal cycle in some other floodplains involves the migration of many species other than those which are 'swamp-adapted' (Chapter 14). On them, the total weight of live fish increases to a maximum during the flood and declines with the ebb through back-migration and mortalities. Welcomme (1979) explains how, in these situations, the timing, duration and amplitude of flooding may in large measure determine the recruitment, growth and survival of fish stocks. Accordingly, natural variations in the flood cycle or man-induced disruptions resulting from flood-control dams and impoundments, can have a profound impact on the fishery. In the *Sudd* system at present, the effect of year-to-year variations in the flood may be less marked than elsewhere, because the large, permanent floodplain takes precedence over the seasonal floodplain in their relative importance to the well-being of most fish stocks. It is possible that, in the future, a marked contraction of shade swamp vegetation would reduce its barrier effect to lateral migration by other *Sudd* fishes. This, in turn, would open the seasonal habitat to them for breeding and nursery purposes but at the same time, expose them to the consequences of annual variations in the flood cycle.

The canal

The canal itself will provide fish habitats and fishing opportunities of two sorts. As a 360 km long river channel, with some backwaters in shipping berths and passing bays, it will become self-stocked with species tolerant of flowing water, for example *Alestes, Hydrocynus, Labeo, Distichodus, Bagrus, Clarotes, Synodontis, Lates* and *Oreochromis*, with *Micralestes* and *Chelaethiops* amongst the smaller fishes. Both Euroconsult (1978) and the Swamp Ecology Survey (1983) reports have recommended that subsistence fishing by methods commensurate with bank protection be allowed. Drift-nets, cast-nets and lines would seem to be appropriate from the methods currently used in the *Sudd* and there may also be opportunities for trapping in outlets from the canal.

The second type of aquatic development relates to the barrier effect of the eastern canal embankment to the drainage of some *khors*, and to the seasonal north-westerly movement of surface water, referred to as creeping flow. It is recognised that inhabited areas and sorghum fields on the east bank liable to inundation will require protection from flooding. Elsewhere, however, water may be allowed to accumulate safely, which could lead to the establishment of extensive patches of 'wet' grassland. These will attract catfishes and already in 1982, along part of the excavated northern section of the canal, useful crops of *Clarias* were caught by canal-side opportunists.

A further possible effect of the canal could be to stimulate aquaculture in the region. This view has been advocated by some authorities, based on the assumption that water for the purpose could be drawn from the canal. However, despite widespread and long-held interests in fish culture throughout the continent, except in some research institutions and heavily capitalised ventures, its development in Africa remains in its infancy. Accordingly, and especially in view of the potential for capture-fisheries in the region, it would seem sensible to restrict such considerations to the possibilities of fish-ranching, by seeding natural accumulations of water along the east bank with tilapias. Added to the self-stocked clariids, they would enhance the opportunities for subsistence fishing.

Conclusions

The currently utilised and potential resources for subsistence and commercial fishing centre on species which, with few exceptions, can find suitable feeding, spawning and rearing grounds in the channels, lakes and vegetation fringes of the permanent system. Of the principal fishery-exploited species few appear to make use of swamp vegetation and seasonal floodplain habitats for breeding. Whilst some shrinkage and readjustment to the areas of swamp and river-flooded grassland will accompany an operational canal, overall this should have a minimal impact on the major fishery resources. If the incidence of river *sudd*-blocks and channel interference by hyacinth increase, however, access for barges to fish collecting centres and for canoes to some fishing grounds, may become impeded. On the other hand swamp contraction may make land access easier in places. There will remain more than adequate resources to meet the demands of seasonal opportunistic fishing in floodplain pools and swamp margins.

Clearly this healthy situation will change if the effect of the canal is coupled with a fall in river discharge towards or beyond the levels of pre 1961. In these circumstances the permanent and seasonal floodplains would be drastically reduced, lakes will be lost and others reduced to a seasonal status. Despite the possibilities of a wider access to the seasonal floodplain by *Sudd* fishes, fishery resources of all types will decline and with them the possibilities for expansion. Welcomme (1979) has drawn attention to the detrimental effects on fish and fisheries in the Chari, Niger and Senegal rivers resulting from flood failures provoked by prolonged periods of drought. In view of the unpredictable nature of climatic changes over the catchment of the upper White Nile and their effects on discharge levels, there is a case for regarding the present constraints to the growth of the *Sudd* fishery (Chapter 14) as helpful in the longer term, however irritating to fishery developers. By preventing an over-commitment to fishing by a growing number of people, the risks of serious socio-economic conflict in the event of diminished resources are reduced.

Finally, the canal itself may directly or indirectly add to fishery resources in the region, and by improving communications lead to a more equitable and wider distribution of products from the *Sudd* fishery.

Effects on wildlife

Introduction

The advent of the canal could affect wildlife in three ways. First, as we have seen when discussing the possible effects on livestock numbers, there will be changes in the habitat; some of the habitat types in the area will be reduced, others increased. Species preferring the habitats such as swamp, which are likely to be reduced, may suffer. Conversely, those preferring woodlands or grasslands could increase. Secondly, several of the species in the area, particularly the tiang and the reedbuck, perform regular and extensive seasonal migrations to which the canal will form a barrier. Some will swim, others may try to do so and drown. Others, unable or unwilling to swim, will become divided into two separate populations on the two sides of the canal. Thirdly, the canal and its associated road will improve communications. Large numbers of vehicles will use the road, and it is safe to assume that many of these will carry guns. Animals coming to drink at the canal, or attempting to cross it, will be particularly vulnerable to this kind of assault as well as to more traditional hunting.

The effects of habitat change

It has been shown earlier in this chapter that the operation of the canal will lead to reductions in the area of permanent swamp and of river-flooded grassland. The extent of these reductions depends, as described above, upon the discharge volume of the river. If it remains at its present high level, then the areas of permanent swamp and river-flooded grassland will be reduced by very much less than if it falls back to its pre-1960 level (see Table 17.1).

The permanent swamp and river-flooded grassland habitats cannot, of course, decrease without other habitats increasing. However, the woodlands and rain-flooded grasslands which will replace them are habitats which already cover large parts of Africa; swamp and seasonal floodplain, on the other hand, are already scarce habitats, and the loss of, say, 5% of the swamp area of Africa would be a more serious matter. There must also be concern lest the expansion of woodland which is likely to take place in response to lowered flood levels may be checked by the increasing demands for wood from an expanding population. However, it remains true that there may be an opportunity for both grassland and woodland animals and birds to expand their range and, perhaps, increase in numbers. Nevertheless, for the reasons given above, the habitat losses in the swamps and river-flooded grasslands will be more significant than the gains in other habitats.

The shoebill storks (about 5,500 at present) that live in the swamps of the Jonglei area (RES, 1983) represent a substantial part of the world population of that species. While this same population might survive at a higher density in a reduced area of swamp, this hope is unlikely to be fulfilled. Pairs of shoebills are very evenly distributed through the available habitat, and a loss of habitat is more likely to lead to the death or emigration of some individuals rather than to an increase in density. Many other water birds may suffer similarly, although by far the greatest concentrations occur at the swamp margins, following the fish that move up as the river flood rises, or exploiting those stranded in pools as it retreats. Whatever river level may be arrived at when the canal operates, there will always be a swamp edge, and this edge will diminish less than will the total area. What is most important to many bird species is the maintenance of the regular rise and fall of the river and the consequent seasonal flooding.

The large mammal species fall into four groups. Some, such as the bushbuck, grey duiker, Lelwel hartebeest and oribi, are animals of the woodlands and the rainfed grasslands, and will probably be unaffected by the habitat changes. A second group, including Mongalla gazelle, giraffe, roan antelope, white-eared kob and zebra, also occur mainly in the rain-flooded grasslands and the woodlands, and although they make some seasonal use of the floodplain, it does not seem that this is essential for them. This group will likewise be relatively little affected by habitat change, although, as described later, the canal may restrict their seasonal movements. A third group, including the elephant, reedbuck, tiang and waterbuck, make extensive use of the *toic* and swamp for at least part of the year. These species could suffer severely from habitat loss but, as described below, interference with their migration is likely to be more important. In some cases, this decline will result from habitat loss alone, but in other cases (in particular the elephant) the swamps act as a dry season refuge from hunters; less swamp will mean more hunting. A fourth group, including the hippopotamus, buffalo, Nile lechwe and sitatunga, live in the swamps or *toic* throughout the year. Any reduction in the area of these vegetation types may well lead to a decline in their associated animal species.

A final effect of the habitat changes induced by the operation of the canal will be, as we have seen, to increase the dry season density of cattle in the *toic*. At present, cattle and wildlife co-exist in the dry season, but with reduced grazing areas, co-existence may change to competition.

Interruption of seasonal migration

As has been described in Chapter 15, a number of species undertake extensive migrations which bring them into the Jonglei area for a part of each year. Others undertake more local migrations. The canal and its embankment will present a major obstacle to these migrations. First they must climb the embankment, then descend to the water, swim across, climb out on the other side, and finally scale and descend a second embankment. Not only will this process require a great deal of effort, but it will also make animals very vulnerable to hunting. The likely effects of the canal on the populations of some of the more important animal species are discussed below.

Elephant migrate eastwards from the swamps at the beginning of the rains. During the wet season they remain in large groups in the swamps. Elephant are well able to swim and will probably try to cross the canal, although they will be very vulnerable when doing so. They could probably survive through-out the year to the west of the canal, but not to the east, where dependence on the canal for drinking water would also render them very vulnerable to hunting.

Giraffe appear to make rather irregular movements during the year, and some undoubtedly cross the line of the canal, but cannot swim, so that two populations will result. This will probably not affect the species unduly. However, giraffe prefer to drink if water is available, and may suffer for lack of it; many died on the Eastern Plains during the dry season of 1980.

Reedbuck are the second most numerous of wild herbivores in the Jonglei area. They spend the wet season in the *Hyparrhenia* grasslands of the Eastern Plain, and the dry season in the river-flooded grasslands, scattered in small groups. They almost certainly need these grasslands to survive the dry season, but are unlikely to be able to swim the canal in any numbers, and their numbers will suffer as a result. The species is particularly vulnerable to hunting, and if they are unable to migrate they will probably become more so.

Mongalla gazelle are migratory, some reaching the *toic*, although most remain in the *Hyparrhenia* grassland throughout the dry season grazing the regrowth, without drinking. They are most unlikely to be able to swim the canal with any regularity, so that two populations may develop, a small one to the west of the canal, and a much larger one to the east. Those to the west of the canal will survive the dry season on the *Hyparrhenia* grasslands which will develop there (see Chapter 16).

Other species which may well be split into two separate populations by the canal are the roan antelope, the zebra, and the ostrich. None of these is likely to swim the canal. As in the other species, those to the west of the canal will also depend upon the river for dry season water; those to the east will have to drink from the canal.

The tiang is the most numerous large mammal species in the Jonglei area, and also the most conspicuous migrant. While the presence of a resident population of some 35,000 animals throughout the year, mainly in the northern wooded parts, shows that the area is not intrinsically unsatisfactory as a year-round habitat, it is most unlikely that the main population of 350,000 animals could survive without migrating. If the species is to continue to maintain high numbers it must almost certainly go on migrating. In trying to do this, the tiang will have to swim the canal. Some probably have experience of swimming smaller watercourses and, as a rule, large-herd forming antelopes swim more commonly than solitary species. The westward crossing of the canal will be made in November–January, when most adult female tiang are nearing the end of their pregnancy. They will be handicapped by this but most will probably survive the stress involved and make the crossing successfully. The return journey in late April or early May will be made when the calves are between one and three months old. Heavy calf mortality is likely at this time. Additionally, the large herds will be exceptionally vulnerable to hunting at both crossing times.

The effects of improved communications

The road along the canal embankment will provide a commanding view over the flat country of the Jonglei area. To some this will provide an outstanding opportunity to view the wildlife of the area, but others will see it as an excellent viewing platform from which to shoot animals. It is unfortunately true that most vehicles travelling in the area carry firearms, for one reason or another, and the lure of free meat leads to their use when opportunity offers. Hunting by traditional means will also be comparatively easy close to the canal because the animals will be concentrated, and will be hampered in their movements by the steep embankments.

These opportunities will be heightened because animals will wish to drink from the canal and, if they are migrating, they will wait close to the bank for an opportunity to cross. The problems for animals of drinking from and crossing the canal will be heightened if human settlement along the banks becomes dense. It must be said, however, that tiang at least would drink from a large pool beside the Range Ecology base camp, under cover of darkness, so that the proximity of people *per se* does not appear to alarm them.

Conclusions

The most significant effect of the canal will be as a barrier to animal migration, and the animal most affected will be the tiang. While there is no reason why this species (or, indeed, any in the area) should not survive the changes brought about by the operation of the canal and by its physical presence, populations will almost certainly be reduced.

The effects resulting from changes in the extent of key habitats (permanent swamp and *toic*) will depend very much on the future behaviour of the river. It must be said, however, that river levels, and therefore swamp and *toic* areas, have fluctuated in the past. The species of the area adapted to changing conditions then and will probably do so in the future.

Increased disruption by an enlarged human population and by a larger number of travellers and, when oil exploration resumes, by all-terrain vehicles and helicopters, will undoubtedly be a major problem. There will be a need for carefully formulated regulations, and also for extensive education in the responsible use of the vast and important resource which is the wildlife of the area.

It should be mentioned here that there have been disturbing but unconfirmed reports of the serious effects on migratory wildlife of the uncompleted canal. This, with its vertical sides and few access points, has apparently proved to be an even more formidable barrier than a fully operational water-filled canal.

18

Local effects in the canal zone

In the previous chapters we have been concerned with the hydrological alterations in the regime of the natural river channels, consequent ecological changes which will be brought about by the operation of the Jonglei Canal, and hence effects on livestock, fisheries and wildlife. These effects will be felt both east and west of the Bahr el Jebel system and to an undetermined degree downstream of it. In this chapter we are concerned mainly but not exclusively with the physical effects of the canal on the people whose permanent villages are in its vicinity; those who live in what has been called the 'Canal Zone', within approximately 30 km either side of it.

Settlement patterns and policies
Present settlement patterns
Air survey (RES, 1983) has shown that the majority of the southern Dinka villages south of Kongor are now located west of the canal alignment. North of this most Dinka settlements are east of the canal, and north again there are large numbers of Gaawar Nuer villages on both sides. Lou Nuer villages are all east of the canal, though some of those shown are not occupied by people who migrate to dry season pastures west of it, but move to the Sobat system and its tributaries.[1] (See Figure 10.3.)

Settlement patterns and the canal
As we have seen (Chapter 10) the location of permanent habitations is dependent on the availability of marginally higher, better drained land. The distribution of permanent settlements will continue to be determined by those topographical features, but to an extent these may be affected by the canal. As we shall see later, areas west of the canal will be subject to less flooding from the river, and the canal itself will act as a barrier against

creeping flow coming in from the east. More flood-free land will therefore be available on that side, and it is likely that population movement will be in that direction, perhaps on a considerable scale. As we shall see later the build-up of water against its eastern bank during the rains may result in the opposite effect. The promise of improved access to trading centres and government services along the canal will be a powerful incentive for people to set up their homesteads and cattle byres close to it. For reasons which will be clear below, while desirable for the purposes of administrative control and the provision of services, too high a level of concentration along the canal could have a whole range of consequences which need to be avoided. Apart from the question of water abstractions, and undesirable musterings of livestock at crossing points at the beginning and end of the dry season, with resultant pressures on grazing required by people living in the vicinity for their animals during the wet months of the year, there could be undue demands on land suitable for cultivation. Contrary to all previous practice, planned settlement will become a necessity, though traditional tenure and common property rights related to the control of territory, described in Chapter 10, will be paramount considerations to be taken into account.

Communications

Effects on navigation

The overriding economic advantage of the Jonglei canal will be a more direct and unimpeded navigable route between Khartoum and Juba, a reduction of about 300 km from the existing passage open to river traffic along the natural channels through the *Sudd* region. This should be a major stimulus for contact and commercial exchange between the two parts of the country.

There will be special opportunities for those who live in the vicinity of the canal itself. Berthing places are to be provided at various points along it and these will lead to the creation of new trading centres. Small ports of this kind will open up communications year round with areas previously ill-served owing to poor road communications. Corresponding growth can be expected in marketing facilities for livestock and fish for export, the supply of consumer goods and the import of grain when needed.

However, for those living away from the canal along the present navigation route – on the Zeraf Island, and west and north of the Bahr el Jebel – there will be negative effects that need to be taken into account. The vegetation of the channels flowing through the *Sudd* is either submerged, fringing, or floating. In Chapter 16 it was suggested that the altered hydrological regime in the natural channels will result in an increase in floating vegetation, notably *Vossia cuspidata*, and this could lead to more frequent blockages, a situation likely to be worsened by the inevitable reduction in river traffic which is bound to take the quicker more economic alternative route along the canal.

Regular river traffic through the area has always been a major factor in keeping the channels clear. The Zeraf river itself used to be navigable through to the main Bahr el Jebel near Shambe after the 'cuts' were made in 1910 and 1913, but there has never been a regular steamer service by that route, the river channels formerly being kept open by the fortnightly inspection movements of the Egyptian Irrigation Department craft and the occasional administrative visit by steamer. Old Fangak was the only significant port of call for these craft, the Zeraf island being served mainly by the regular steamer schedule to Wath Kec on the White Nile. The Zeraf had in fact been blocked for a number of years before 1979, but was partly reopened in 1982 during oil prospecting surveys. Consideration needs to be given to at least an essential minimum level of traffic along the White Nile and Bahr el Jebel to meet the needs of the area in trade and services, mainly medical and veterinary, and simultaneously help keep the natural channels open. In this regard it is probable that the commercial interest of oil companies may require them to dredge these channels, as they were already doing downstream of Malakal in 1982.

Given a return to discharges nearer to the pre-1961 mean, the diversion, in years of low flows, of substantial amounts of water down the canal could lead to the natural channels of the river becoming too shallow for commercial river traffic and fishery barges (see Chapter 17) thus adding to their navigational problems and again the danger of major vegetation blockages – a vicious circle. That being so, a minimum daily input into the Bahr el Jebel will be necessary, one figure suggested being 35 million m^3 per day (SES, 1983).

It is worth adding that navigational advantages between Kosti and Juba brought about by the more direct canal route will be much reduced if attention is not also given to keeping open the Juba-Bor reach, which has frequently been subject to extreme navigational difficulties in the dry season.

Effects on road transport

The construction of roads in the Jonglei area has always presented almost insuperable problems. There are formidable difficulties of terrain and soils – *gilgai* formation with cracks, hollows, small mounds or hummocks and holes (see Appendix 1) that make surfacing hard to construct or maintain during the dry months of the year. Many depressions require extensive earth embankment, easily eroded or washed away in the rains, when the surface in any case becomes sticky, rutted and usually impassable to motor traffic. For the most part roads and tracks have to be cleared of vegetation, repaired or completely remade annually after the rains cease. Road work requires either very intensive labour inputs or the use of expensive grading machinery, and there is no readily available surfacing material for improvement, though laterite has been used to good effect on the Bor–Juba stretch. The provision of a satisfactory network of roads, linking the berthing points to administrative centres and essential medical and veterinary services, is a major priority if the best use is to be made of the opportunities the canal will bring to the area.

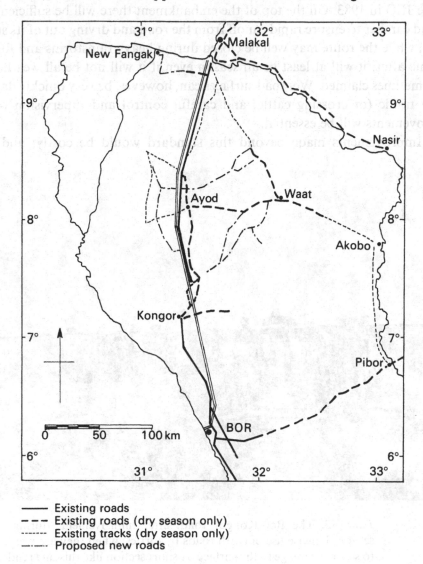

——— Existing roads
– – – Existing roads (dry season only)
------- Existing tracks (dry season only)
—·—·· Proposed new roads

Figure 18.1. Status of roads and proposed extensions 1983.

However, the canal banks will provide the base for a main north–south road route, which is at present closed annually for three or four months of the year owing to rain and wet muddy surfaces. The route to be followed is along the east bank from the Sobat mouth as far as the latitude of Kongor, where it is planned that it should cross over to join the recently improved Bor–Kongor road, for it is on that side that most people live from this point southwards (see Figure 18.1). Failure by the government, through the PJTC, to secure the necessary finance for this and at least two other bridges (at the head and tail of the canal) was the subject of bitter public debate at the Bor conference held by the JEO in 1983. On the top of the embankment there will be sufficient slope and camber to ensure rapid runoff from the road and drying out of its surface, so, while the route may well be closed during heavy rainstorms and for some time after, it will at least be all season even if it will not be all weather as is sometimes claimed. Wet road surfaces can, however, be very quickly damaged by traffic (or crossing cattle) and careful control and supervision of road movements will be essential.

Improvements made beyond this standard would be costly, and to be

Plate 32. The Bor–Kongor road, after improvement, during the wet season. Unwise use of earth roads by lorries, when they are wet, can lead to serious damage to the surface. A short section like this one renders large lengths of road effectively useless. Photo: Michael Lock.

justified in terms of through long-distance traffic, would require equivalent surface standards north of the canal outfall to Kosti. High river levels could cause flooding back up the watercourses which the present road north of Malakal crosses, calling for bridging or diversion. It is also questionable whether investment in such costly improvements would not create problems of road–river competition, a matter that will need close attention since river services in the Sudan have not for a long time been run on a viable or efficient basis.

Water availability and quality

One of the principal local advantages claimed for the canal is that it will provide a permanent water supply through areas that are virtually waterless during the dry months of the year. It has already been suggested in this chapter that too much reliance on this source of supply could lead to an undesirable degree of concentration of settlement along the canal causing a variety of local problems. Indeed, the Range Ecology Survey, in its recommendations for future development (RES, 1983), made detailed proposals for a domestic water supply network, the objective being not only to promote more even exploitation of cultivation and grassland resources, but specifically to avoid over concentration of population along the canal.

A further premise of this plan, which is described in greater detail in the next chapter, is that, since the water of the Bahr el Jebel does not currently satisfy the World Health Organisation's minimum purity standards, this will also apply to the canal. Consequently, plans for domestic use of its water must entail treatment installations if finance from international sources is to be forthcoming. In fact, detailed plans for the abstraction of water specifically for domestic consumption along the whole of the canal do not appear to have been made, except the gravity-fed supplies mentioned later, which are for multiple purposes.

However, no international organisation calls for minimum water quality standards for livestock, and the canal may prove more suitable as a source of supply for animals. There are problems here too. The first is the damaging effect of their hoofs by regular movement up and down the canal banks, especially after rain. The vegetation mat suggested earlier might not stand up to constant use, unlike specific canal crossing places which would be used only twice, or at most four times, a year. Planned watering points appeared to be required with, if necessary, some degree of reinforcement of the banks to prevent erosion and facilitate the movement of livestock. A second problem could be the degeneration of available pasture (see Chapter 16) if high grazing densities are maintained along the canal for too many weeks in the year. This may be parly offset by the improved pasture conditions on the eastern side of the canal, where drinking water may also be available outside the canal for at least part of the dry season, but probably too muddy for use during the rains. On the west bank, as far as the crossover point near Kongor at any rate, there is likely to be another problem. Livestock, to water from the canal, will have to cross the main road, with inevitable damage to its surface, particularly in wet conditions, though in those circumstances for most of the rainy season livestock would fortunately be watering at pools away from the vicinity.

Proposals which may at least solve some of these problems have been submitted to the JEO for the construction of offtake structures between km 40 and km 309 in areas where the canal crosses depressions of watercourses and where gravity abstraction would be assured (Associated Consultants, 1981). These will be piped through the banks for domestic water, with simple treatment plant, for livestock supply and in some cases wildlife, as well as for small-scale irrigation schemes. It was essential that these pipes or valves should be installed prior to the completion of the canal embankment, though this is no longer everywhere feasible (RES, 1983) and virtually impossible without excessive expenditure once the canal is in operation. The proposals are imaginative and important and their implementation should have very high priority if the rehabilitation and continuation of the project ever takes place. Abstraction in this way (with pipes ranging from 0.35 m to 0.76 m in diameter) is only possible where the canal water levels will pass above the surrounding ground to permit gravity flow. Similar schemes where gravity abstraction is not possible would probably require diesel pumps, which would raise problems of maintenance and cost. Indeed, in these parts of the canal it may prove more reliable and economic to provide ground-water supplies, not only more widely distributed as suggested earlier and, under the plan described later in Chapter 19, but, where needed, close to the canal. It may even be that at certain points seepage from the canal bed will feed shallow aquifers nearby and the possibility of wells to exploit them.

Health and disease

Water supplies from the canal, as well as surface supplies throughout the region, raise questions of public health. There is evidence to suggest that there may be an increase in schistosomiasis (bilharzia), a disease whose vectors, small aquatic snails, particularly of the genus *Bulinus*, thrive in irrigation canals, where there are usually concentrations of people. This is another reason for the need to plan the dispersal of the population, though human bilharzia is transmitted when people wade in contaminated water which is likely to happen in any case. A greater incidence of guinea worm may also occur, since it is transmitted through water used for drinking in which people with infected legs have walked. This underlines the need, not only for treated domestic supplies, but improved health services, and education and extension work in simple preventive measures (see Chapter 19).

Effects of the canal on groundwater sources

Except in the area north of Mogogh where a high degree of salinity occurs, there is more groundwater available in the Jonglei area than is likely to be required in the immediate future. It is, however, necessary to consider very briefly the possible effects of the operation of the canal on groundwater in reducing areas of permanent swamp and seasonally inundated floodplain west of it. Data on boreholes and shallow wells are available (JIT, 1954 and RES, 1983), but not much is known about the mode of recharge of the aquifers, so that predictions can be no more than very tentative. Comparison of recent groundwater levels with those recorded in the 1950s, and especially before the increased river flooding of the 1960s, indicate a significant rise in the water table close to the swamp and floodplain. Little or no change appears to have taken place well away from river-flooded land, and – as a generalisation – the farther east, the less the change in the water table. From these comparative observations, the Range Ecology Survey indicates that only marginal changes in level might be expected if the canal operation coincides with continued high levels of White Nile inflow into the *Sudd* region. If river flows return to discharges closer to those pertaining before 1961 (e.g. the 1905–61 mean), the groundwater will presumably gradually sink to pre-1961 levels, a drop slightly increased by the drainage effect of the canal's operation (RES, 1983). This is really to say little more than that the operation of the canal will not cause decreases in the level of groundwater in areas close to the present edge of the floodplain much more than would be the case if the areas of flooding return naturally to pre-1961 conditions.

The canal itself is in effect a new river channel, though closely confined within its banks and the area of its water surface (about 20 km^2) is insignificant compared with the area of swamp and floodplain. Its depth will be less than 5 metres below ground level for most of its length, well above the more permeable strata present at deeper levels. There are, however, places where the canal alignment runs through sections of sandier more permeable soils nearer the surface, for the most part where it intercepts the beds of watercourses which cross its alignment from east to west. Here the effect may be to provide perennial rather than seasonal recharge for the shallower aquifers in these places and hence supply promising sources of water in the form of shallow open wells which will hold out through the dry season.

Flooding against the east bank of the canal

One of the more significant local effects of the canal will be to impound water in the rainy season on its eastern side, water derived not only from accumulations of local precipitation but also the build-up of water (see Appendix 3) coming in from the east and south. This has always been predictable, but was actually observed where the canal embankment had been completed before the cessation of construction work in 1983, the area of inundation in some places extending several kilometres east of the canal and persisting well into the dry season. This will have a number of localised effects, many of them beneficial provided that the right steps are taken as soon as construction work is resumed and not left, as it was in 1983, an unresolved and therefore controversial question.

The most obvious beneficial effect will be the flooding, which, as described in Chapter 16, will tend to produce nutritive and palatable grasses similar to those in the beds of seasonal watercourses to be found in the Eastern Plain, e.g. *Echinochloa* spp., which are of such great importance in the dry months of the year. Such newly created dry season grazing areas along the east side of the canal should go some way to compensate for losses in river-flooded pastures brought about by the operation of the canal. According to recent estimates (RES, 1983) something between 500 and 1000 km^2 of dry season grazing may be provided in this manner, thus partially offsetting the losses in riverain pasture, though the period during which it can continue to sustain livestock may vary from year to year. As in the case of river-flooded grasslands, its grazing value will presumably diminish as the dry season proceeds, but its nutrient status may rise in time since dissolved material will be carried in and deposited.

Yet such flooding in the vicinity of the canal will also create problems; it will cause difficulties of access to the canal embankment, making daily communication between villages and waterfront difficult and, more seriously, impeding major human and livestock crossings at crucial times of the year. Uncontrolled, it may also cause damaging inundation of permanent habitations and their adjacent cultivations; indeed, it had already caused considerable hardship in this way before November 1983 when the matter was raised at the JEO's Annual Co-ordinating Conference at Bor. The failure of the PJTC to acknowledge the serious nature of this problem caused much public concern. The resentment thus generated could be partially allayed if remedial measures were undertaken such as the building of embankments or causeways across the area of flooding leading to villages, crossing points and berthing sites along the canal.

Landing places may be expected to develop along what will become the direct river route between the northern and southern Sudan with important local and inland trade implications. Causeways will therefore also be needed to link with berthing places on the east side of the canal and to serve the lateral lines of road communication leading to the main road along the eastern canal bank. These feeder roads are an essential part of the infrastructure required for the future development of the region.

The ponding of local precipitation and creeping flow against the east bank of the canal also calls for the construction of a system of low dykes or bunds parallel to its eastern margin and, according to local topographical features, at appropriate distances between 3 and 10 km from it. Besides protecting the villages, such measures would create an additional band of *Echinochloa*-dominated grassland, thus further reducing dry season grazing pressures, while also providing an extra line of drinking water for livestock and wildlife through the region for at least part of the year. Such protective dykes would also reduce the size, if not the number, of lateral embankments needed. An important consideration in siting these dykes would be the traditional seasonal migration routes across the line of the canal, discussed below.

The possibilities of cross-drainage by siphoning water under the canal to relieve the ponding effect to the east of it have been mentioned in Chapter 2, and might be economically justified if such flooding proves excessive and cannot be adequately controlled. But such measures would scarcely be viable simply to reproduce grassland conditions as they are now to the west of it. Apart from *toic* type grazing grasses immediately east of the canal bank, areas of prolonged *Hyparrhenia* regrowth may well increase at the margins, partly compensating reductions on the other side. The difficulty here is that for those then living west of the canal this would necessitate four crossings of livestock in the course of the year, which apart from logistical problems, would increase the stress to which livestock are subjected. Mention has been made elsewhere (Chapter 17) of the potential for fisheries in areas along the east side of the canal where the ponding effect is deepest.

The political situation has precluded the monitoring of these flooding effects from the ground since November 1983, but monitoring by satellite imagery would provide records of great value for the development of the canal area in the future, even if the canal is never completed.

Effects of the canal on the seasonal grazing cycle: crossing points

The absolute need for the bulk of the population and their animals to move with the seasons according to the availability of different pasture types and of water has been made clear in previous chapters. Cattle must of necessity move to the *toic* or river-flooded grasslands during the three or four

driest months of the year. Equally, rain-flooded pasture serves an intermediate but relatively short purpose in the grazing cycle at the end of the rains
when regrowth appears after burning and again when the rain first falls again
during April or May (see Chapter 10). Topography and geographical
distribution of land types dictates an east–west axis for this seasonal movement; and, for most of the people in the canal zone, the movement crosses its
alignment.

When the rains cease, most southern Dinka first drive their cattle eastwards
to take advantage of the regrowth of rain-flooded grasslands, particularly
Hyparrhenia rufa, before moving them to the riverain pastures in the west. As
we have seen this is an immensely valuable supplementary grazing practice
which relieves pressure not only on the floodplain or *toic*, but also on wet
season pastures that may have been overcropped towards the end of the rains.
It is a practice that needs to be developed if riverain pasture is reduced by the
effects of the canal. The obstruction of this movement by the canal will vary in
this part of the Jonglei area. Many of the people in the northern parts already
live east of the canal alignment (see Figure 10.3) and can make use of the
grasses in the Eastern Plain (Dinka: *aying*) before crossing the canal alignment to reach the *toic*. This will involve two crossings each year. In the south
the Bor Gok live east or south-east of the canal, and many will not need to
cross it anyway. Many of the Twic Dinka and most of the Bor Athoic,
however, now have their permanent wet season settlements west of the canal
line. The same applies to many Gaawar Nuer farther north, though they may
gain more extensive areas of intermediate pasture land west of the canal if
river-flooded grasslands recede – a return to conditions as they were before
the high river levels of the 1960s onwards.

It is also the case that many Dinka spending most of the year west of the
canal line do not now move their cattle as far into the Eastern Plain as they
did before, for fear of Murle raids. Indeed, as we have seen, in many cases
water supply points have been allowed to fall into disrepair and even wet-
season village sites in the more easterly and exposed areas have been
abandoned for this reason (see Figure 11.1). We must assume, however, that
this unacceptable threat from the east will in the end be removed. Given any
diminution of river-flooded grasslands, Dinka in these areas will be bound to
make use of these eastern rain-flooded intermediate lands once again and
more extensively and for longer periods. Some may return to abandoned
village sites, but those whose villages remain west of the canal may have to
cross it with their herds, not twice but four times in the year. The risks
involved and labour inputs required in effecting these crossings in relation to
the viability of subsistence pastoral activities are serious considerations;

viewed as factors in the development of a commercially orientated livestock industry they could be critical. Nevertheless, cattle are regularly swum across the main river near Bor (and across rivers elsewhere in Africa) without ill-effect. In the canal zone the total annual numbers of cattle crossings have been estimated at 800,000, though the stabilisation of stock movements due to new grazing resources created by the accumulation of flood waters east of the canal itself may reduce that number by some 100,000 (RES, 1983). These figures were calculated from aerial surveys carried out in one year and may be expected to vary in future according to the circumstances of the herds. They demonstrate the magnitude of the problem of herd movement across the canal, and it must be remembered that the people themselves, perhaps as many as 250,000, plus at least 100,000 goats and sheep, will also need to cross. The wet season distribution of livestock is indicated in Figure 10.3 which shows the incidence of dwelling huts and hence areas of settlement on the higher ground within the Jonglei area to which all animals are taken at the height of the rains. Figure 12.1 shows the distribution of cattle according to three seasons – mid-wet, early dry and late dry respectively – and thus the movement of cattle populations.

The scale of these crossings is likely to create very great logistical problems; to meet these seasonal requirements four bridges have been planned, of which two would be at the extreme southern and northern ends of the canal and therefore of rather less use as far as the main movements of humans and livestock are concerned. Twelve motorised ferries, roughly one every 22.5 km, have been proposed, though the problems of operating and maintaining machine-driven transport systems in this remote part of the Sudan are all too obvious, even though such difficulties may be alleviated by the improved communications provided by the canal itself. The total rejection by the government of the idea of hand-operated ferries, common enough in other parts of the southern Sudan and, indeed, many parts of Africa, has not been fully explained. Such ferries have not caused any insuperable obstruction or hazards to navigation; the manual operation of far less costly equipment at more frequent but controlled points along the canal could provide a simpler and more reliable service.

The locations of bridges and ferries (see Figure 12.1) were proposed after the analysis of cattle distribution data from aerial surveys (JEO 1982), although a number of the proposals were subsequently ignored by the PJTC. So few crossing points will, however, undoubtedly lead to conflict over traditional migration routes and grazing rights, with consequent inter-sectional friction. Movements of livestock to and from rain-flooded pastures vary from year to year and place to place according to prevailing climatic conditions. Likewise, the timing or movements into the *toic* are determined by water levels in the perennial river channels and the speed and extent of their fall, and the convergence of the herds on dry season pasture occurs gradually as the season advances. Nevertheless, while movement over the bridges need present no problem, any great concentration of livestock and their owners waiting to cross at 12 motorised ferry points could raise immeasurable difficulties of disease control as well as the problem of making available adequate fodder without denuding the vicinity of every blade of grass. The mustering of humans and livestock in numbers far in excess of those to be found in the largest of cattle camps will also create problems of hygiene and health. The inevitable periodical breakdown of mechanised ferries could also create crises unless, as will probably happen, cattle become accustomed to swimming the canal.

Kadouf (1977) has drawn particular attention to the nature of land tenure in the Southern Dinka area, observations which apply equally to the Nuer (Howell, 1954), covering all three categories of pasture and hence well defined and exclusive rights of usage common to tribal units and segments within their recognised territories (see Chapter 10). These also include rights in water points and pools, and fishing rights – a vital economic asset, since many inland watercourses and pools are prolific in fish, and are easily caught as the waters drop. Such resources are often used by others but only with the express permission of the owners; friction and conflict occur if these rights are infringed. Territory extends in naturally defined segments running west to include riverain pasture and east to incorporate the larger tracts of rainfed grasslands. Unless the owners of these corridors each have their appropriate crossing points, it will be inevitable that more people will need to cross the territories of others, often making use of their water supply sources, and at times their fishing resources, on the way, with a consequent conflict of interests and confrontation. Since in these circumstances more crossing points than planned will be required, it seems likely that people will simply swim at least the healthy adult cattle across the canal; calves, smaller stock, women, old people, children, and the paraphernalia that is normally carried to the cattle camps, being taken across by canoe. Nuer are less familiar with this method of crossing, and may need instruction, but are likely to learn quickly enough if necessity so dictates. Few obstacles will deter the Nilote from moving his cattle when needed.

Swimming cattle across would appear to be feasible, certainly in that part of the canal where the banks are not steep, i.e. from km 40 to km 310, where the slope is to be 1:8. Nevertheless, uncontrolled crossings of this kind could be the cause of some erosion of the canal banks, though it is likely that a vegetation mat will develop and stabilise them, particularly if the canal levels remain relatively constant. The protection of canal banks at crossing points – the provision of reinforced ramps, for example – might be required if fluctuating levels result from the method of operation recommended in Chapter 20. The alternative is adequate numbers of the more elaborately designed swimming points advocated by consultants (see Chapter 2). Given the high costs, the virtue of these is debatable, particularly if velocities in the canal fluctuate, resulting in variable points of landing downstream by the cattle. Priority should be given to the most cost-effective methods, as well as those capable of simple adaptation and easy maintenance by the people.

It is likely that a canoe hire industry will develop in the area to assist the crossing of cattle and to ferry people, calves and smaller stock over the canal. Mahogany required for these craft is not available locally, and it may be that the provision of suitably designed boats of steel or fibreglass to perform all these functions would be a service reasonably provided by government, and certainly less costly than motorised ferries.

The build-up of water on the eastern side of the canal will cause great difficulties for livestock movement, particularly in the early part of the dry season. Water and excessively muddy conditions on that side may therefore also require the construction of causeways from the canal embankment to a point beyond the flooded area (RES, 1983).

The problem of canal crossings, for some people and their livestock perhaps four separate crossings each year, will undoubtedly be difficult to resolve. Without satisfactory solutions the canal will prove to be a damaging obstacle to the maintenance and development of the pastoral component of the economy, which as we have seen is wholly dependent on mobility to make use of seasonally variable resources.

Note

1. A rather different picture of population distribution is shown in El Sammani (1985) and Abdel Magid (1985) based on the work of Watson, incorrectly suggesting a rather lower density east of the canal line, particularly in the Lou Nuer area. This presumably arises from the difference in sampling methods.

Part VI

Rural development and the future

19

Rural development, 1974–83: a decade of unfulfilled promise

Introduction

During the brief interval between the end of the first civil war and the outbreak of hostilities again in November 1983, the southern part of the Jonglei area at any rate had received vastly more attention to its economic needs than it had in the previous 20 years, and, in terms of capital transfer, technical co-operation and volume of expatriate expertise, more than in the whole of the 57 years of the Condominium Government.

It is not the purpose of this book to evaluate the results of these activities, but it is worth recounting the main thrust of investigation, trial and project implementation during that period, to highlight some of the difficulties encountered, and in doing so to identify those areas of the development process that will need the closest and most immediate attention when peace is restored and work on the canal can begin again. It is worth adding that most measures required will be development priorities whether the canal is completed or not, and would have been so had the project never been conceived. Except perhaps in their location, there has never been any clear distinction between the nature of development needs in general and those that relate more specifically to any disadvantages which the physical presence of the canal or its hydrological operation may cause.

As a prelude to the account which follows, which is not a record of consistent success, it is perhaps fair to draw attention again to the immense technical and other difficulties in such unpredictable climatic conditions and with so many unpromising environmental features. Canal or no canal, solutions to the problems created by these circumstances call for extensive and very long-term research, experiment and trial. The time available to the Jonglei Executive Organ, donor agencies that assisted in its programmes, and others working on development measures initiated independently by regional and provincial authorities, turned out to be all too short.

The work of the Jonglei Executive Organ

The National Council for the Development Projects for the Jonglei Canal Area was established by Presidential Decree in October, 1974, (Republican Order No 284) to investigate the impact of the project locally, to devise methods of maximising its benefits and minimising any adverse effects, and to plan and implement schemes for 'integrated economic and social development in the area'. Its Executive Organ (JEO) was created to put these objectives into effect and to secure funds from international and bilateral donors for the purpose. Matters concerning the civil engineering works, including design, alignment, construction and future operation of the canal remained exclusively the responsibility of the Permanent Joint Technical Commission for Nile Waters. (For terms of reference of these two bodies see Notes 1 and 2 to Chapter 2).

The head office of the JEO was, and still is, in Khartoum, with subsidiary offices in Malakal for forwarding materials and in Juba for project liaison with the regional government authorities, its main base for field operations being at Bor. The justification for maintaining its headquarters in Khartoum, over 1,000 km from the centre of the Jonglei area and the scene of action, was largely dictated by the perceived need to be close to the planning and financial decision-making departments of the central government and, even more important, to have access to the representatives, resident or visiting, of the donor agencies from which funds might be expected (Garang, 1980). Its precise powers in the exercise of its functions do not appear to have been very clearly defined. As its Commissioner pointed out in 1981 the mandate implicit in its stated objectives was sufficiently broad to parallel many activities already projected or undertaken by government at all levels. This led many people to assume that the JEO was a substitute for other government departments, when in fact it could only supplement their services (Awuol, 1981).

Co-operation and co-ordination on so diverse and ill-defined a scale, with few explicit powers and controls, were bound to be difficult. Moreover, relations between the JEO and the donor agencies involved with field projects in its own programme were often closer than those between the JEO and national and regional ministries. Indeed, donor agencies were at one time advised by the JEO to form their own direct links with regional and provincial authorities, thus bypassing one of the purposes for which the JEO had been created (Garang, 1980).

These difficulties in co-ordination, compounded by distance and the intractable problems presented by the physical environment, are mentioned here to give some perspective in judging the success of the JEO programme. As an organisation it also lacked staff, particularly its own professional advisory personnel in a whole range of disciplines necessary to deal with closely interrelated problems. There are lessons for the future to be learned from both the deficiencies in resources available to the organisation and the particular difficulties it had to face, but in parenthesis it is worth remarking that the most striking weakness appears to have been the lack of JEO (or National Council) representation, especially southern representation, on the Permanent Joint Technical Commission – a point discussed further in the next chapter.

The JEO's early programme was concentrated on finding the finance for, commissioning and then co-ordinating a series of studies on the potential effects of the canal. The outcome of at least some of these studies has formed a major part of the data base upon which this book has rested. These studies were to be the foundation of future development initiatives. Not surprisingly, however, political pressure increased on the JEO to justify its existence by translating plans into positive action. In 1980, the then Commissioner, the late Mading de Garang, had to admit 'in truth the Executive Organ, although it has been in existence for five years, has little to show the people of the area'. A year later, his successor, James Ajith Awuol, remarked that 'the Canal Zone lags far behind in development and its citizens do not appreciate long scientific research which does not provide them benefits they firmly believe they should receive'. While this was not entirely correct in all respects (many other areas in southern Sudan were in receipt of far less development assistance even at that stage), it was a reasonable expression of political opinion. Attempts were therefore made to inaugurate a more closely integrated programme, drawing together the projects funded and staffed by different donor agencies. To this end the first of the JEO's Annual Co-ordination Conferences was held in 1980.

Aid sources and donor agencies

The principal development projects in the Jonglei area were financed and executed by United Nations agencies (UNDP, UNCDF, UNOPE) with FAO responsible for the two main schemes in the field: the *Sudd* Fisheries Project and the Kongor Rural Development Project. The UNDP was also responsible for financing technical advice to the JEO over a period of nearly five years.

The two other major donors in the sphere of rural development over this period were the EEC and the Netherlands Government. The former was primarily responsible for studies of the engineering aspects of the canal and its related structures, and also of the natural systems to be affected by it. The Dutch Government's role was mainly in fields of development that were equally pertinent whether or not the canal was ever built. The significance of their work forms a major part of this chapter.

Social services

The JEO made two attempts to plan for community service development, in an area which, until 1979, had been characterised for 20 years by an almost total lack of even the most basic health and education facilities. The first of these was outlined in a document entitled: 'Proposals for Mid-Term Programme and Crash Programme for the Development of Livestock and Socio-Economic Services in the Jonglei Canal Area' (JEO, 1979). This early programme was based on the concept of model villages, in which a cluster of buildings for the provision of social services would be constructed around a central core, on the assumption that this would attract people from surrounding communities to come and settle there. This approach to community service development was confined to the 'Canal Zone'.

Compensation for those households to be displaced by the canal construction were an early priority. Those people living in other parts of the Jonglei area, Nuer and Dinka to the north and west of the Bahr el Jebel for example, would also be adversely affected by the operation of the canal, but no community services were proposed for that part of the area for them. Their interests were not taken into account in the JEO's development planning process, just as they were excluded from the research programmes of technical consultants. The needs of the people below the canal outfall also appear to have been largely ignored, though attention was drawn to the possibility of adverse effects downstream by El Sammani, one of the JEO's contemporary advisers on these matters, who identified 'an estimated 90,000 that shall be directly affected to the north of the Sobat Mouth' (El Sammani, 1985, p. 60).

The second attempt, 'A Comprehensive Plan for the Development of Community Services in the Jonglei Canal Area' (JEO, 1983), was mainly based on field experience acquired by a UNDP consultant on Regional Planning and Development Management, Steven Lawry, in 1981, and had the benefit of being able to draw on the lessons learned from the JEO's previous activities. In attempting to complete this programme, the JEO, like other agencies in the area, had come up against many unforeseen difficulties. Reliable building contractors with experience of local conditions were not available, at least at an acceptable cost, with the result that the JEO felt itself compelled to undertake construction tasks for which it was ill equipped. Lack of managerial skills and trained construction workers, the high cost and difficulty of obtaining imported materials and the poor quality of local ones, when combined with a short building season of less than six months and untested assumptions about the interest of the local people in these projects, inevitably led to an unsuccessful programme. The comprehensive plan took a realistic look at these failures, as did Jonathan Jenness (pers. comm.), in his final report as technical adviser to the JEO; the JEO was to stop building schools for which there were no teachers, and leave the whole matter of educational policy and planning to the regional education ministry.

There was some limited success in the field of public health. At a time when the Bor Provincial Hospital had no water supply, to say nothing of inadequate staff, no equipment and no drugs, every little helped. The CCI construction camp at the Sobat mouth ran a daily clinic, its first interest being the health of its staff, but freely and willingly giving treatment to a wider clientele. They also assisted in the setting up of a human health and nutrition survey, with clinical improvement as the ultimate objective, particularly amongst the Shilluk of the area. The JEO/UNDP integrated rural development project centred on Kongor included the construction of a small health centre, which opened in 1983 and was staffed for a year by a British volunteer doctor and nurse, with rather limited success in the short period available. The Dutch Government, as part of their many activities based in Bor District, constructed a training centre at Baidit 25 km north of Bor for primary health care staff from the entire province. Although finished in 1981, a number of administrative problems evidently intervened, and the centre was not commissioned. Nevertheless, village dispensaries throughout Jonglei province were equipped with basic drugs under this programme, and a simple record-keeping system was initiated.

Between 1972 and 1984, then, there was not much tangible or enduring progress in the field of social services. There remained a chronic shortage of trained manpower; and there was inadequate co-ordination between the JEO and the regional ministries of Health and Education. There were encouraging signs in one field only, that of building design, an essential requirement for social services. The Dutch-funded Malek Appropriate Technology Centre was successfully erecting ferro-cement buildings, with promising results. This technique solves a major problem of soil movement, since foundations are not required, materials are durable and resistant to termite attack, and the method is easily learned, while costs are relatively low.

Communications

Perhaps no part of the life of the Jonglei Area resembled less in 1974 its former state of 17 years previously than the communications system. Records recovered from Bor[1] reveal that, in the late 1950s, a telegraph network still kept all the administrative centres, both in the area and beyond, in daily contact; regular steamers arrived on time, bringing passengers, mail, fresh food, cereals, building materials and fuel; an extensive network of roads corresponded on the ground with the layout shown on the maps, and these roads were, each year, both protected from abuse in the wet season by closing them after heavy rain and maintained to the standard that local materials and manual labour would permit.

The ravages of the civil war, the climate, the floods and the effects of two decades of economic and administrative neglect meant that, after the signature of the Addis Ababa Agreement in 1972, not one of these systems was effectively in operation. Many roads were by then scarcely discernible; the telegraph lines had been turned into ornaments; the arrival of a steamer in Bor could not be predicted with an accuracy greater than two or three weeks. This was still essentially the situation in 1980.

Road communications then began to improve rapidly. With the financial assistance of the Dutch Government, an all-season road with a firm laterite surface, the only material available within reasonable distance, was constructed between Juba and Bor in 1979 and 1980. This joined an excellent road provided by the same donors, that in turn linked Juba with the Uganda border at Nimule. In 1981 and 1982, the road was continued northwards for the 120 km from Bor to Kongor. The total lack of suitable surfacing materials for this stretch of the road meant that it was not quite an all-weather link, but it was, without doubt, an immensely improved and important artery for potential development. Rightly judging that capital expenditure on the roads would be useless without attention to the recurrent maintenance costs, the Dutch government also financed (through Ilaco as Consultants and De Groot as contractors) a road maintenance unit which operated from Bor between 1981 and 1984, a good example of imaginative aid policy.

In addition to these major road construction projects, subsidiary dry-weather roads to the major court centres in Bor District were constructed, under Ilaco supervision, as part of their programme, Bor Area Development Activities, in 1982 and 1983. Road development in Bor and Kongor Districts had, by 1983, just about returned to the point where it had been left 28 years previously. Meanwhile the main road link between Malakal and Juba was improving at about the same rate, and the road on the eastern earth embankment of the Jonglei Canal had progressed 80 km southwards from the

Sobat Mouth between 1978 and 1981, had advanced to 150 km the following year, and was at 260 km by the end of 1983. At this point a dry season connection with the Dutch road to Kongor was a matter of some 15 kilometres. Although the movement of troops to counter the rising threat of civil unrest was the main motive behind its provision, one major obstacle to road transport in the area was also at least temporarily lifted in 1983 by the construction of a pontoon bridge across the Sobat river.

River transport, meanwhile, was also improving, though not quite so rapidly. New barges and diesel-driven pusher units were provided by the West German Government under an agreement with the state-operated River Transport Corporation; CCI, the canal-building consortium, and Chevron, prospecting for oil, operated their own large barges between Kosti and the *Sudd* region. Chevron had also increased the scope of their enterprise to the point of dredging the riverbed in some places north of Malakal, and reopening blocked channels in others within the *Sudd* itself; the Juba Boatyard constructed and sold, partly to private buyers but mainly to development aid agencies, small ferro-cement barges. Together, these measures did not much improve passenger traffic on the river, but they were just starting to have an important effect on the economy as a whole, when they were interrupted once more by the outbreak of civil war in 1983.

Other, perhaps more fanciful, developments in the field of communications included the construction, in both Bor and Malakal, in 1978–9, of a satellite communications receiving dish. Five years later, the one in Bor at least had yet to be brought into service.

Agricultural development

Claims that the area has great potential for agricultural development were put to the test at Bor and at Pengko between 1976 and 1983, in a project funded by the Government of The Netherlands, in partnership with the Southern Regional Ministry of Agriculture, and executed by the Dutch consulting company, Ilaco. Initially called the Pengko Pilot Project (PPP), this enterprise had no connection with the potential future Jonglei Canal, nor, at first, was there any real dialogue with the JEO, except on logistical matters (the PPP, unlike the JEO, was always well equipped with spare parts, fuel and other essentials). The results of this project are summarised below, and bear a message that is far from welcome. The study revealed that the prospects for economically viable production of cereals on a large scale in the area are small. This was the outcome of well organised and practical research, but perhaps because too many hopes had been placed on this form of development, the results of the trials were regarded with quite unjustified scepticism in official circles.

Large-scale mechanised agriculture – sorghum and maize

The purpose of the farm of the Pengko Pilot Project, located just east of Bor, was to collect reliable data on which a decision regarding the technical and economic feasibility of a large-scale mechanised, and possibly irrigated, crop production scheme in the Pengko Plain could be based. The project started in 1976 with the clearing of a large area of woodland, just south-east of Bor. Of this, 109 hectares could be irrigated, and 200 used for rainfed crops. Initially, it was intended to try sorghum and maize. The first trials on small plots gave encouraging results; up to 4 tonnes/ha of sorghum were obtained. Top yields of maize were slightly less. However, problems arose when these crops were grown on a large scale and production was heavily mechanised. Not more than 500 kg/ha of sorghum could then be harvested, even less than yields obtained by local farmers. Maize production was even worse. The most important reasons for this were hydrological. The soil within a field was not homogeneous; sandy topsoils alternated with heavy clays, interspersed with truncated termite mounds. This led to micro-topographical differences within a field which prevented proper drainage, leading to serious waterlogging problems and, in turn, to poor plant development and a luxuriant growth of weeds.

Even planting on ridges was not sufficient to protect the plants from flood damage. To try to remedy this, cambered beds were constructed, with the highest point in the middle and drains on either side. This system did indeed increase sorghum yields, but only to 1,000 kg/ha, still a low figure. This could be attributed to a number of factors: rainwater did not penetrate into the soil immediately, but flowed off the surface of the cambered beds into the drains, thus rendering the crop very susceptible to dry spells. This could be partly overcome by regular loosening of the soil. The crop was badly damaged by birds (weavers and pigeons), though this could perhaps be reduced by planting late-maturing varieties. Other problems encountered in the large-scale trials included the limited period suitable for mechanised cultivation. The soil proved too hard at the end of the dry season and the fields to be inaccessible to tractors once it was thoroughly wet. Thus many tractors are needed so that cultivation can be carried out rapidly in a relatively brief period in the cropping season, and this results in an uneconomic use of plant, which would be largely idle for the rest of the season. The cambered beds meant that the size and the accessibility of the plots were reduced, because of the drains. The now undulating topography led to uneven ripening, making mechanical harvesting impossible, and because the crop was ready for harvest before the end of the wet season, drying of the grain was difficult.

On the basis of these trials it was concluded that a very complex agronomic system would be required to cultivate sorghum on the Pengko Plain including protection against overland flooding, a comprehensive drainage network to remove excess water (perhaps even including pumps) and a system of cambered beds to ensure adequate drainage within the fields. A preliminary economic analysis on the basis of a farm of 1,000 hectares showed that annual yields of at least 4.5 tonnes per hectare would be needed just to cover recurrent expenditure, excluding the costs of capital development. As such yields were never obtained even in small-scale trials at the pilot farm, it was concluded that there was no scope for fully mechanised production of sorghum or maize.

Trials with rice

Rice was also included in these experiments. Some very good yields of up to 8 tonnes per hectare were obtained on trial plots. Again, however, problems appeared when the operation was scaled up. Machinery bogged down in wet spots in the fields, delaying or even rendering impossible the application of herbicides and fertilisers, and later the harvesting of the crop. Germination was very poor on wet and low spots. Weeds constituted a serious problem, especially sedges (*Cyperus* spp.) and grasses (*Echinochloa* spp.). Herbicide (propanil) gave satisfactory control in the trial plots, but application to large fields was difficult for the reasons mentioned above. Hand weeding was not feasible owing to a lack of manpower. Delayed harvesting led to losses of over-ripe grain, and lower grain quality due to sun-cracks. Lack of proper drying facilities exposed the harvested grain to occasional showers when it was taken outside for drying; wetted grains germinated. Birds were a particular problem in dry season crops, because these matured in April when there is not much other food available for the birds.[2] Notwithstanding these problems, some large fields gave yields of up to 4 tonnes of paddy per hectare.

At the same time, experiments were started in the Pengko Plain, 40 km east of Bor. In the first two years very low yields were obtained, but the third year gave much better results. This was probably due to the decomposition of the turf, during which the soil nitrogen is monopolised by the bacterial decomposers and is unavailable to plants. However, application of even high doses of nitrogenous fertiliser resulted in yellowish, sickly plants, so that nitrogen deficiency is not on its own an adequate explanation. Other problems

440   *The Jonglei Canal*

encountered were weeds, damage by grazing wild animals, and periodic drought when there was no reliable source of water for irrigation.

The experiments with rice led to the conclusion that irrigation is a prerequisite for successful rice cultivation, and that from a technical point of view, reasonable yields can be obtained. However, the internal rate of return is insufficient to justify large-scale fully mechanised and irrigated rice production.

Apart from economic considerations, there were other constraints. The Pengko Plain is an uninhabited area, and it was therefore difficult to recruit staff. Security was also a problem because of fear of Murle raiders from the east; the labourers were only prepared to work in the Pengko Plain if they had a police escort.

However, the most serious limiting factor is probably a logistic one. Such large-scale projects depend on inputs which have to come from abroad. Apart from the need for foreign exchange, the lines of communication are immensely long and difficult. The Pilot Project could obtain the necessary inputs through a well-organised communication system with Nairobi, Khartoum, and Europe, but, while the canal may improve the services when foreign donor support is withdrawn, such systems could prove difficult to maintain.

The Jonglei Irrigation Project

Shortly after the decision was made to embark on the construction of the canal, an ambitious 84,000 hectare irrigation scheme alongside the canal due west of Kongor was planned. This was based on the earlier canal alignment beginning at Jonglei. The area was later surveyed, and with the aid of the Dutch Government a pilot scheme was proposed near Nyany, subsequently the Range Ecology Team's study site. Despite its ambitious scale and consequent political appeal as a strategy for economic 'transformation' rather than the 'improvement' approach (Garang, 1981; Yong, 1976), the project appears to have been dropped from the development agenda. After the realignment of the canal in 1978, no alternative site was proposed or surveyed (see also Chapter 13 p. 326).

Agricultural development in Bor and Kongor Districts

Using the simplest of approaches of small-scale replicated trials and the introduction to individual farmers of new tools and techniques, progress was just starting to be made to improve local crop production in the canal area when civil unrest halted all further trials during the growing season of 1984.

Pasture management

An important development was the construction of flood control dykes, to protect people's homesteads from river-flooding. Between Jalle and Kongor, in the 1970s, people had, on their own initiative, constructed by hand a dyke which normally protected them from the river flood-waters coming from the west. In 1979–80, the PJTC and CCI provided machinery to raise and strengthen this hand-built dyke and, two years later, the FAO project at Kongor extended it. The high floods of 1983 breached even this dyke. Apart from the greater security that these dykes gave to the people living behind them, they provided incidentally useful information on the factors limiting the distribution of pasture types in the area (see Chapter 8). This made it possible to predict what other bunding and water-harvesting enterprises might do.

Most of the grasslands of the Jonglei area are rain flooded, but, as explained in previous chapters, they only have a seasonal pattern of use at present. These rain-flooded pastures are of high quality early in the rains, but later dry out and become unpalatable. A second opportunity for grazing these pastures comes at the beginning of the dry season, when burning can stimulate a flush of new growth if there is sufficient residual soil moisture. At both of these times, drinking water may be the main factor that limits exploitation by livestock. Even if water were available, it should be stressed that there is no possibility of using these, the *Hyparrhenia* grasslands in particular, throughout the year (see Chapter 7).

The Eastern Plain is drained by a network of very shallow watercourses running in a northerly direction, obvious from the air by virtue of their slightly different vegetation, but very much less apparent from the ground. As we have already noted, the Range Ecology Survey report suggests that low bunds should be constructed across these to impound water. This could produce areas of better, longer lasting pasture, and the hole from which earth is excavated to create a dam wall towards the middle would act as a *hafir*. This suggestion was tested by the FAO Kongor project (McDermott & Ngor, 1984). Little improvement in grass growth or change in species composition could be detected, but this was hardly to be expected within the first year of the trial. On the other hand, the hafir provided a valuable centre for more prolonged use of the *Hyparrhenia* grassland. The value of these bunds and hafirs would be increased if they were coupled with a more systematic approach to burning, with fires being started early enough to ensure the soil moisture necessary for regrowth, yet not so extensive as to be able to damage the economically important supplies of thatching grass. Unfortunately, these important pasture development initiatives, like similar ones conducted at the same time by BADA[3] at Pengko, 120 km to the south of the Kongor experiments, were interrupted by the civil war, just as they were gathering momentum. Other forms of pasture management in the area, such as haymaking and silage production, have been much discussed (see Chapter 17), experimented with both by the JIT earlier and more recently at the government experimental ranch at Mafao near Juba, but have not been put to the test by practical application in the circumstances of the existing pastoral economy.

Livestock development

In the decade leading up to 1983, gradual progress was being made in the field of livestock development. Weekly auctions were held in the larger administrative centres (Bor, Kongor, Ayod and Malakal), and records of the transactions were kept (see Appendix 8). A rough check was kept on animal health through carcase quality inspections at Bor and Malakal slaughter-houses. But it was in the provision of veterinary services that most progress was made. Veterinarians formed part of the teams of three foreign-financed projects, active variously between 1979 and 1983 (BADA at Bor, the Range Ecology Survey team at Nyany, and the FAO project at Kongor) and collaboration between them and the regional veterinary service meant that a limited, though highly popular, clinical service was provided, particularly in the southern part of the Canal zone, and that some progress was made in providing vaccination cover against some of the major bovine diseases.

A project to improve the productivity of local cattle under conditions of intensive management, was started at Bor in 1981; it met with a number of difficulties and, despite good management, was a long way from being economically viable, or, more importantly, from having any impact on the traditional livestock husbandry practices of the Dinka.

A complicated proposal for a beef fattening and breeding enterprise was also considered on the initiative of CCI at Sobat Mouth, with financial assistance from the French Government; this too made no impact on the traditional system.

Fisheries development

The FAO *Sudd* Fisheries Project was based at Bor between late 1980 and early 1984. The aim of the project was to make better use of the fish resources of the *Sudd* region, by encouraging greater commercialisation of the catch. To this end, a project boat made regular visits to buying centres which were established in the centre of the swamps, and where sun-dried fish were purchased for onward sale in Bor and thereafter also in Juba. Tentative steps towards the setting up of fishermen's co-operatives were also made. Both this latter enterprise and the promotion of the sun-dried fish trade, showed signs of stimulating local interest in the fisheries economy. There were, nevertheless, many practical difficulties, though they were of the sort which could be overcome in the long run. Fisheries development in the *Sudd* clearly has very great potential which could be realised when circumstances permit.

Forestry and the development of woodland resources

Reference has been made (Chapter 8) to the decline in areas of woodland which followed the increase in flooding from 1961 onwards. It has also been noted that reduced flooding which will result from the operation of the canal may lead to the regeneration of woodland, particularly *Acacia seyal*, in many parts of the area, though other factors, especially fire, are also important. Pressure on available woodlands for fuel and timber is in any event likely to remain heavy, notably in areas where there have been increased concentrations of human settlement following floods from the west and threats of Murle aggression from the east (see Figure 11.1). The retreat of these floods as a result of the canal's operation may offer opportunities for wider population dispersal, but despite rational planning of settlement there may well be other areas in the canal zone where over-concentration occurs. The rapid depletion of timber stocks may therefore continue unless steps are taken to halt it. Plans and proposals for community forestry in the Jonglei area were in preparation by FAO (Poulsen, 1983), but these were also disrupted by the outbreak of hostilities in 1983. Observations on the potential for forestry are given in Appendix 10.

Water development

The present transhumant nature of the rural livelihood is deter-mined as much by water supply needs as it is by the seasonal availability of nutritive grazing. Water is, of course, available for much of the year in natural depressions, *khor*-beds and pools, the location of which is one of the factors that dictates the distribution of wet season settlements, just as it does the use of the grasslands. Prior to the early 1970s, the great majority of the people of the area, and of course also their livestock, were dependent throughout most of the year on these natural sources of water. This was not universally so, for in the 1940s and early 1950s an ambitious programme of construction of hand-dug, brick-lined wells had been undertaken, particularly in what is now Kongor and Bor Districts. The intention of these was twofold: to improve the quality of water for health reasons, and to enable people who were obliged to leave their homesteads in the dry season for want of water, to return to them

early in order to start cultivating, thereby improving agricultural productivity. This programme led to the construction of some 40 wells, the majority of
which were still functioning adequately 35 years later.

A subsequent programme, a decade later, saw the sinking of deep boreholes
with diesel powered pumps in the major court centres (Kongor, Duk Payuel
and Ayod, for example). The output of these was obviously vastly superior to
that of shallow open wells, but operation and maintenance difficulties made
them a much less reliable source of supply. By the mid-1970s, with scarcely 50
man-made water points serving an area of some 70,000 km^2, plans to improve
and extend water supplies were an obvious priority.

In 1978, the JEO began a programme of water development, in conjunction
with the Regional Ministry of Rural Water Development, that was based on
the drilling of deep boreholes and their equipment with diesel-operated
pumps. The chronic fuel shortage at that time, lack of spare parts and of
transport to facilitate servicing and maintenance, demonstrated the local
impracticability and high cost of this form of supply. Many of the boreholes
quickly became non-operational (RES, 1983) and the UNCDF, the source of
finance, cut off its support in 1982. The JEO then turned its attention to *hafirs*
and a programme of excavation began in 1982. In the two seasons up to 1983
when work ceased, 18 tanks of this kind had been dug in the four main
districts of the Jonglei area.

Despite its hastily conceived and subsequently unfortunate experiences in
the practical aspects of water development, the JEO also had the foresight to
commission an overall water development plan for the Jonglei Canal area,
which formed a part of the Range Ecology Survey. While this plan was
nearing completion, BADA, the Dutch-funded development project at Bor,
produced an outline water development plan for that District and for the
neighbouring districts of Pibor, outside the Canal zone to the east, and Waat
(see Figure 10.1). By the end of 1983, 35 boreholes had been drilled and
equipped with handpumps, after all the development agencies involved in the
area had come to a tacit agreement to abandon developments based on the
use of diesel-operated pumps. Despite this welcome acceleration of water
development activity, there were at the end of 1983 still only just over 80
functional man-made water points in the whole of the four administrative
districts through which the Jonglei canal will pass.

The overall plan was based on investigations of the depth and productivity
of the aquifers, and of the water quality in the different water-bearing strata.
This was related to the distribution and population density of the people of
the area, and their anticipated water consumption needs. A technical evaluation of the reliability and cost effectiveness of different types of drilling and

hand-pumping equipment was another feature of the plan. The plan was deliberately designed to provide water for people, but not in large enough quantities for livestock, with the intention that the transhumant pastoral economy, undoubtedly the optimal use of the land and its resources under present-day conditions, should not be disrupted (RES 1983, vol. 6).

This same concern, and also a desire to avoid the risks of grassland abuse that have bedevilled livestock water-development projects in virtually every other pastoral area in Africa, together led to the most politically contentious part of the plan's recommendations (see also Chapter 18). The main argument is as follows: a completed canal, full of water, if allowed to be the only source of water-development in the area, would become a very forceful attraction for the concentration of human settlement along its length; this would be bad for the political and natural resource economies of the area (human and animal health hazards, grassland abuse, conflicts over traditional rights); these forces can only be overcome by counteraction, in other words by providing water in a well-dispersed network throughout the area to satisfy people's domestic needs *in situ*, or at least as close as would be economically feasible. Consequently, the plan (RES, 1983) proposed to make as little use of the canal's water for humans as possible. This was not, in the political context, a welcome proposal, since one of the major claims made of the canal, in terms of its benefits to the local population, was that it would bring water to an otherwise seasonally waterless area. However, there is a further reason for doubting the universal benefit of the canal as a source of domestic supply; the water, if untreated, is not going to be of acceptable quality, by World Health Organisation standards, for human consumption. Of course, people will drink the water in the canal, come what may; but no government or external donor could ever plan to finance a programme that was liable to cause a deterioration in human health. Treatment, however simple in design, is relatively expensive and not always practical.

Nevertheless, the plan produced as part of the Range Ecology Survey acknowledges and incorporates proposals made previously, under contract to the JEO, to abstract water under gravity from the canal (Associated Consultants, 1981). As outlined in Chapter 2, there are three distinct reaches of the canal. The first 40 km southwards from the outfall at Sobat Mouth, will have a level largely controlled by the regime of the Sobat river, and will rise seasonally above surrounding ground level. The next reach, from km 40 to km 310, has a fall whereby the water in the canal will be slightly above the level of the surrounding land particularly where it crosses depressions or watercourses, with a head of between 0.5 and 1.5 m, or rather more than that if the canal were to be operated at 25 million m³ per day. Gravity abstraction is therefore possible within this reach, and the proposal mentioned above was for abstraction points between km 55 and km 193, which were to include simple treatment works for domestic supply, some offtakes for small-scale irrigation schemes and in some cases, the release of water down existing *khor*-beds otherwise denied their natural flow forever by the damming effect of the canal. As mentioned earlier, works for this purpose needed to be undertaken before the banks of the canal were completed and certainly before it comes into operation. This is a prime consideration to be taken into account if construction, including very necessary repairs to the banks so far completed, is resumed.

Conclusions

It is clear from this brief but by no means comprehensive account of some of the development activities in the area, that the JEO and other development agencies had made considerable efforts to meet basic needs in Bor and Kongor Districts, with some but not comparable enterprise in other districts of the Jonglei area.

It is not difficult to see why the JEO, with restricted authority and powers of co-ordination, inadequate staff, and with such immense difficulties of communication and supply, failed to achieve all its targets. It would appear from the records that by 1983 the annual co-ordinating conferences convened by the JEO were just beginning to act as a dynamic and positive forum for the sharing of knowledge and experience. There were at least the signs of a breakthrough in the solution of some of the seemingly intractable problems posed by the physical, and to some extent, the human environment. There was, none the less, a long way to go; and if canal-related development is to proceed successfully at some future date, it must be accompanied by a radical strengthening of the role of JEO, or some successor body, if it is to achieve its stated objectives and those of its parent body, the National Council. This is discussed more fully in Chapter 20.

The final impression which emerges from the experience of this relatively brief period is that development solutions to the particularly difficult circumstances of the environment will require many more years of patient trial and investigation. Patience may be difficult too; the demand for immediate and positive action will be urgent and exacting in circumstances that require rehabilitation measures in a situation far worse than even the most extreme form of drainage scheme could create.

It must be remembered, however, that with the possible exception of rice production and fisheries development, neither the JEO nor other agencies working in the area had by 1983 identified large-scale cost-effective agricultural schemes, or even smaller more traditionally orientated improvements demonstrably superior to systems operative in the finely balanced and often very unstable subsistence economy. The prospects and the challenge are there, but any new programme of development must begin where others were forced to leave off and must be preceded by a thorough examination of the experience gained in this interlude in the violent and damaging history of the area in the last 33 years.

NOTES

1. A complete set of the Bor District Monthly and Annual reports from the 1950s through the early 1960s were transferred from Bor to Juba by the Southern Records Office in 1981 and 1982. They were being kept by the Regional Ministry of Culture and Information in Juba at the time the Regional Government was abolished. As far as is known they are still being kept by the Southern Records Office there.

2. Damage caused by birds is one of the major constraints on crop production in the area. This is an important fact to be borne in mind when planning irrigation projects operative in the dry season. Because of the food and water so provided, *Quelea* and other bird pests may well congregate on irrigation schemes at a time of year when they would have normally left the area.

3. 'BADA is the component of the activities which falls under the Dutch Government's recent "Programmatic Development Approach" '.....in effect a methodology to ensure that donor funds are spent in a development enhancing and poverty alleviating manner. Crucial elements are: flexible funding, long-term commitment, focus on problems instead of once-and-for-all technical interventions, heavy reliance on participation, institutional development and involvement of local people, development of a balance of countervailing powers between government agencies and (rural) people instead of a one-sided strengthening of Government institutions'. (Rölling, 1981).

20

Conclusions: looking to the future

There is an almost total dearth of verifiable information about what is happening in the Jonglei area since the outbreak of the new civil war in 1983. Communications by land and river have been severed. The civil war appears to be centred in the Nilotic region and hostilities more intensive than during the earlier conflict. Food shortages were reported in 1987 to be acute in most areas, and there are severe famine conditions in others. There have been massive movements of people escaping from the vicissitudes of war. Since 1983 there has been a steadily increasing flow of refugees across the Ethiopian border. 110,000 were reported to be in Ilubabor Province by July 1986 and some 6,000 in south-western Ethiopia. The population of Bor town had declined to virtually nothing, whereas Malakal, at the northern end of the Jonglei area, swelled by fugitives from the clashes between government forces and those of the SPLA, had grown to about 80,000, more than double its earlier population. All canal-related development work had ceased by 1984. The southern Sudan had become one of the major disaster areas of the world.

Future social and economic trends

The repercussions of this renewed outbreak of civil disturbance are likely to be infinitely more damaging to the inhabitants of the Jonglei area than any possible adverse effects from the canal, even if the canal brought no counter-balancing benefits. The future impact of the project, if indeed it is ever completed, must be seen in this perspective. It is as yet impossible to assess the consequences of renewed hostilities on the rural economy of the area; so much will depend on the length and continued vehemence of the struggle, as well as the political terms under which rapprochement between north and south is in the end achieved. The longer-term socio-economic implications are equally difficult to predict and will depend upon the nature and extent of rehabilitation measures in a situation where a new start will be necessary and in circumstances even more difficult than in 1972.

It is, however, worth reflecting on the nature of some of the social and economic trends apparent before the present conflict began, and in particular to identify some of the causes of the incipient changes already detectable. As we have seen in Chapter 11, the processes of diversification in the subsistence economy, gradual as they may at first have been, started more than fifty years ago with administrative measures that served to create the beginnings of a cattle market, the growth of the grain trade, at that stage limited wage earning opportunities, and thus the first move towards the evolution of a cash economy. This was followed, after the Second World War, by further diversification in the form of migrant labour, a process stimulated by the growing opportunities for more lucrative wage earnings. This movement was further encouraged by new chances to purchase cattle that could be used to advantage in the bridewealth transactions of marriage. All these interdependent economic processes coupled with some educational advance began to impose certain stresses on the foundations of the social system.

The floods and the intensification of hostilities in the Jonglei area during the 1960s tended to increase the temporary population movement to the northern towns. After the restoration of peace and the retreat of the worst of the floods, migrant labour continued to expand even further, while urban local employment opportunities also grew in Malakal, Juba, and, especially for the southern Dinka, in Bor. Nevertheless, the acquisition of cattle for the purpose of marriage remained an important economic and social target for the male population, and for the younger unmarried men who formed the bulk of the migrant labour force a totally dominant priority, intensified, perhaps, by cattle losses sustained during the height of the floods. Despite the growth in demand for consumer goods and social services and the need for grain, the search for cattle remained paramount. Quite apart from their direct role in the economy as one relatively reliable source of food, the exchange of cattle continued to be the means of achieving both the continuity of the lineage and the image the individual projected for himself in posterity, as well as the means of creating a network of kin relationships that marriage through bridewealth distribution brings and with it a measure of economic security that a high risk environment requires.

Owing to the hazards imposed by the environment, the Jonglei area has nearly always been a net importer of cereals. The disruption of the grain trade by the current struggle in circumstances in which many people are either unable to cultivate at all or are short of labour has meant that present conditions approaching universal famine were inevitable. What are the prospects for the future? An attempt has been made in Chapter 11 to trace the economic and social consequences of the combination of the earlier civil war and floods in the 1960s. There is some evidence that the severity of losses in livestock, movement of large segments of the population, reduction in land suitable for cultivation, and the neglect of dry season water installations created, for a while, food shortages beyond the redistributive capacity of the kinship system. A variety of external influences and experiences, as well as local disasters, had had their effects on this aspect of the social system. These effects differed in degree in different parts of the area, but some contraction of kinship ties and perceptions of kinship obligation were discernible, though perhaps the survival of the system at all was more remarkable than the signs of erosion within it.

How far then were the factors leading to modifications in the network created by marriage, and the emergence of a more individualistic attitude towards cattle acquired by purchase, directly due to natural economic hardships arising from high river levels and severe flooding, or how far due to the consequences of civil disorder? This is a question that must remain unanswered; the processes are complex, the causes interrelated and there are in any event no quantitative data available on which a comparative judgement could be made. It is, however, worth remarking, that these processes of social change appear to have been more evident in those areas more directly exposed to hostilities than those subjected largely or solely to the economic effects of floods. This was the case among the Eastern Jikany who were very directly affected by fighting during the first civil war. Moreover, afterwards, the proximity of the international frontier, and the support given by Ethiopia to the rebels in that area in retaliation for the Sudan's support of Ethiopian dissidents elsewhere, made it virtually impossible for the government either to disarm the people or effectively intervene to prevent feuding. It is here that modifications in the social system have been most marked.

The same processes have not been so evident west of the Nile where people were protected from hostile intrusions by the very floods that reduced their herds and ruined their crops. Less evidence is available from the Nuer of the central part of the Jonglei area, though it seems to be the case that similar modifications to the conventional marriage system, and with them doubtless at least temporary contractions in the kinship network, were forced upon them because for a period they had lost most of their cattle, and had been compelled to turn to fishing and hippo hunting for a livelihood. This applied particularly to those whose homes were on the Zeraf island or close to the Zeraf river. South of them, however, the Dinka – particularly of Bor District – were not only subjected to the damaging encroachment of floods, but, being accessible and close to the main line of communication, also bore the impact of engagements between government and rebel forces and, a decade later, from the raids of the Murle, armed by the government for the purpose of opposing the rebel forces. Here too, especially in the rapidly growing town of Bor, there had been signs of changing attitudes to the main tenets of the kinship system and a tendency to ignore some of the collective responsibilities, more individualistic perceptions of the ownership and disposal of cattle purchased from earnings, and an increase in the incidence of marriages engineered by elopement rather than through conventional and elaborate forms of marriage negotiation.

It would be misleading to attempt to isolate any one factor as bringing about these social changes; they are part of multiple processes and pressures – what Lawry (1982) has referred to as 'endogenous change' – and complex internal and external factors. It does seem likely, however, that the circumstances of total disruption, great distress and near economic collapse engendered by the present war, however much the political cause may be supported by the Nilotic population of the area, may lead to further erosion of a system which is of immense advantage in a subsistence economy so vulnerable to the variable hazards of the physical environment. We do not attempt here to argue in favour of the social system of the Nilotes *per se* or their conventional models of economic relations sustained by the kinship pattern. This is not a matter for external judgement. Change will in any event accelerate, not only as the result and aftermath of the present civil war in which political rather than ecological factors have brought about terrible and devastating conditions of real hunger and all the evils that go with it. It may also accelerate if the policy of 'transformation' through the introduction of large-scale agricultural schemes rather than 'improvement' of the existing economy, as advocated in the more optimistic days when the Sudan was viewed as the potential 'breadbasket of the Middle East', proves a less conjectural target (Garang, 1981). It is, however, fact not conjecture to say that the need for a system which provides a measure of food security will not be less but greater in the circumstances of tension and economic stress that may be expected to continue long after the present civil war is over. The need will persist unless, and until, greater certainty of food sufficiency can be assured. That, apart from measures to avoid any possibility of major decimation of the herds by disease such as have occurred in the past, must rest on new and assured methods of crop production proved to be technically feasible and economically viable.

As we have seen from the previous chapter the solution to the latter problem has so far largely escaped those who were able for a few years to concern themselves with investigation and trials, essential preliminaries to sound development in difficult ecological conditions. Research has now once again been interrupted by war. No alternative solution lies, either, in the processes of rapid urbanisation such as occurred in Bor town and continues to occur for different reasons in Malakal. Beyond offering employment opportunities, which for the educated elements of the population is a growing need, and some entrepreneurial openings for those able to compete with external trading communities which have the advantage of capital, supply networks, and transport, these processes have added nothing to the productivity of the Jonglei area. Continued migrant labour and reliance on remittances in one form or another from a massive but endlessly changing absentee element in the population is no answer either.

Canal operation and management

Weaknesses in the powers and structure of the Jonglei Executive Organ, the handicaps to which it has been subjected, and the lack of co-ordination between departments of government and donor agencies concerned with the development of the area have been mentioned in the previous chapter. Moreover, there have inevitably been very divergent opinions between the two statutory government bodies concerned; the one, the PJTC, charged with the design, construction and downstream water benefits of the canal, as well as with the wider national and international issues of Nile control; the other, the JEO, with the interests of the people of the area through which the canal passes and the development of their resources. The composition of these bodies, with no mutual representation, sharpened these differences which increased as the canal advanced. Too little attention was paid to local anxieties or the execution, or even design, of some of the measures proposed for the alleviation of damage likely to be created by the canal. Relations between the two bodies were far from harmonious, a situation which tended to accentuate differences between north and south at a time when a wholly collaborative approach was essential. If, in the end, the project is completed and put into operation, it follows that a reconstituted and much strengthened Jonglei Executive Organ – something perhaps in the nature of a canal authority will be called for. It is not within the scope of this book to suggest the precise constitution of such a body or the nature of its relationship with other departments of government. There are precedents and parastatal models in other parts of Africa and the world. The range of interests and functions required of it will, however, be apparent from the experience related in previous chapters, for example:

Discharge regulation

This ensures optimal discharges through the canal with the aim of passing the maximum quantity of water within its capacity while minimising damage to the local environment and related interests. Regulation to meet both requirements is not impossible, as was demonstrated in the case of the earlier (Equatorial Nile) project under the 'Revised Operation' (see Chapter 1), though in the present case, with no control upstream, discharges would have to be adjusted annually according to the natural flows of the river. In this connection, as part of their mathematical study of the potential effects of the canal on areas of flooding, Sutcliffe & Parks (1982) experimented with routing different flows in order to study the likely effects of varying discharge operating rules on the areas of floodplain seasonally covered and uncovered. In one test, instead of a steady diversion of 20 million m^3 per day, flows of 15 million m^3 per day were diverted down the theoretical canal in the wet season (between May and October) and 25 million m^3 per day were diverted in the dry season (between November and April). In another test exactly the reverse was simulated, with high flow passed down the canal in the wet season and low flow in the dry. The results of these tests are presented in summary form in Table 16.1. They show that flows can be manipulated without appreciable change in the net benefit downstream of the canal's operation. Most important to local interests is the indication that, if low flows are diverted in the wet season, the area of *toic* lost is half that which would be lost under a steady flow regime. The total annual evaporation from the *Sudd* is approximately proportional to the average flooded area; consequently the seasonal variation in diversions down the canal may be predicted to have little effect on evaporative losses from the *Sudd* and, in turn, on the total water benefit in terms of additional water made available for irrigation purposes downstream.

A simple hydrological model of the kind described is certainly not the perfect and ultimate way of predicting the effects of varying the throughput. But the implications are extremely significant and suggest the need for imaginative operating regulations if the economic interests of the people of the Jonglei area are to be taken into proper account. There are guidelines here for any future canal authority; those who control the sluice gates at Bor will, to a large degree, control the environmental condition of the *Sudd* region.

Maintenance of navigation in the Bahr el Jebel

Disadvantages to navigational requirements in this part of the Jonglei area have been discussed in Chapter 18. Navigational access to river ports such as Adok and Shambe must be assured, as well as the interests of fishermen and the fishing industry, actual and potential. This will require a minimum safe discharge down the natural channels (SES, 1983), and is a requirement which will have to be reconciled – probably with some difficulty – with the seasonal control of discharges suggested above.

Measures to alleviate obstacles to the grazing cycle, and local flooding effects of the canal

These effects have been outlined in Chapter 18. Any canal authority would need to be directly concerned with the design, planning and execution of all measures to facilitate the crossing of the canal, including water supply at crossing points and the control of movements and pasture requirements in their vicinity. It would also need to be responsible for planning and assistance for the resettlement of those displaced by the excavation of the canal itself or by flooding east of it, and measures to maximise the benefits of the latter as well as alleviate disadvantages. Planned settlement of the canal zone, bearing in mind local considerations of tenure and recognised migration routes, will be a major need and a complex task.

Monitoring and research

Apart from local development programmes and related research which will be needed whether the canal is completed or not, many local effects have been predicted but only actual experience of the impact of the canal will reveal their extent. Monitoring of all aspects of the local impact will therefore be a major responsibility for any body concerned with the welfare of the area. This will involve the collection and collation of all data relevant to local economic and social interests. Among other things:

(a) Survey and evaluation of the effect of the canal's operation on areas of river flooding each year, using the predictions outlined in Chapter 16, will be necessary as a point of reference until real data have been accumulated. Apart from hydrological measurements which it is assumed will continue to be available and the re-establishment of any measuring stations which may have been disrupted by the present disturbances, a wider network of reliable rain gauging points are clearly needed for this area. Images also need to be obtained and analysed from the best available satellite based earth resources monitoring system (LANDSAT or its successors) thus indicating the extent of flood areas and seasonal and annual changes.

(b) Demographic data are unreliable though difficult to obtain (see Appendix 6). A comprehensive programme in this field will be required if the effects of the canal on human interests are to be correctly assessed and evaluated. It will also be needed in connection with any resettlement within the Canal Zone if it is to be sensibly and equitably planned. These data would include the standard requirements of a census that must logically precede any form of development or remedial measure, including, for example, family size and composition, education, and household structure with particular reference to relationships between rural and urban communities, migrant wage earners, urban employment and trading. The current lack of quantitative data will be more than apparent from Chapter 11.

(c) Recent economic data, particularly concerning the production and import of grain over sufficient periods to indicate trends, have also been conspicuously lacking in the Jonglei area, though this has been partly due to the relatively short period between 1955 and the present during which research and recording have been feasible. Some indication of market trends in livestock have been discernible, but long-term registers need to be established. Household surveys over a wider range of samples in different parts of the Jonglei area and over greater periods of time are needed. Given the extreme seasonality of Nilotic nutritional levels, variations from year to year and between sections within a community, more research is required in this subject, despite the difficulties of dietary survey among transhumant peoples.

(d) The Range Ecology Survey led to the conclusion that a research programme covering the quality and consumption patterns of water and the effect of new supplies on human settlement distribution as well as the annual agricultural calendar will be needed, not only in relation to general development plans in the area but to the optimum use of the canal as a source of supply without causing undesirable concentrations along its banks (RES, 1983).

(e) Two effects of the canal are likely to be a serious contraction in the amount of dry season grazing and a general, though perhaps marginal, reduction in the movement of people with their livestock across the line of the canal, thus diminishing the way they exploit the seasonally variable grassland resources throughout the year. These effects need careful and continuous review to cover all the basic factors of production outlined in Chapter 12, with special emphasis on comparing these effects on the different performance of Dinka and Nuer cattle, the changing status of the market economy, and possible changes in prevalent livestock diseases, whether or not vaccination and control programmes are successfully implemented. The effects of the canal on the distribution, seasonal migrations and overall numbers of the three species of domestic livestock will need to be monitored, at least every few years, by the system of low level aerial survey already advocated.

(f) Fisheries, a very promising line of development but far from fully exploited, will be the least affected by the operation of the canal. Monitoring effects on the movement and breeding habits of fish will nonetheless be important, as well as on navigation in the smaller river channels used for the collection and transport of fish.

(g) Wildlife will be deeply affected (Chapter 17) particularly because of the difficulties and risks of canal crossings and greater pressures on reduced river-flooded grasslands which they share with domestic livestock. The discharge regulation suggested above will therefore favour wildlife as well as domestic herds. Separate investigations and aerial surveys will be required at three stages: before the canal is commissioned; at the time water is first passed through it, and after it has been in operation for some time – these to determine the effects on each of the major herbivore species.

(h) Crop production will not be affected by the operation of the canal. Indeed, the relief from river flooding west of it will open up larger areas for cultivation. Gravity abstraction, feasible in some sections of the canal, will offer opportunities for at least small-scale dry season irrigation schemes. Agricultural research and experimental programmes (such as those initiated by the JIT and later Ilaco and BADA, but subsequently discontinued) in both crop production and animal husbandry are matters of general development concern in the area, whether the canal is completed or not. If it is, any canal authority must either be directly responsible or empowered to co-ordinate the activities in this regard of other government bodies concerned. Long-term research, experiment and trials are essential in this difficult environment, and the absence of consistent and sustained programmes of this kind have been one of the principal obstacles to development in the Jonglei area.

(i) The effect of the canal on public health also requires constant attention, and, while this is a matter for action by local medical services, a canal authority should be required to monitor any spread of canal related diseases. For example, bilharzia was not prevalent in the Jonglei area up to the 1950s even though the hosts (snails) for the parasites had been identified (JIT, 1954, p. 251). It is now widespread, carried by migrants returning to the area from the north. Not only will the canal improve communications and hence movement from north to south thus increasing this threat, but even small-scale irrigation schemes from it would magnify the problem. Controls, monitoring, medical research and education are therefore matters of great importance and priority.

It will be apparent from previous chapters that research, and indeed measures intended to develop the area in anticipation of the canal, in the period 1972–83 tended to be concentrated on Bor and Kongor Districts. This imbalance needs to be redressed and greater attention will need to be paid to areas along the canal north of these districts (RES, 1983). Moreover, this will have to extend farther to include areas downstream of the outlet of the canal, on the Zeraf island and west and north of the Bahr el Jebel. These latter areas will not share the advantages or disadvantages of the physical presence of the canal but will be deeply affected by its ecological effects on pasture, fisheries resources and probably navigation. All too little is known about effects downstream of the canal outlet. Any future organisation, the JEO or its successor 'authority' will need to focus on these parts of the Jonglei area too.

Such an authority may have many other canal related concerns and interests to promote and protect. It will have a vital role to play, in collaboration with the Permanent Joint Technical Commission, in the optimum control of the Nile flow through the canal and the natural channels, reciprocal representation being highly desirable. It will also be concerned as mediator in disputes arising over economic interests related to the operation of the canal, and with positive development measures to take best advantage of any local benefits it may offer and to offset adverse effects in what is still one of the most deprived areas of Africa.

Further upper White Nile water conservation drainage schemes
Jonglei Stage II

The Jonglei Canal Stage I is in limbo and will remain so until an acceptable formula for peace between north and south becomes more than a pious hope. It therefore seems unlikely that plans for the enlargement of the Jonglei scheme – Stage II – will advance very rapidly. It is, however, necessary to mention in brief the nature of this extension as well as the other canal drainage schemes which are part of ultimate plans to make use of White Nile water, in these latter cases outside the Bahr el Jebel system (see Figure 20.1).

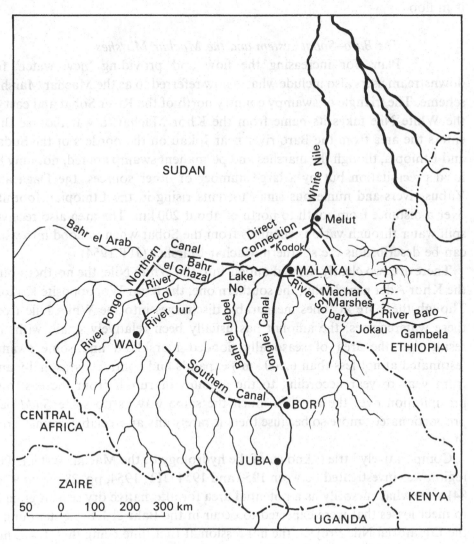

Figure 20.1. Further drainage and conservation schemes.

Jonglei Stage II would entail doubling the present planned capacity – from 25 millions m³ per day to 50 – and a target of approximately double the benefits. This would be achieved by excavating a second canal either close and parallel to or some distance away from the present canal. This second stage would inevitably require control and annual storage upstream, mainly in Lake Albert (Mobutu), and, if the equatorial lakes drop to levels below those that have prevailed during the last 25 years, some measure of additional control and storage in Lake Victoria. Discharges could thus be controlled and timed to reduce seasonal fluctuations in lake outflows and modify backwater effects at the junction of the Sobat with the White Nile when the former river is in flood.

The Baro–Sobat system and the Machar Marshes

Plans for increasing the flow and providing 'new water' for downstream users also include what is now referred to as the Machar Marshes scheme. The triangle of swampy country north of the River Sobat and east of the White Nile takes its name from the Khor Machar, a watercourse that enters the area from the Baro river near Jokau on the borders of the Sudan and Ethiopia, though the marshes and permanent swamp are fed, not only by local precipitation but by a large number of other sources, the Daga and Yabus rivers and numerous small torrents rising in the Ethiopian foothills over a distance from south to north of about 200 km. The area also receives spill water through various channels from the Sobat when in flood into what can be described as the southern Machar Marshes (JIT 1954).

There are two clearly defined outlets into the White Nile; the northern one, the Khor Adar near Melut, the southern one, the Khor Wol opposite Kodok. Though there are at times measurable discharges into the White Nile from these watercourses, the amount has usually been relatively small, with the result that the area of seasonally flooded marshland and some swamp (estimated at not less than 6,500 km²) expands and contracts seasonally and from year to year according to the amount of runoff from the east and precipitation over the area, in much the same way as does the *Sudd* but, proportionately, more so because there is rarely any appreciable drainage out of it.

Comparatively little is known of the hydrology of the Machar system. The region was investigated between 1950 and 1954 (JIT, 1954, pp. 28–9 and 971–84) somewhat cursorily as a potential area for alternative dry season grazing to meet losses that were expected to occur in the Bahr el Jebel system under the Equatorial Nile Project, the impression at that time being that it was not fully exploited by the Nuer inhabitants (the Eastern Jikany) from the south and east, or the Paloich and Dunjol Dinka from the west (see also El Hemry & Eagleson, 1980).

Plans envisage a canal with an off-take downstream of Jokau on the River Baro, just within Sudan territory, and an outlet above Melut on the White Nile, a distance of some 300 km. Annual benefits are estimated at 4.4 km³ – not far short of those predicted on the average for each phase of the Jonglei Project I – but in the absence of detailed investigation this figure can be no more than a very tentative calculation. It may be noted in passing that the alternative might be a dam on the River Baro near Gambela, a project that was mooted earlier and might prove more cost effective since, if feasible at all, a reservoir would be created of 25 km³ capacity. This would avoid the principal disadvantage of the Machar drainage scheme that 'in low years the losses are much reduced and the economy, in water, which can be made is also reduced perhaps to the point of vanishing altogether' (Hurst, *et al.*, 1947: Nile Basin Volume VII).

Drainage of the Bahr el Ghazal system

A similar large-scale drainage scheme is in mind for the Bahr el Ghazal system, which, apart from local precipitation, is fed by numerous rivers and their tributaries rising from the Sudan–Zaïre watershed. Large amounts of water flow into this region – the combined annual mean discharge of the main Bahr el Ghazal tributaries (Lol, Pongo, Jur, Bahr el Arab) is approximately 13.5 km³ – but most of this water never reaches the Nile, being taken up by evaporation and transpiration in the seasonally inundated floodplains and permanent swamps in which many of these rivers terminate. The annual mean discharge of the Bahr el Ghazal at Lake No is not much more than 0.6 km³ and a good deal of that comes from spill from the Bahr el Jebel below Adok.

The Bahr el Ghazal scheme would consist of a canal, 425 km in length, running from the vicinity of Wau, first north and then east, to the Ghazal–Jebel junction at Lake No. Unfortunately the water benefit of this would be much reduced by the backwater effect up the Bahr el Jebel, impeding the flow in the natural channel and causing wasteful spill once more into the swamps on either side. To avoid this an extension canal is considered necessary running in a direct line from Lake No to Melut ('the Direct Connection'), an additional distance of 225 km. The gross annual benefit is expected to be 5.1 km³. Further savings are tentatively suggested by the provision of a 'southern-going' canal, tapping rivers on the eastern side of the Ghazal basin and conveying the water to the head of the Jonglei canal over a distance of about 300 km. Another plan appears to envisage a 'north-going' canal, in this case 840 km in length and joining the White Nile near Malakal (Chan & Eagleson, 1980).

Details of the basic hydrology of these areas and of the engineering aspects have not been published, but costs alone may delay their implementation indefinitely. The sequence is not decided, but the general plan appears to envisage the completion of the Machar and Ghazal schemes before the second stage of the Jonglei Project. This latter, to be fully viable and feasible, would require not only the substantial costs of upstream engineering works, but the successful outcome of delicate political negotiations with Uganda and Zaïre over Lake Albert, and with Tanzania, Kenya, Rwanda and Burundi too, if any modification to levels in Lake Victoria were needed (see Chapter 3).

Internally, little or nothing has yet been done to assess the impact of these schemes on local interests (but see El Hemry & Eagleson, 1980). For the Ghazal and Machar schemes much the same effects as Jonglei Stage I could be expected; massive ecological changes and losses in dry season pastures and fisheries, existing and potential. On the credit side there could be a greater degree of predictability as far as flooding is concerned, but no guarantee against even further reductions in vital local resources in years of low inflow. Navigation westwards into the Bahr el Ghazal basin, or eastwards towards the Ethiopian border if the Machar Canal was the chosen alternative to a dam on the Baro river, might be expected to improve.

Jonglei Stage II, however, could multiply the adverse effects in lost pastoral and fisheries resources to unacceptable levels without any further improvement in counterbalancing assets such as north–south communications. Those local benefits would already have been achieved under Stage I. Moreover, the shattering effects on pastoral management of the need to cross two canals – four or in some cases eight crossings – each year are manifest. There are other unanswered questions related to all three schemes in aggregate; for example a combined additional annual mean discharge estimated at 19.1 km³ at Melut (ARE, EMWP, 1981) could result in disastrous flooding on both sides of the White Nile and for a long way downstream (see also Chapter 16). Protective measures, or alternative livelihood schemes, would be necessary, the feasibility of which has not apparently been examined or the potential additional costs taken into account.

These observations are made in parenthesis to the main subject of this book which does not extend beyond the Jonglei Canal Stage I. It can be argued, though not with much conviction, that the best development of the waters of the Nile must be to use them where they can give the optimum economic return. It has, for example, been suggested that some form of joint collaborative agreement might be reached to utilise Egyptian water quotas on Sudanese land, a combination likely to be less demanding in terms of water, or alternatively to trade water for land (Whittington & Haynes, 1985). Similar arguments could be deployed in support of the use of the waters of the Bahr el Jebel, the Machar Marshes or the Bahr el Ghazal in other parts of the Nile basin, present local usage being vital but scarcely as viable in strictly economic terms. The speculative geopolitics of this form of reasoning is no part of this book; the scale of the damaging environmental impact on the region will be referred to again later.

The local impact of the canal if it is not completed
Throughout this book we have made the tacit assumption that, whatever the causes of the civil war, or the period it will last, the canal will one day be completed. With the growth of population in Egypt and further demands for irrigation expansion in the northern Sudan, more water will in the long run be needed even if economies achieved by the reuse of drainage water in Egypt could meet requirements in the shorter term. Pressures for its completion will mount as the years go by, as they doubtless will for the other schemes described above. The possibility that the canal might not be completed, at any rate in the foreseeable future, cannot, however, be lightly dismissed. Political circumstances may continue to preclude further construction work.

Even if peace and public security are restored, the costs of completion, including the repair of eroded sectors, and the use of other types of machinery should the bucket wheel be permanently immobilised, may well exceed the original outlay on excavation alone. At present, up to 260 km from the Sobat mouth, the abandoned canal is a vast trench which catches water during the rains. It is true that recent reports suggest that the retention capacity of the canal bed has proved less than expected, partly because of the relatively limited precipitation between its banks and partly perhaps because the surface layers of clay may have been ruptured in places during excavation. Nevertheless, this trench and its embankments are obstacles to human and animal movement, while on the eastern side they already restrict the flow of rain water and creeping flow which bank up against it, according to some reports for considerable distances. Cattle have also been lost when trying to negotiate the slopes. Large numbers of wild animals – particularly tiang – have broken limbs or been drowned, especially at the southern end where the banks had not been levelled off before construction ceased and drainage ditches had been constructed in advance of the main earthworks. People have to drive their cattle across the line of the canal along their traditional east–west migration routes, a difficult task if access is impeded by heavy flooding east of it, and hazardous if conditions at the bottom of the canal are muddy. Some may have to take them southwards to skirt the canal at its unfinished end, a movement which may not only cause additional stress to the livestock from travel over long distances and poor grazing *en route*, but cause political friction if, for example, numbers of Gaawar Nuer cattle enter Dinka territory on the way and cause unwelcome pressures on their resources. Flooding east of the canal will already have caused difficulties to wet season occupation of permanent villages close to its banks, but on the credit side there may be a substantial but varying supply of dry season grazing for quite large numbers of livestock.

Given that economic circumstances, even after the restoration of peace in the area, are such that the canal is permanently abandoned as a project, certain relatively modest engineering measures could be taken to modify detrimental effects and maximise potential benefits. For example, earth ramps could be built by excavation across the canal from bank to bank, with drainage works beneath them if that was necessary. Such solid earthen bridges would serve as crossing points at traditional migration routes for the people and their livestock, while catering also for seasonal wildlife movement if they were constructed at relatively short intervals – preferably less than 10 km apart – along the length of the canal. The flooding of permanent habitations could be relieved by bunding at an angle to the canal bank, with points of entry for the water into the canal itself, at the same time increasing the levels of stored water in it. There might be difficulties of abstraction, particularly for domestic purposes, requiring pumped supply and filtration plant. Health could otherwise be endangered; vectors of bilharzia could thrive in such conditions for example, though the risks might be no greater than those found in *hafirs* or natural pools. If the grass species composition in areas seasonally inundated east of the canal improve in the way predicted (see Chapter 16) a more elaborate network of dykes to impound the flood waters could increase this beneficial effect. Erosion of the canal banks will undoubtedly take place, but, if maintained, the eastern bank could still be used as a road route to link with the Kongor–Juba trunk route completed in 1982.

Political perspectives
The construction of the Jonglei canal and development measures for the area around it have become issues in the current confrontation between north and south in the Sudan. The indisputable fact that the canal is designed primarily for the benefit of users downstream of the Southern Region, whatever the coincidental advantages it may bring locally, had for long been a source of anxiety and resentment for the people of the Jonglei area. Lack of consultation at the very outset, the seeming inability of the

National Council for the Development Projects for the Jonglei Area, a body specifically created to allay these fears, to represent local interests forcefully enough, and the growing realisation of the inadequacies of the powers and resources allocated to its Executive Organ were the cause of mounting concern among rural peoples and their educated kinsmen and leaders. To this was added disappointment over the lack of rapid material progress in canal-related development plans. The reality that results on the ground must inevitably be preceded by long-drawn survey, research and trial in so unpromising a physical environment is not an argument that has much appeal for a local population whose expectations have been aroused by repeated claims of approaching material advancement. The fact that the development of the Jonglei area had received relatively more attention in this regard, albeit very unevenly distributed, than other neglected and impoverished parts of the southern Sudan was unfortunately no consolation.

It would be a mistake to suggest that the canal and its effects are central to the present struggle which stems from other fears and grievances, but it has become to the people of the 'South' something of a symbol of what to them is a one-sided exploitation of their territory and resources. To some, too, the improvement in communications, both river and road, which the canal would offer represents a threat, rather than an asset, that may facilitate political penetration and domination by the north and the perpetuation of what they regard as infringements of human rights in a country of varied ethnic origin, culture and religious belief. One of the first actions of the renewed conflict was an attack on a subsidiary construction camp of the engineering contractors, CCI, on 5th November, 1983. A second attack was made on the company's main base near the Sobat Mouth in February, 1984. The company then withdrew and all construction work came to a halt. The very costly and complex bucket wheel excavator (see Plate 6) was abandoned at the point it had then reached. Since then there have been conflicting reports of its fate, but whether or not it has been damaged by SPLA forces or by others, three years of neglect in the southern Sudanese climate will already mean that it will be difficult, immensely costly or even impossible to restore to working order.

From the point of view of the southern Sudan, the initial decision in 1974 to proceed with the present canal project was made without reference to southern representatives. This was only two years after the Addis Ababa Agreement under which the southern provinces were granted a degree of regional autonomy over their internal affairs; following 17 years of total confrontation and war, the circumstances were sensitive to unilateral action of this kind. Confusion also arose between the widespread damaging effects of the earlier canal scheme under the Equatorial Nile Project predicted by the

Jonglei Investigation Team in 1954, as has been explained in Chapter 1, and the comparatively less extensive disadvantages of the new canal. The demonstrations which occurred in Juba in 1974 as a protest against the project were sparked off by this as well as the rumour that plans for the canal included resettlement of some two million Egyptian peasant farmers along the line of the canal. Such settlement plans on that scale, though reported in the Egyptian press, may have been largely imaginary, and in any event most unlikely to succeed, given that financial and other inducements have tended to draw Egyptian agricultural labour into the cities or to oil-rich neighbours. The incentives to tempt them into the desolation of the *Sudd* region are lacking; the probably marginal economic viability of large-scale agricultural irrigation schemes there, even if technically feasible, is hardly likely to generate attractive wages. The pull of the potential oil industry might in the long run prove a greater incentive, though the experience of Middle East oil countries is that migrant labour tends to congregate at the 'downstream' (refining and shipping) end where the jobs are, not at the source.

Nevertheless, the Integration Charter of 1982 between Egypt and the Sudan revived once more latent fears of some form of substantial alien presence in the canal area, since it gave Egyptian citizens the right to buy and occupy land in the Sudan. This raised in the minds of the southerners the vision of Egyptian companies and businessmen purchasing land for agricultural schemes along the canal, thus opening the way, if not for Egyptian labour, for considerable, if indirect, external economic and political influence in the Jonglei area and the Southern Region. It is therefore evident that the cancellation of this Charter, among other treaties and protocols, has been high on the agenda of many southern Sudanese political organisations, notably the SPLM, as well as a renewal of development plans for the Jonglei area, adequate provision for full participation in political and economic decision-making related to those plans, and more particularly in the operation of the canal regime as it affects southern interests. These are strictly political aspirations, and apart from the repeal of *Shari'a* law and total reform of the Sudan's constitution, SPLM objectives include an equitable share in future economic benefits derived from natural resources potential within their territory, the first being oil, a factor unrelated to the canal; the second water which the canal will release.

The discovery of oil in the Southern Region has changed the development prospects of both the country as a whole and the immediate Jonglei area. On the basis of exploration and drilling which had been completed by the end of 1983 it is known that there are substantial oil deposits in the Bentiu (Western Nuer), the Adar area (Maban) and in Jonglei Province itself. There may also be potential for natural gas. The outbreak of civil war has halted further exploration and all plans for oil extraction. The international oil surplus and the dramatic fall in oil prices in 1985–6 has raised further doubts about the time schedule in exploiting the Sudan's oil reserves, and these factors, as much as the re-establishment of peace and stability, will determine when the Sudan's oil will begin to flow. But the oil potential within the Jonglei area or areas adjacent to it is indisputable and is bound to be a crucial factor in the future development of the region.

Oil became a point of contention between the central and regional governments of the Sudan before 1983. The Addis Ababa Agreement of 1972 reserved all oil and mineral rights of the Southern Region to the central government. Subsequently, the regional government was not consulted or involved in the negotiation of concessions with the multinational oil companies, and the potential share in revenue that could accrue to the Southern Region from oil-related activities within its territory was seen to be reduced first by the decision to build the nation's first oil refinery outside it and secondly, by the later decision to pump oil out of the region by pipeline direct to the Red Sea. Between 1980–5 the central government was also involved in proposals to redefine regional boundaries so that the Bentiu and Adar fields would be placed outside southern regional jurisdiction.

It is for these reasons that various organised southern Sudanese political groups are now pressing for a re-examination of the nation's oil policy as part of the resolution to the current civil war. The most powerful group, the SPLM/SPLA, includes many persons from the Jonglei area and has been able to establish its control over large parts of the Southern Region. Since it is their military activity which effectively stopped all oil exploration and extraction, as well as construction of the canal, their role in the renegotiation of the distribution of potential oil revenues may be crucial. The annual budget of the Southern Region as a whole will have a legitimate claim to an equitable portion of oil revenues. What will be available for use in the Jonglei canal area is a matter for internal negotiation. But as there will be no substantial revenues generated in the area by the Jonglei canal itself, the financing of administration, and social services and development in the Jonglei area will most likely come from, or be guaranteed by, future oil revenues.

The demand for some kind of 'water rent' is also part of southern aspirations, and is not without justification since, as we have seen, water, albeit fluctuating and uncontrolled, is crucial to the animal husbandry and fisheries components of the present local economy, as well as being important to longer-term local irrigation potential. This is likely to be an open debate in terms of international riparian rights; the abstraction of water upstream is usually the source of dispute with downstream users, whereas this is the reverse. Moreover, the water largely derives from the East African catchment, not from the *Sudd* itself and brings into the debate the claims of upstream states such as Uganda. Nevertheless, it is the view of an appreciable body of informed opinion that since the additional water transferred downstream will be lost for the purposes of local development, even if such potential is likely to remain untapped at present, it is reasonable that the extraction should be matched by a transfer of funds from Egypt and the Sudan's central budget related to the economic benefits of the increase in irrigated agriculture. It is held that, at a minimum, this rent should be set to cover a yearly sum of funds to be applied to canal area and associated regional development; to monitoring the impact of the canal; projects for the amelioration of harmful effects; and research to identify viable alternative livelihood schemes (Jenness, 1985). Moreover, while the technically feasible and economically viable use of water for irrigation purposes within the Jonglei area has yet to be demonstrated, it is not ruled out forever and a fair quota must at least be notionally reserved for the purpose.

Whatever the manifest needs of the growing Egyptian population, this is not water than can be logically claimed as part of Egypt's 'historic' or 'acquired rights'. Unlike its predecessor of 1929, the Nile Waters Agreement of 1959 between the Republic of the Sudan and the United Arab Republic does not contain any provision, even by inference, safeguarding the interests, in this case of the inhabitants of the Bahr el Jebel, Bahr el Zeraf, Bahr el Ghazal and the Sobat River areas (all expressly cited in the Agreement), if measures are taken to prevent 'losses and increase the yield of the river for use in agricultural expansion in the two Republics'. The fact of the matter is that water evaporated in the *Sudd* region is not a total loss; it has its vital local value in the subsistence economy and has done from time immemorial. It also has potential for future development in the Jonglei area. In this context there are fundamentally different perceptions of riparian rights. For downstream users, water saved from evaporation by major drainage or diversion works financed by them is 'new' or 'additional' water; it is water saved and theirs by right. For those who live in the Jonglei area it is not water lost for, as we have seen, in the process of seasonal inundation of the floodplain valuable economic assets in pasture and fisheries are created, though it is less easy to argue that this natural process is the most economical use of available water resources.

In Chapter 3 a supply and demand balance sheet for the Nile is projected on the basis of hydrological data available at the time. That chapter was written before the recent fall in the overall annual discharge due to low rainfall in the

Ethiopian catchment of the Blue Nile system. Even if evidence of long-term climatic change is as yet inconclusive, for Egypt and the Sudan this unexpected shortfall is a dramatic reminder of the periodic if irregular variations in Nile flows which have occurred in the past. Deficiencies in the Blue Nile will undoubtedly focus attention again on the resources of the White Nile, including, perhaps, once more the storage potential of the great lakes of East Africa, and the key role of a diversion canal through the *Sudd*.

Just as the governments of Egypt and the Sudan must concede that what to them are losses of water by evaporation in the *Sudd* are not necessarily losses to the people who live in its vicinity, so the latter must recognise the dire need for more water of those who live downstream. Arguments over these issues are likely to continue, but within certain limits are capable of equitable solution. It would not appear that in the present tragic confrontation between north and south in the Sudan, the SPLM and other less conspicuous representatives of southern opinion are fundamentally opposed to Stage I of the Jonglei Canal as a project. Resentment stems largely from their exclusion from the decision-making processes leading to the canal's design, alignment and execution and the uncertainty of an adequate say in its future operation.

Solutions to the other projected drainage schemes described very briefly above are likely to be much more difficult. It has been the purpose of this book to survey objectively the whole range of considerations that must be taken into account if the present canal proceeds. By extension those considerations apply with greater force to these other projected schemes since in aggregate they could be catastrophically damaging to the environment of a large proportion of the southern region, and likely to be wholly unacceptable unless unquestionably viable alternative livelihood schemes can be devised. But these are long-term not immediate issues. It must be accepted that in the negotiation of all forms of river control the aim must be towards 'the benefit of *all* the inhabitants of the Nile Valley' (JIT, 1954), even if those objectives are difficult to achieve without substantial compromise.

APPENDICES

Appendix 1

SOILS

Introduction

The soils of the Jonglei area were studied by the JIT (1954, pp. 97–126), and subsequently by South Dakota University (1978), El Hassen *et al.* (1978) and Ilaco (1979, 1981*c*). South Dakota's study was based on satellite imagery; Ilaco's studies were confined to the southern end of the area. All studies were summarised, for the southern part of the area, by Euroconsult (1981).

Structure and classification

The soils of the Jonglei area have developed on recent alluvium. This is a fine-grained and well-weathered material and it is not surprising that the soils are generally clay-rich and poor in nutrients. During the dry season they crack deeply but the first heavy rains cause the soil to swell, sealing the cracks and producing a very impermeable surface on which water stands for long periods. When wet, the soils are slippery and incapable of bearing any load; when dry they are extremely hard and virtually impossible to cultivate.

The soils generally contain much clay and coarse sand, with relatively little silt and fine sand. This bimodal size distribution, in which the clay particles fill the spaces between the sand grains, makes for a very impermeable soil. The uppermost 10–30 cm of the soil tends to be much richer in sand than the deeper layers. The reasons for this are uncertain. One possibility is that alternate wetting and drying of the soil, leading to alternating reduction and oxidation, has led to destruction of clay by the process of ferrolysis

(Brinkman, 1963). Another, perhaps more likely, was originally pointed out by Stigand (1918). He observed how constant trampling by animals and men leads to the separation of clay and sand so that paths through flooded areas quickly become very sandy, as the clay particles are carried away by the water. Euroconsult (1981) found that hollows were richer in clay than hummocks; this is the pattern which would be expected to develop if the process observed by Stigand continued for a long time.

The presence of the superficial sandy layer has led to problems in their classification. The shrinking and swelling of the soils when dried and wetted, as well as their high clay content, are typical of vertisols. The official definition of a vertisol requires that no horizon should have less than 30% clay – and the surface layers often have less clay than this. Such soils would correctly be referred to as alfisols. There is also considerable variation over very short distances, correlated with microtopography, with more-or-less typical vertisols in hollows and alfisols on the hummocks.

Nutrient status

This brief account is based on results published by Ilaco (1979) and on analyses of 17 samples from the botanical study sites in the Nyany area (RES, 1983). The general picture is of soils which are low in most of the essential plant nutrients.

Nitrogen content is low, as is organic carbon; this suggests a low organic matter content. Nitrogen in vegetation is lost each year in fires, although much is probably translocated to the roots early in the dry season.

Phosphorus is very low both in total and in available forms. It is not lost in fires, and could well become concentrated in the cattle camps, where ash from burned dung accumulates.

Potassium is present in moderate amounts and is less likely than either nitrogen or phosphorus to be limiting to plant growth. Calcium and magnesium are also in moderate supply; neither is likely to be limiting, but Euroconsult (1981) have pointed out that low calcium : magnesium ratios, such as pertain, tend to produce soils with poor structural properties. Sodium, which exceeds 10% of the exchangeable cations in some of the subsoils, can produce poorly structured soils.

Minor elements analysed include copper, manganese, zinc and cobalt. All except manganese were on the low side.

Special features

Three soil-related features are characteristic of the region and pose particular problems or possibilities. These are briefly discussed below.

Gilgai

This is an Australian term used to describe a regular pattern of low mounds and depressions occurring on fairly level ground. It is associated with seasonally shrinking and swelling soils, and there is certainly a causal connection between the soil behaviour and the pattern although the mechanisms are not well understood. The difference in level between the tops of the mounds and the bottoms of the depression is often of the order of 50 cm, but this may be less in the darker soils of the eastern plains and more in sites where termites build mounds preferentially on the hummocks and accentuate the pattern. Patterns of large amplitude have been described, with their associated vegetation patterns, in the Bor area (Ilaco, 1981*c*). Aerial survey shows them to be widespread over the study area.

Gilgai patterning is important for several reasons. It makes the levelling of land for irrigation projects very difficult; for crops such as rice, precise levelling is important so that even flooding and even crop development is achieved. The levelling of roads and airstrips is also complicated by the pattern. Furthermore, it is said that the pattern can re-establish itself after levelling if seasonal wetting and drying continues.

Hard tussocks

A feature of many parts of the area, particularly those which are seasonally shallowly flooded, is an abundance of small mounds up to 30 cm in diameter, with almost vertical sides and with grass growing from them. During the dry season they become extremely hard. Essentially these appear to be grass tussocks in which soil has accumulated. They are probably the result of the worm activity. Castings are made preferentially within the tussocks, perhaps because the soil there is better aerated and richer in organic matter. Animals tend not to tread on the tussocks, so that castings produced between them are destroyed, and the soil between is compacted. Although ants and termites often work within the tussocks, they are unlikely to be the primary building agents.

Hard tussocks hamper land clearance, and they also make cross-country driving a slow and damaging experience.

Calcrete

Calcrete (referred to as calcareous nodules by Ilaco (1979*a*)), consists of stone-like nodules occurring in layers in the soil profile, usually at some depth. These nodules are characteristic of soils which are seasonally wetted and dried. In the south of the area they are generally small and not very abundant, but they may be commoner elsewhere and future soil surveys should note their occurrence, as stone is so scarce in the region that they could substitute for aggregate in building.

Appendix 2

CLIMATE

This appendix gives a brief description of the climate of the Jonglei Region, to summarise what is known of its origins and the factors controlling it, and to say something about the effects of the climate on the environment and on the life of the people of the region. It is not based on particularly satisfactory data, for only at two stations, both at the edge of the region (Bor and Malakal), are long and continuing series of rainfall records available; three others, two of them more central (Tonga and Shambe), have ceased recording but provide long series of readings starting in the first decade of this century. Others were interrupted for varying lengths of time, and many did not start until about 1950 and then only lasted for a few years. Temperature records are even scarcer, and records of other parameters must be extrapolated from outside the region.

Rainfall

There is a clear division between the wet season and the dry season. Although rain can fall in any month of the year, showers between December and March are light and are rarely of any significance to plants, animals or men. The main wet season lasts from May to October; April and November are transitional months, often with one or two violent storms and heavy rain of short duration. While it is possible to demonstrate a slight tendency for mean annual rainfall to decrease from south to north (Bor 904 mm; Malakal 789 mm), the fluctuations from one year to the next greatly exceed this.

There is a tendency for a break to occur in the rainy season, in which little rain may fall for three to five weeks. This tendency appears to decrease from south to north. Thus at Bor, the wettest month is separated from the next wettest by an interval of at least one month in 70% of years. At Malakal, such an interval occurs in only 30% of years. This may demonstrate a trend towards a bimodal rainfall pattern, as is expected towards the equator. Such a break in the rains is, because of its unpredictability, often catastrophic for agriculture. It is this unpredictability in the timing of the rain that is a major factor that leads to the apparently paradoxical situation in which people of the Bor area import much sorghum from the lower rainfall areas around Renk, much farther north but with a more reliable rainfall distribution.

Much of the rain that falls during the wet season does so in brief and often very local heavy storms. At Nyany in the 12 months from June 1981 to May 1982, 76% of the rain fell in 25 days in 26 storms. During these storms, the rate of precipitation greatly exceeds the rate at which water can soak into the soil (the infiltration rate). Water accumulates on the surface, and, because infiltration rates are very slow, forms a sheet of water which can flow slowly down the very gentle slope of the land, usually northwestwards. This phenomenon, known as creeping flow, is discussed further in Appendix 3 below.

Origins of the rainfall

The Jonglei area has a tropical summer-rainfall climate in which rainfall is associated with the passage of the Inter-Tropical Convergence Zone (ICTZ). The ICTZ follows the latitude at which the sun is directly overhead at

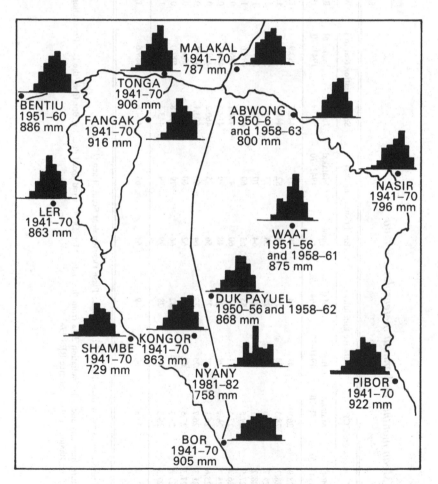

Figure A.2.1. Histograms of monthly rainfall for 14 stations in and around the *Sudd*. Average annual totals are given below the station name, and refer to the indicated period. Where means are based on years between 1950 and 1964, they are likely to be higher than those derived from the period 1941–70.

Table A.2.1 *Mean monthly and annual meteorological data*

Month	Temperature (°C) Bor 1941–70	Temperature (°C) Malakal 1941–70	Relative humidity (%) Bor 1941–70	Relative humidity (%) Malakal 1941–70	Bright sunshine (%) Juba *	Bright sunshine (%) Malakal 1941–70	Solar radiation (cal/m²) Juba *	Solar radiation (cal/m²) Malakal 1941–70	Windspeed (m/s) Bor 1977–81	Windspeed (m/s) Malakal 1950–53	Evaporation (Eo) (mm) Bor 1941–70	Evaporation (Eo) (mm) Malakal 1941–70
January	28.0	26.9	41	24	78	83	442.5	480.9	1.60	3.89	217	192
February	28.7	28.3	40	21	69	82	448.9	527.2	1.61	3.98	190	188
March	29.7	30.5	47	24	58	73	445.7	530.4	1.63	3.35	202	205
April	28.7	31.1	60	36	54	70	432.5	533.7	1.50	2.77	186	192
May	27.5	29.5	68	55	63	60	479.7	495.6	1.17	2.50	183	170
June	26.7	27.5	71	66	60	45	456.1	446.9	0.94	2.46	139	138
July	25.7	26.1	78	74	48	39	412.6	440.1	0.75	2.64	140	152
August	25.6	26.1	77	77	56	45	460.2	469.4	0.82	1.97	140	133
September	25.3	26.7	75	73	64	50	503.1	480.3	0.78	1.74	150	138
October	27.3	27.5	68	66	62	60	474.2	484.0	0.86	1.83	177	152
November	27.8	27.3	57	42	66	79	455.8	490.0	0.90	2.39	189	165
December	27.3	26.5	46	28	79	89	443.5	496.2	1.97	3.98	217	183
MEAN TOTAL	27.4	27.8	60	49	64	65	454.6	489.6	1.18	2.77	2,130	2,008

Note: * Period of record not known, data derived from Pengko Plain Development Study, Ilaco (1981c)

Source: Temperature, humidity and solar radiation data from the Sudan Met. Dept., wind speed data for Bor from the Bor Production Farm and Ilaco (1981c), wind speed data for Malakal from Bhalotra (1964). (Source RES 1983)

midday, but tends to lag behind the sun, so that the wettest month is usually about a month after that in which the sun was overhead at midday.

Recent work (Hills, 1979) suggests strongly that most of the moist rain-bearing air masses that affect the Jonglei area originate in the Indian Ocean. Only in July and August, at low levels, is there evidence of penetration of Atlantic air masses from the west. This conflicts with earlier views about the Jonglei climate (JIT, 1954).

Temperature

The Jonglei climate is typically tropical in that diurnal temperature fluctuations greatly exceed the annual fluctuation in monthly mean temperatures.

At Nyany in 1981–2 the greatest daily ranges (*c.* 20 °C) occurred in the dry season (February) and the lowest (*c.* 10 °C) in the wet season (August). This agrees with earlier accounts (Ireland in Tothill, 1948).

Relative humidity and evaporation

Relative humidity is closely linked to temperature. The lowest humidities occur in the dry season, when the mean monthly RH at 0800 h at Malakal is 23% in February, and the highest during the wet season, when it reaches 88% at 0800 h at Bor in August (Ireland in Tothill, 1948).

Evaporation at Bor and Malakal exceeds precipitation in almost all months of the year. The exceptions are at the height of the wet season and only in August at Malakal is the difference significant.

Sunshine and solar radiation

Figures are only available from Malakal and Juba, the latter outside the study area. In both sites the annual average percentage of the potential sunshine is 64 or 65%, but at Malakal there is a marked peak of over 80% in December–January and a minimum of 39% in July. At Juba no month exceeds 80% or is less than 48%, paralleling the trend, shown above for rainfall distribution, for better-marked seasonality in the north of the area. In both sites the solar radiation peaks do not coincide with the peaks in percentage bright sunshine, probably because of the prevalence of dust (*harmattan*) and smoke hazes at the height of the dry season.

Summary

The general picture of the Jonglei climate, then, is of a markedly seasonal climate with very considerable variations annually. These variations manifest themselves mainly in the rainfall, which varies from year to year, from place-to-place (both from year-to-year and in a single year), and in its distribution through the year. When the vagaries of local climate are seen in conjunction with the variations in the flood regime produced by East African rainfall, the problems posed for both agriculture and for animal husbandry are very large.

Appendix 3

CREEPING FLOW

Overland or sheet flooding – referred to as 'creeping flow' by the Jonglei Investigation Team and in common use ever since – is characteristic of the rain-fed grasslands of the Jonglei area. It was defined as 'the slow movement of large bodies of water across a plain which slopes very gently and is almost impermeable' (JIT, 1947), impermeable that is after saturation of the soil by rainfall in the wet season (see Appendix 2). It is a flood that 'creeps in the sense that it is 'water advancing over a flat plain without any defined channels'. It can come from afar and appear unannounced by any concurrent precipitation of rainfall in the immediate vicinity. It is not a hydrologically or hydraulically unique phenomenon, for it is no more than a manifestation of runoff in particular topographical circumstances, but it is a frequent, though unpredictable, occurrence in the Jonglei area during the rains, does great damage to crops, and causes discomfort to people and livestock. It also washes away roads unless a very high outlay is made in providing adequate culverts. However it has a natural beneficial function in increasing the retention of moisture in the soil after the rains have ceased (see Chapter 7), producing more abundant regrowth after burning of perennial grasses (notably *Hyparrhenia rufa*). Controlled by dykes or bunding it may offer potential for improved pasture or the production of crops. It occurs in the very flat parts of the grassland plains – the Eastern Plain (see Glossary) being a particular example – where distinct drainage channels with clear-cut banks or regular floodplains have not developed, partly because the velocity of the flow is impeded by the density of the vegetation it produces.

It is sometimes claimed that creeping flow is due solely to the incidence of heavy rain falling on these plains. However, given the upper limits of rainfall that is precipitated in that region in the space of a few hours in the circumstances of a moving storm and the relatively rapid evaporation which usually follows a downpour, sheet flooding over a wide area must often in part be augmented by spill from the perennial rivers in the region. The JIT suggested that creeping flow is usually the combined product of rainfall accumulations and spill from the main drainage channels, noting on the one hand that 'overland flooding ... on the Eastern Plain was worst when rainfall was heaviest', and on the other that 'river-spill, or torrents runoff, cannot travel far unless assisted on its way by heavy rain' (JIT, 1954).

The contribution of spill water from the larger rivers and watercourses – the distinction is hard to define except in respect of size of the channel and the period and amount of discharge – is borne out by the presence of fish, which are usually found moving with these creeping floods. While the lungfish (*Protopterus aethiopicus*) is capable of aestivating in a mucus-lined burrow in dry mud, and there are other species that can breath air and thus tolerate foul waters of low oxygen content, other fish require open water to survive. It seems likely therefore that overland floods bearing quantities of fish are at least partly composed of spill waters from perennial sources.

It has also been suggested that some of the flooding in the southern part of the Eastern Plain comes from runoff from the Imatong mountains and adjacent hills. However, east of Pengko the slope is north-easterly (Figure 10.4) and the main tributaries of the Pibor river, the Lotilla, Veveno and below them the Geni, intersect nearly all floods from the south, though at times the flow must be beyond their capacity and therefore continues northwards. Some spill water comes from the Bahr el Jebel in years of high flow via the Atem system, but suggestions that water from this source reaches the Eastern Plain from upstream of Bor are incorrect since the river is there contained within its incised trough (see Chapter 5), with no significant eastern breaches in the form of depressions or watercourses.

Creeping flow is rarely more than a few centimetres deep, though greater depths have been reported. For example, flow depths of 0.50 m were recorded at the Pengko sub-station in 1979, although this ceased altogether after the rains had ended (Ilaco, 1981). There may, however, be some confusion here between creeping flow moving across country, unconfined by noticeable rises or banks on either side, and a shallow yet not wholly unrestricted channel where depths of this magnitude might be expected. Pawson, in his account of floods in the vicinity of Waat in 1947, speaks of one of them advancing about 50 km in the space of two weeks over a front of over 30 km with a 'definite flow of water about a foot deep and with a current strong enough to make a ripple behind a man's leg when he stood in it'. This form of moving flood contributes to, but is clearly different from, virtually static flooding that occurs in deeper depressions, the depth and duration of their waters dependent on the rate of infiltration and subsequent evaporation, and, incidentally, producing different, longer-lasting pasture grasses.

Creeping flow advancing in a northerly direction over the Eastern Plain and down the line of broad longitudinal drainage systems (see JIT, 1954, Figure D 27, which shows these systems, though these are inevitably less distinct than the map suggests) eventually reaches the more distinct channels of the Khors Fullus and Nyanding which drain into the Sobat, and the Khor Atar which joins the White Nile. The flow is here often impeded by the alluvial banks of these watercourses, which are slightly higher than the surrounding land, or by reverse flow up them from the Sobat when that river is high. Severe local flooding can occur in consequence, though except for widely separated pools this area usually dries out rapidly after the rains have ceased and is notoriously lacking in water and palatable grazing in the dry months of the year. It is for this reason that many of the Lou Nuer who occupy this territory are forced to migrate northwards to the Sobat or the lower parts of its southern tributaries or, in the case of some sections, to move westwards to share the *toics* of the Gaawar Nuer or the southern Dinka during the dry season.

The distinction between floods caused by spill from a river into its immediate floodplain (*toic*), runoff moving overland due either to spill from similar sources farther up the slope, or accumulations of rainfall following heavy storms, is therefore hard to make. Much of the Jonglei area is covered with water, visible from the air, at the height of the rains. But the causes of creeping flow have been only intermittently researched and quantitative data are almost entirely lacking. As we have seen, the JIT suggested dual sources from river spill or rainstorms in varying amounts, but spill in one locality is after all due to precipitation farther up the catchment. In the Pengko Plain (part of the southern half of the Eastern Plain) 'rainfall is believed to be the only source of creeping flow' (Ilaco, 1981), an observation which may be true of relatively short-lived floods over limited areas.

The extent of flooding in the wet season over the Eastern Plain has been exemplified rather than precisely delineated in the much reproduced map in the Jonglei Team report (JIT, 1954, Figure D28), but it has rarely been pointed out that this was the outcome of only two reconnaissance flights on successive days in 1951 supplemented by local enquiry. There is no reason to believe that the same degree or location of flooding would occur every year; nor was the map intended to indicate the limits of flooding. Seasonal variations in all forms of flooding are demonstrated in Figure 10.5. Information for these maps was obtained by air survey at three different times of the year, but in different years (1980–1) and at a time when river-flooded areas were vastly much greater than in the 1950s. Again, surveys carried out in other years might reveal a different pattern, but the comparative extent of mid-wet, early and late dry season flooding is clearly demonstrated, even though the sources of water are not indicated.

Creeping flow is yet another unpredictable hazard in an unstable environment as well as a useful though variable source of green regrowth pastures. It is also important in the context of the effects of the canal. Much of the flow will be intercepted, with consequent ponding of water against the eastern canal bank with both beneficial and harmful effects as described in Chapter 18. The harmful effects can be remedied, but since there is no plan to siphon accumulations of rainfall or creeping flow under the canal (except at the Khor Atar), rain-flooded grasslands west of it, which may be expected to increase as seasonally river-flooded *toic* recedes, will be less frequently and deeply flooded and yield less abundant regrowth after the rains.

Creeping flow offers opportunities of controlled use by appropriate banking for supplementary irrigation of rain-grown crops, especially rice, or the production of improved pasture for dry season use. A good deal of thought has been given to water entrapment and harvesting in this way, as well as to the flood protection of crops (JIT, 1954; SDIT, 1955; Ilaco, 1981; McDermott & Ngor, 1984). Trials have not so far been markedly successful; the incidence of creeping flow has been found to vary greatly from place to place and year to year, thus making uncertain the location of schemes likely to yield consistently favourable results. The potential for this form of development (see also Chapter 16) is none the less very much a priority for further and sustained investigation and trial in the future.

Appendix 4

SUDD DIVERSION: EARLIER SCHEMES AND CANAL ALIGNMENTS (see chapter 1)

Garstin's general plans for the control of the upper White Nile first put forward at the beginning of the century were followed by successive proposals with many variations. All required training works through or round the *Sudd*, and with or without Lake Albert Dam as a major engineering component. For the reduction of water losses in the *Sudd*, two alternative methods were constantly in mind: either some kind of diversion channel excavated outside but close to the natural river channels, or well to the east of them; or alternatively banking of the natural channels, or a combination of both canalisation and embanking.

It is difficult to identify the precise dates of the initiation of these various plans, and more particularly their evolution and abandonment, for very nearly all possibilities were under consideration at the same time, with a large number of factors to be considered – technical feasibility, method and costs of construction, relative savings in water and hence cost : benefit.

Garstin's initial idea had been to remodel the banks of the Bahr el Jebel; the first diversion scheme, sometimes referred to as the 'Bor-White Nile Project' was suggested to him later by J. S. Beresford, Inspector General of Irrigation in India. This was to be a canal on a direct line from Bor to the Sobat Mouth, ironically – considering the wealth of other possibilities debated in the interim – the closest to the present Jonglei Canal and its alignment. The principal obstacle at that time was that it would have required dredging from both ends, a slow and costly process. The use of land machinery – drag lines – was not considered.

The next important proposal was the Bor–Zeraf project, which would have consisted of a canal from Bor to a point on the Bahr el Zeraf some 170 km upstream from its mouth. Remodelling of the winding Zeraf channel would have been a costly undertaking, so the next modification was to be a continuation of the canal in a straight line to the White Nile, but with a reduced section and a connecting link with the natural channel of the Zeraf to make use of its spare carrying capacity. There were economic advantages to this plan, since cuts could have been made from the natural channels of both the Bahr el Jebel and the Bahr el Zeraf to give easy access to dredging machinery.

The next variation (1923) was the 'Jonglei Canal Project', the first of many variations, starting near the small village of that name close to the Atem channel. It was found that the Atem would provide sufficient water for the first stage, but the greater discharge required for the next stage, after completion of the Lake Albert Dam, would need an extension of the canal to a point above Bor or alternatively embanking the Atem up to its take-off point from the Bahr el Jebel. The exit would again have been near the Zeraf mouth, the alignment corresponding to what became known as Line VI.

The alternative use of drag lines along a more direct route overland would have been more economical, even with the railway which was to follow along the bank of the canal as it proceeded. But drag lines require dry conditions, and could therefore only be used during the dry season for about four months in the year. Moreover, a lot of the land at the southern end was found to be below river levels, and given high discharges might be under water for long periods, and hence delay construction.

The major alternative to these canal, or part canal, schemes, harking back to Garstin's first ideas, was the 'Remodelling of the Bahr el Jebel'. The main channel was thought to be relatively stable, but an increase in slope and hence velocity caused by straightening the numerous bends would have caused erosion, led to new bends and a return to the natural meanderings of the river unless very expensive protection works were included in the plan. Alternatively the channel would have had to be enlarged along its whole length, involving more excavation work than a canal along the more direct route.

This in turn led to examination of the 'Double Embankment Project', banks on both sides of the river and averaging about 1.7 km apart. Within these limits the natural channel would be allowed to take its own course for about 600 km with its natural bends, the banks being about 450 km in length. The hydraulic predictability of such a scheme was open to doubt, and another variation, the 'Pharaonic Project' was considered. This idea was derived from the method employed by the first Pharaoh Menes, in which the west side of the Nile in Egypt was banked, while the east was left open as a form of flood escape. On the unproven assumption that more water was lost west of the Bahr el Jebel, the scheme involved banking that side from Mongalla to Lake No, with about 100 km of embanking up the Bahr el Ghazal to avoid losses from ponding. Surveys had also revealed a faster rise in ground levels on the eastern side of the Jebel and Zeraf rivers so that although there would be much increase in flooding it could be contained by increased spill into the eastern part of the *Sudd*. This was, in concept, also a flood protection measure – an important consideration until the Albert Dam had been built with adequate capacity to control this contingency. The sequence in mind was therefore: Pharaonic Scheme, Albert Dam, and then by embanking the east side, the Double Embankment Scheme.

Meanwhile attention had been given to the possibilities of diverting water from the Bahr el Jebel into the Pibor–Sobat system. During the exceptionally high flood of 1917–18 spillage of considerable amounts of water east of the Nile and above Bor had been reported. It was therefore thought that a relatively inexpensive connection might be made between the Nile and the Pibor River by way of the River Veveno and thence to take advantage of the wide bed of the Sobat, which in the dry season – the timely period – is virtually empty. However, subsequent survey work revealed that the intervening plain, about 80 km across, was over a metre above the highest recorded river levels in the Bahr el Jebel. No such connection in 1918 could have occurred. Moreover the Veveno was found to have no well defined channel in its upper reaches and to be too tortuous in its lower reach to be much use as a channel. A canal would therefore have to be very deep to carry water through, and a barrage on the Bahr el Jebel would be required. Such a canal would be some 220 km long from Gemmeiza to Pibor Post; and since the carrying capacity of the Pibor proved to be much less than supposed, a relief channel from Akobo about 130 km long would also be required to carry the water to the Sobat. The saving would average about 2 km^3 in the 'Timely' season. The cost : benefit was clearly unfavourable and the idea was dropped in 1932.

At no time during the debate on these various schemes, with many modifications to each, except after the inconvenient intervention of the Sudan Government in the late twenties, were the local effects or the impact on the interests of the inhabitants of the area considered at all.

Appendix 5

AERIAL SURVEY TECHNIQUES (see Introduction)

During the Range Ecology Survey, three low altitude aerial surveys of the study area were carried out. The dates were November–December 1979 (early dry season), late March 1980 (late dry season), and early September 1981 (mid-wet season). The surveys used the Systematic Reconnaissance Flight method, described in detail by Norton-Griffiths (1981).

A grid system was superimposed on to the 67,900 km^2 of the area, each cell of the grid being 10 × 10 km. East–west lines were flown through the centre of each of these squares, in a four-seater aircraft, at a height of 300 ft and a speed of approximately 160 km per hour. An accurate radio navigation system – an essential aid in the featureless terrain – ensured that the aircraft remained within 0.5 km of a predetermined track, and a radar altimeter allowed very precise height control. Flying was done in the morning to take advantage of the better weather at this time of day, but at the altitude used (300 ft) cloud was not a problem.

Two rear-seat observers each counted all the animals and other objects that fell within a band delimited by rods attached to aircraft's wing struts. If the aircraft remains at a constant height – not difficult in the level terrain of the area – the strip counted retains a constant width on the ground. In these surveys the strip width was 145 m on each side of the aircraft, so that just under 3% of each 10 × 10 km grid cell was surveyed. Large herds of cattle and other animals were photographed, and observer estimates checked later by reference to the photographs.

The front-seat observer recorded various attributes of the physical environment such as vegetation type, land use, surface water, extent of burning, and grass height and greenness. Records were made within each 10 km grid unit, and also within minute-long sub-units within these. He also took oblique colour photographs (about 1,500 on each survey), which were used in the preparation of vegetation maps. Once during each sub-unit, he also recorded aircraft height so that any necessary correction resulting from variations in altitude could be applied to the strip counts.

The minute-long sub-units were then analysed statistically to give estimates of total numbers of animals and other objects, with standard errors (a measure of precision of the estimates). The sub-units were also divided into various groups (strata) on the basis of criteria such as vegetation type, and estimates of numbers with standard errors made on the stratified samples. Descriptions of the statistical methods used, and the theory underlying them, can be found in Jolly (1969), and Sokal & Rolf (1973).

Appendix 6 (see Introduction to PART III)

HUMAN POPULATION

Neither *de facto* (a record of actual physical presence of people on a given day) nor *de jure* (according to usual place of residence) census methods are suited to highly mobile transhumant societies such as the Dinka and Nuer, though the latter method can be tested by sampling.

Figures for population according to tribal groups within districts were estimated in 1953 by the Jonglei Investigation Team as a matter of expediency by the then standard method of applying a multiplier, in this case 4.5, to names recorded on lists of taxpayers drawn up by the administration. Manifestly such a method is open to a high degree of error and figures were published with all the reservations which were apparent at the time (JIT, 1954, pp. 228–31), but they were at least indicative of the relative size of tribal groups. A further estimate was made by the Southern Development Investigation Team a year later, based on rather more up-to-date tax-payer figures and interim information supplied by the Department of Statistics who had by then begun the National Census (completed in 1956). A multiplier of 5 was used in this instance, though pilot census samples elsewhere were already beginning to indicate that an even higher multiplier might be more accurate. Comparison between these and earlier population figures (e.g. 1930) to indicate population trends (see Kelly, 1985) is likely to be very misleading.

It should be noted that the total area which would have been affected by the Equatorial Nile Project as calculated in 1954 was much larger than that likely to be affected by the present canal. The National Census of 1956 was conducted on a more professional basis, but even so figures for highly mobile Nilotic communities are open to considerable doubt. There is no reason to believe that subsequent estimates have been any more accurate. Various figures are given below and are quoted direct from the Range Ecology Survey report (RES, 1983) except that the JIT estimates have been adjusted to correspond more closely with tribal groups living within the Districts cited, and SDIT estimates have been included as well as figures from the National Census of 1956 as they apply to the area.

Various figures are given below (Table A.6.1) without comment, except to say that Ilaco tested their calculations by *de facto* methods on a sample basis and their figures could be more accurate than the rest. However, their various samples each prompted the use of a different multiplier which has inevitably led to considerable differences between their 1975, 1977 and 1979 figures. Unfortunately their work was confined to Bor District.

Human population figures for the 'Jonglei Area' (see Figure 10.1) include only those people east of the Bahr el Jebel and south of the Sobat, and incorporate the Lou Nuer, only about one-third of whom will be directly affected by the canal. Cattle population figures west of the Nile (i.e. those cattle that are driven to the Nile *toic* in the dry season) are given in Appendix 8, but rather illogically neither the JEO nor any other agency recently operating within the Jonglei Area present human population figures for the owners of those cattle. These, using SDIT (1955) figures and including the Ruweng Dinka just north of Lake No, amount to 178,025, though many of them herd their cattle along inland water courses and around lakes outside the Jonglei area. Using the same source, the Shilluk number 100,620, and the Dunjol, Paloic and Abialang Dinka (north of Malakal) 32,915.

It would be misleading to interpret the increase in total figures given in recent years over those recorded in the 1950s as indicating a growth in human population. Indeed, there is some reason to believe that the population has declined between 1955 and 1983 owing to epidemic diseases and mortality during the civil war and the high floods of the 1960s. It needs to be stressed that no population statistics or estimates at any period are of sufficient accuracy to indicate trends.

Table A.6.1

Source	Census year	District (present boundaries)					Total
		Bor	Kongor	Ayod	Fangak	Waat (Lou)	
JIT, 1954	1952	58,139	65,453	36,040	65,038	(67,275)	291,945
SDIT, 1955	1954	67,905	77,300	47,135	73,725	(74,750)	340,815
National Census	1956	– 130,620 –		– 108,331 –		(103,638)	342,589
JEO, 1976a	1955/6 & 1973	48,620	81,000	– 97,948 –		(53,638)	281,206
Ilaco, 1975	1973	49,700 ⎫	Excluding Bor town				
Ilaco, 1977	1976	57,000 ⎬					
Ilaco, 1979b	1979	44,990 ⎭					
World Fertility Survey	1973/79	95,127	92,895	58,925	66,697	(68,404)	382,048
JEO, 1983	1979/81	92,044	91,112	56,757	66,634	(51,889)	358,436

Appendix 7

BRIDEWEALTH: THE DISTRIBUTION OF CATTLE IN MARRIAGE (see Chapter 10)

Marriage is not a single event, but a series of negotiations, each followed by the transfer of cattle and accompanied by ceremonies, ritual and sacrifice, often taking several years before a union is secure. Although many gifts, usually sheep, goats, tobacco, grain, spears and cloth, and more recently imported commodities, are made by the relatives of the bride and bridegroom to each other, it is the transfer of cattle in bridewealth that is essential, ensuring the legal status of the relationship and securing a basis for good and close relations between the two groups of kin. Marriage is more than simply the creation of a relationship between two people. Kin from the families of both individuals are intimately involved, those on the groom's side providing cattle and those on the bride's side receiving them; the bride's relatives expect to get their shares of bridewealth from people who stand in an equivalent relationship to the groom as they themselves stand to the bride. Marriage thus involves the formation of a series of new social ties involving many different people. Rules of exogamy serve to spread the web of relationship widely, bringing previously unrelated people into relations of affinity. As time passes and children are born to the union these links become stronger.

Cattle are claimed in marriage by the bride's kin according to convention-ally recognised rights. Among the Nuer, for example, on the bride's father's side, her father, her brothers and half-brothers, her father's brothers and sisters, and in most cases her father's father, father's mother and father's mother's brother all have the right to claim cattle from the family of the groom. Similar rights are claimed by persons standing in equivalent relation-ships to the bride on her mother's side. If one of these relatives is dead, which is often the case, the right is claimed by his or her heir. Nilotes have models which can be quoted to indicate both the conventional pattern of distribution and the composition in age and sex of the cattle in each portion claimed by individuals standing in a particular relationship to the bride. It is important to stress, however, that such quotations are perceived as a point of reference from which marriage negotiations can proceed rather than immutable precepts.

Marriage negotiations, 'the invocation of cattle', are a trial of wits for both parties, requiring persistence, knowledge of the state of the herds belonging to the groom's relatives, usually in detail, skill in citing precedent, and in the case of the groom's kin, a great deal of patience. Nuer sometimes speak of marriage settlements as a battle, *kur*, while among the Southern Dinka one observer has described them as 'a strange type of combat' (Kadouf, 1977, p.

77). Settlements reached over the number of cattle acceptable to the bride's family take into account the current state of the herds, for they may have been much reduced by disease, and the perception of what the model suggests they may reasonably expect, though – again as a point of reference rather than an inflexible rule – there has usually been a recognised minimum claim in cattle which must be met if the legality of a union is to be established.

While there may be a flexible minimum there is, however, no upper limit, and among the Dinka in particular the social and political importance of the bride's father is a factor which tends to increase the amount of bridewealth demanded well above the average or the numbers normally quoted as the model, especially if, as is often the case, there are several contenders for the girl's hand. Deferred payments are, within certain limits, acceptable to the bride's immediate family, especially if promises to pay are supported by prospects of cattle being acquired by the groom's family on the marriage of one of his sisters. Indeed, Nuer regard the cattle transferred to the bride's relatives on her mother's side to some extent as deferred payment standing over from the marriage of her mother (Evans-Pritchard, 1951, p. 78).

Although the principles and objectives of the bridewealth system are the same for Dinka and Nuer, there are many differences in detail from area to area. The extent of marriage relationships and the notions of reciprocity involved are, for example, highlighted by the Western Dinka practice of 'reverse' transfers (*arueth*) of cattle on a considerable scale made by the wife's kinsmen to those of her husband when the marriage is regarded as fulfilled and likely to remain stable, though this is not customary among Dinka or Nuer of the Jonglei area.

The extent to which a single marriage creates a web of kinship among several lineages is illustrated in the central Nuer example below:

Zeraf and Western Nuer Versions

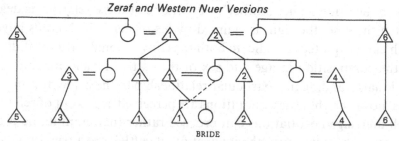

BRIDE

Lineage 1 = Bride's father and agnatic kinsmen.
Lineage 2 = Bride's mother's father and agnatic kinsmen.
Lineage 3 = Bride's paternal aunt's husband and agnatic kinsmen.
Lineage 4 = Bride's maternal aunt's husband and agnatic kinsmen.
Lineage 5 = Bride's father's maternal uncle and agnatic kinsmen.
Lineage 6 = Bride's mother's maternal uncle and agnatic kinsmen.

(From: Howell, 1954)

If one examines this diagram and realises that similar networks of relationship are also created, for example, on the marriage of all the bride or bridegroom's brothers and sisters, to say nothing of their less immediate kin, the extent to which reciprocal social and economic obligations are formed and symbolised by the transfer of cattle will be appreciated. The bridewealth system results not only in the circulation of cattle, a vital economic resource making bridewealth transfer a massive process of economic distribution around the region, but establishes relationships which, as we have seen, in times of hardship in one area allow people to benefit from the assistance of those better provided in another.

Appendix 8

LIVESTOCK POPULATIONS (see Chapter 12)

Populations

For total livestock populations in 1982 within the Jonglei Area, as delineated by the Range Ecology Survey, see Table 12.1, p. 281.

Table A. 8.1 *1982 Livestock populations by district, Jonglei Province*

District		mid-wet	early dry	late dry
Ayod	Cattle	81,896	89,164	83,843
(12,446 km²)	Sheep/goat	10,426	15,956	15,508
Waat	Cattle	71,239	47,056	99,272
(5,074 km²)	Sheep/goat	16,341	4,265	22,895
Fangak	Cattle	44,010	34,204	65,223
(14,060 km²)	Sheep/goat	12,701	11,068	24,077
Bor	Cattle	36,980	61,513	63,333
(5,474 km²)	Sheep/goat	10,920	19,785	28,101
Kongor	Cattle	103,429	118,115	249,030
(15,712 km²)	Sheep/goat	13,896	22,902	29,109
Total	Cattle	337,554	350,052	560,701
	Sheep/goat	64,284	73,976	119,690

Source: RES, 1983

Sales

Cattle sales figures from within the Jonglei Area are only available from Bor and Kongor auctions.

Table A. 8.2. *Categories and total numbers of cattle sold in Bor 1974–9*

	Bulls		Cows		Calves		
Year	Number sold	% of total	Number sold	% of total	Number sold	% of total	Total
1974	1,166	44	1,037	39	454	17	2,657
1975	2,322	61	778	21	680	18	3,780
1976	2,700	58	828	18	1,152	24	4,680
1977	2,646	56	783	17	1,323	27	4,752
1978	2,214	52	729	17	1,269	31	4,212
1979	3,503	59	1,453	25	971	16	5,927

Source: Ilaco, 1981

It is more difficult to give precise sales figures for Kongor. During 115 days of sales over the period 1980–2, 755 cattle were sold (RES, 1983). By extrapolation, which does not take into account seasonal fluctuations in sales and changes between the three years, the figures arrived at from the Range Ecology Survey total 1,560 cattle sold per annum. Calculating small stock sales (total 329) in the same manner, approximately 624 sheep and goats were sold over the same period. During 1980 and 1981 sheep and goats were tabulated separately in the auction records, and over this period goats represented 96.9% of small stock sales. Between May 1982 and April 1983 3,141 cattle, 1,926 goats and 146 sheep were sold at Kongor (McDermott, J., 1984, p. 10). Very few cattle sales are conducted in the cattle camp because the official auction receipt is security against accusations of cattle theft, which may follow an informal economic transaction.

Herd structure

Given the drastic impact of flooding and civil war on the herds and flocks of the area, it is instructive to consider the various figures for herd age and sex structure, in order to assess the extent of recovery through the 1970s and early 1980s. Unfortunately the statistical data collected are mainly restricted to the southern, Dinka, parts of the Jonglei Area, and many places severely affected by flooding, Zeraf island for example, have not been visited by researchers.

The ratio between newborn males and females in cattle is approximately one to one. However, most pastoralists will maintain a much greater proportion of female animals in their herds, disposing of male animals, save those used in breeding, in sacrifice, slaughter, exchange and sale. There is considerable variation in the literature concerning estimates of sex distribution in average or typical herds. For example, female animals comprised 67.6% of herds studies by the Hunting team in the Western Sudan in 1973 (Hunting Report, 1974, p. 20). Elsewhere Demiruren (1974) gives an estimate of 75% females as the average for African nomads, whilst Dyson-Hudson N. and R. (1970) report that amongst the Karamojong the proportions range from 49% to 85% (Dahl & Hjort, 1976, p. 31). Amongst the Nilotes of the Jonglei area the sex distribution recorded in populations was as in Table A.8.3.

These figures seem to be within the range of what is regarded as normal for African pastoralists, with the possible exception of Nyany, which seems to have contained a low proportion of males.

Dahl & Hjort (1976, p. 44), generalising from a wide variety of studies of pastoralists in the Sahel and East Africa, give a 'tentative' female age distribution in nomadic cattle herds:

Table A. 8.3.

	Duk Padiet	Kongor	Baidit	Bor	Nyany	Nuer*
	1976	1977	1979	1980	1982	1980
Sample size	1,837	451	720		1,277	762
Females	73.3%	74.3%	73.6%	72.6%	78.8%	66.0%
Males	26.7%	25.7%	26.4%	27.5%	21.2%	34.0%

Notes: *Nuer herds were studied near Ayod and Woi.

Source: RES, 1983.

Table A. 8.4.

Calves	0–7 months	13.5%
Heifers	7 months–3 years 6 months	25.5%
Cows	over 3 years 6 months	61.0%

Dinka and Nuer female populations in the Jonglei area work out at:

Table A. 8.5. *Duk Padiet 1976 (Dinka)*

	Observed	Predicted	Chi2
Calves	218	181.71	7.25
Heifers	230	343.23	37.35
Milking cows	517		
Dry cows	381	821.06*	7.21
Total	1,346		51.81

Note: * All cows
Source: Calculated from Payne & El Amin 1977, p. 70.

These proportions are for populations, rather than for individual herds. The data are difficult to interpret, being fairly crude – for one thing the criteria used to define the different age categories vary from report to report. Despite the lack of precision, as well as the broad span of years in which the various studies were carried out, it is interesting that two populations, Param and

Table A. 8.6. *Param (nr. Kongor) 1977 (Dinka)*

Calves	56	45.23	2.57
Heifers	132	85.43	25.39
Milking cows	91		
Dry cows	56	204.35*	16.09
Total	335		44.05

Note: * All cows
Source: Calculated from Payne & el Amin 1977, p. 71.

Table A. 8.7. *Baidit 1979 (Dinka)*

Calves <1 year	125	71.55	39.93
Heifers	116	135.15	2.71
Cows	289	323.30	3.64
Total	530		46.28

Source: Calculated from RES 1983, p. 45.

Table A. 8.8. *Ayod and Woi 1980/81 (Nuer)*

Calves <7 months	52	68.04	3.78
Calves 7 m – 2 yrs	50		
Heifers >2 years	72	128.52*	0.33
Cows	330	307.44	1.66
Total	504		5.77

Note: * Calves 7 m–2 yrs and heifers >2 years
Source: Calculated from RES, 1983, p. 102.

Nyany (Tables A.8.6 and A.8.9), contained less cows than might have been expected, perhaps indicative of earlier difficulties (but whether these were the effects of flooding or of localised epidemic is unknown). In addition, the Nyany herd contained a higher proportion of females to males than any of the other samples. Only one population, Duk Padiet in 1976 (Table A.8.5), appears to have been in decline. The Nuer populations from Ayod and Woi (Table A.8.8) were consistent with Dahl and Hjort's generalised 'normal' female herd to a significant extent. The age criteria used in this sample appears to match that used by Dahl and Hjort reasonably closely, and so one can suggest that this population has a normal, in a real, as well as in a statistical sense, age distribution. This implies that in these two Nuer areas, at least, the long-term effects of the mid-1960s flooding have not been as drastic as might have been feared. The figures do not tell. however, whether or not herds were of sufficient size to satisfy people's social and economic needs.

Table A. 8.9. *Nyany 1982 (Dinka)*

Calves <7 months	65	135.95	37.03
Calves 7 m–2 yrs	157		
Heifers >2 years	425	256.79*	110.19
Cows	517	614.27	15.40
Total	1,007		167.62

Note: * Calves 7 m–2 yrs and heifers >2 years
Source: Calculated from RES 1983

Appendix 9

THE RELATIONSHIP BETWEEN SWAMP INFLOWS AND OUTFLOWS (see Chapter 16)

In order to model the hydrological behaviour of the *Sudd* with the Jonglei Canal in operation, relations between inflows measured at Mongalla and outflows deduced by subtracting Sobat flows from flows measured at Malakal, are required. The results of regression of monthly and quarterly outflows on inflows with various lags are given in Table A.9.1. The regressions were carried out in both linear form $q_t = aQ_{t-3} + b$ and logarithmic form ln $q_t = c$ ln $Q_{t-3} + d$ where ln is natural logarithm.
The equation selected is

$$\ln q_t = 3.928 + 0.4110 \ln Q_{t-3} \tag{1}$$

which is equivalent to

$$q_t = 50.8(Q_{t-3})^{0.4110} \tag{2}$$

where inflows and outflows are in $m^3 \times 10^6$/month. This may be tabulated in $m^3 \times 10^6$/month and in $m^3 \times 10^6$/day (Table A.9.2.):
This equation implies that outflow exceeds inflow at low flows, and as this is an unrealistic extrapolation another equation is required for prediction of low flows but cannot be derived directly as it lies below the range of measurements. Accepting that spilling is responsible for transmission losses between inflow and outflow, a reasonable relation for steady conditions would pass through the origin at 45°, and the simplest equation with this property and a realistic shape is

$$q = Q + cQ^2 \tag{3}$$

and only one value of c ensures that this curve meets

$$\ln q = 3.928 + 0.4110 \ln Q \tag{4}$$

without a discontinuity of gradient. The curve

$$q = Q - 0.0002144Q^2 \tag{5}$$

joins the regression equation in this way at $Q = 1,729$, $q = 1,088$ and provides the values tabulated (Table A.9.3). The joint curves provide a reasonable fit to the lower range of measurements.

Because the Canal offtake will be at Bor, the relationship between flows at Mongalla and at Bor latitude was needed; this was derived by multiplying the intervening flood area, which was related to Mongalla flow after detailed survey (JIT, 1954; Table 192), by average evaporation less rainfall to give the loss. This provides the relationship between flows at Mongalla (Q_M) and Bor (Q_B)

Table A. 9.1. *Regression analysis for Sudd inflows (Q) and outflows (q) (1916–72 omitting 1963–6)*

No var	Name	Coeff	seb	t	R²	see	Const.
Linear – monthly:							
1	S	0.031	0.001	30.88	60.2	110	896
1	Q_t	0.1741	0.0077	22.54	44.6	201	813
1	Q_{t-1}	0.1817	0.0075	24.24	48.2	194	795
1	Q_{t-2}	0.1927	0.0071	27.05	53.7	184	769
1	Q_{t-3}	0.1976	0.0070	28.25	55.9	180	758
1	Q_{t-4}	0.1992	0.0070	28.41	56.1	179	755
1	Q_{t-5}	0.2022	0.0069	29.16	57.4	176	748
Logarithmic – monthly:							
1	S	0.2806	0.0095	29.67	58.3	0.127	4.530
1	Q_t	0.3286	0.0156	21.01	41.2	0.151	4.563
1	Q_{t-1}	0.3449	0.0152	22.76	45.1	0.146	4.437
1	Q_{t-2}	0.3716	0.0142	26.12	52.0	0.136	4.232
1	Q_{t-3}	0.3837	0.0138	27.76	55.0	0.132	4.139
1	Q_{t-4}	0.3880	0.0138	28.26	55.9	0.131	4.105
1	Q_{t-5}	0.3909	0.0137	28.55	56.4	0.130	4.084
2	S	0.1952	0.0231	8.47	59.5	0.125	4.298
	Q_{t-3}	0.1312	0.0326	4.03			
Linear – 3-monthly means:							
1	S	0.0313	0.0015	20.35	66.3	151	889
1	Q_t	0.182	0.0127	14.33	49.3	185	793
1	Q_{t-3}	0.212	0.0107	19.76	65.0	154	722
2	S	0.0185	0.0042	4.40	67.8	147	804
	Q_{t-3}	0.0939	0.0288	3.26			
Logarithmic – 3-monthly means:							
1	S	0.287	0.0147	19.52	64.4	0.112	4.472
1	Q_t	0.3407	0.0259	13.15	45.0	0.140	4.470
1	Q_{t-3}	0.4110	0.0211	19.47	64.3	0.112	3.928
2	S	0.1556	0.0410	3.80	66.4	0.109	4.123
	Q_{t-3}	0.2010	0.0590	3.41			

Notes: S is storage; q_t is outflow in month t; Q_t is inflow in month t.

Coeff is the regression coefficient b, in an equation of the form $q = bQ + c$; *seb* is the standard error of estimate of b; t is the Student's t statistic; R^2 is the coefficient of determination; *see* is the standard error of estimate of the relationship; Const. is the intercept c.

$$\ln Q_B = 0.9854 \ln Q_M + 0.096863 \tag{6}$$

which may be inserted in the relationship between Mongalla inflow and swamp outflow

$$\ln q_t = 3.928 + 0.411 \ln Q_{M,t-3} \tag{7}$$

to give the equivalent relation between Bor inflow (lagged by three months) and outflow

$$\ln q_t = 3.888 + 0.4171 \ln Q_{B,t-3} \tag{8}$$

The parabola which passes through the origin at 45° and joins this equation without discontinuity of gradient is

Table A. 9.2.

Q m³ × 10⁶/month	q	Q m³ × 10⁶/day	q
500	653	16.7	21.8
1,000	869	33.3	29.0
1,500	1,026	50	34.2
2,000	1,155	66.7	38.5
2,500	1,266	83.3	42.2
3,000	1,365	100	45.5
4,000	1,536	133.3	51.2
5,000	1,683	166.7	56.1

Table A. 9.3.

Q (m³ × 10⁶/month)	q	Q (m³ × 10⁶/day)	q
0	0	0	0
200	191	6.7	6.4
400	366	13.3	12.2
600	523	20	17.4
800	663	26.7	22.1
1,000	786	33.3	26.2
1,200	891	40	29.7
1,400	980	46.7	32.7
1,600	1,051	53.3	35.0

$$q = Q_B - 0.0002127Q_B^2 \text{ for } Q_B < 1732 \tag{9}$$

Equation (6) may be used to predict flows past Bor from Mongalla flows, and Equations (8) and (9) can be used to predict swamp outflows from Bor flows.

Appendix 10

POSSIBILITIES FOR FORESTRY

The shortage of wood for building has been highlighted in Chapter 8. What are the possibilities for improving the situation? There are three: better use of the local species; planting within the area; and increased planting outside the area with a view to using the improved transport facilities provided by the canal. Let us examine these in turn.

Better use of local species

The local species most used for construction are *Acacia seyal* and *Balanites aegyptiaca* (both trees); in the south of the area, poles of *Zizyphus* spp. and *Piliostigma thonningii* (small trees or shrubs) are sought after for the main frames, and the stiff, hard, twiggy branches of *Securinega virosa* (a shrub) are used for infilling the frame prior to applying the mud plaster which forms the bulk of the wall thickness. *Balanites* and *Acacia seyal* are particularly resistant to seasonal waterlogging, and therefore form the bulk of the woody vegetation within the area – neither produces a particularly strong or termite-resistant timber, and they are probably used mainly because there is nothing better.

Balanites has a long tap root and is said to be difficult to transplant, being best sown at stake. Protection from fire and grazing would be needed for the young plants and establishment of plantations of any size would certainly be very difficult. On the credit side, a carefully managed plantation would produce a crop of fruits as well as timber, and also some fodder from the fallen leaves and fruits.

Acacia seyal has been used in the past as a timber and fuelwood tree. The Forestry Department (Sudan Government, 1953) has recommended clear-felling when stands are full sized but not over-mature, leaving a few trees to act as seed bearers. After felling, the area should be burned as early as possible; this is said to promote regeneration of *Acacia seyal* and *Zizyphus*. Clearance of very mature stands tends to produce grassland. They also suggested ways of extending the life of *Acacia seyal* poles, which are normally quickly destroyed by the wood-boring beetle *Sinoxylon senegalense*. Recommendations were felling, barking, and then soaking in water for 14 days. After this the poles were to be dried in air for 28 days and then immersed in creosote, heated to 230 °F (110 °C) and allowed to cool in the liquid for 18 hours. At a simpler level, they point out that the borer only attacks on the underside of logs, so that stacking vertically produces less attack, and removing the bark discourages it further.

Plantations within the area

The major limiting factors to the growth of trees in the area are seasonal waterlogging, flooding, and the cracking clay soils. Only a few species can withstand these conditions. The alternative is to look for better-drained sites, but these are, by and large, already being exploited by the local people for cultivation and dwelling sites.

Suitable planting sites might be found on the Duk Ridge and in the Zeraf Jebels area, where there are higher 'islands' of better-drained soil which might provide a planting site for introduced species such as neem (*Azadirachta indica*) – which does well at Bor – and cassia (*Cassia siamea*). A few *Eucalyptus* species were planted at Bor by ILACO, and their performance should be assessed when opportunity offers.

The banks of the Canal might provide useful better-drained planting sites. The actual water margins could with advantage be planted with bamboo, as have the banks of the Nile between Sobat Mouth and Malakal. This serves the dual function of stabilising the banks and providing a valuable source of poles. The raised banks could also provide planting sites, although the bank with a road along it would be less suitable as, for trees to grow, water must percolate into the bank, whereas this is undesirable for road maintenance. Some tree growth, however, would help greatly in stabilising the bank. The bank without a road, if made as broad as is possible without compromising its containing function, could support a long linear plantation of *Cassia siamea* or *Azadirachta indica*, which would profit from the extra depth of soil, the better drainage, and the nutrients in the freshly exposed soil.

Plantations outside the area

Arguably this would be the best way of providing for the needs of the people for building timber, but not for fuel. There are trial plantations of both neem and cassia in the Mongalla area. Both grow rapidly and provide good straight poles. Both also coppice well (i.e. they regrow from the stumps when cut). Extension of these plantations, on the better-drained soils of the ironstone plateau, with transport of the poles into the area by lorry and boat, might go a long way to providing for the needs of most of the population for building timber.

Education

Any method of improving timber supplies will require the co-operation of the local people, because plantations will have to be protected from fire and grazing animals. Plantations will also have to be tended in their early stages and supervised at the harvest stage. All this will require a substantial input of educational and public relations work – which should ideally be undertaken by the JEO.

EPILOGUE

Exceptional local rainfall and the disastrous floods of 1988 in Khartoum, combined with the unexpected return to a high peak discharge in the Blue Nile, have occurred since this book went for press. Additional water from the Jonglei Canal would have added to the back-up in the White Nile above the junction, and the events also call into question the combined discharges of all projected schemes for increasing White Nile flows (pp. 394 and 462) should they coincide with a very high peak in the Blue Nile. These events and reflections are further reminders of the need for overall Nile control.

REFERENCES

Abdel Bagi, Subaei, M.A., Olsson, L. and Sadiq Nasir Osman (1976). *Labour Migration in the Jonglei Area*. Khartoum: Jonglei Socio-Economic Research Team.

Abdel Ghaffar, M. Ahmad (1976). *Anthropology and Development Planning in the Sudan: the Case of the Jonglei Project*. Khartoum: Economic and Social Research Council.

Abdel Magid, Yahia (1975). *Problems encountered in River Basin Development. The Case of the Nile Basin*. Budapest: Proceedings of the UN Interbasin Development Conference.

Abdel Magid, Yahia (1982). *Conservation Projects of the Nile and Irrigation Development in the Sudan*. Paper presented to the Royal Geographical Society Conference on the Impact of the Jonglei Canal in the Sudan, London, Oct. 1982. (*See also* Howell, 1983).

Abdel Magid, Yahia (1985). The Jonglei Canal: a conservation project of the Nile. In *Large-scale Water Transfers: Emerging Environmental and Social Experiences*, ed. G. N. Golubev & A. K. Biswas — pp. 85–101. Oxford: Tycooley.

Abou al-Atta & Abd al-Azim (1976). *Long-Range Planning in the Sphere of Irrigation and Drainage*. Cairo: Conference on Long-Range Planning and Regional Integration [in Arabic].

Adams, M. (1983). Nile Water: a Crisis postponed? *Economic Development and Cultural Change* **31**, 639–43.

Adams, W. Y. (1977). *Nubia: Corridor to Africa*. London: Allen Lane.

Adamson, D. & Williams, F. (1980). Structural geology, tectonics and control of drainage in the Nile Basin. In *The Sahara and the Nile. Quaternary Environments and Prehistoric Occupation in Northern Africa*, ed. M. A. J. Williams & H. Faure, pp. 225–52. Rotterdam: A. A. Balkema.

Anderson, J. (1944). The periodicity and duration of oestrus in Zebu and grade cattle. *Journal of Agricultural Science*, **34**, 57–68.

Anon (1931). *Lou Nuer*. E.N.D. [Eastern Nuer District] 66.A.1 (Nasir files).

ARE (Arab Republic of Egypt) Ministry of Irrigation (1981). *Water Master Plan* (EWMP), 17 volumes (UNDP/EGY/73/024) March 1981.

Arkell, A. J. (1955). *A History of the Sudan*. London: Athlone Press.

Associated Consultants (1981). *Small-Scale Abstraction of Water from Jonglei Canal*. Khartoum: Jonglei Executive Organ.

Associated Consultants (1982). *Pipe Offtakes and Associated Structures*. Khartoum: Jonglei Executive Organ.

Awuol, J. A. (1981) (Commissioner, JEO). *Address to Jonglei Province Development Conference*. Khartoum.

Awuol, J. A. (1982). *The Role of the Executive Organ, National Council for the Development Projects for the Jonglei Canal Area*. Paper presented to the Royal Geographical Society Conference on the Impact of the Jonglei Canal in the Sudan, London, October 1982.

BADA (1984*a*). *Progress Report No. 13*. Kingdom of the Netherlands, DGIS, The Hague.

BADA (1984*b*). *Progress Report No. 14*. Kingdom of the Netherlands, DGIS, The Hague.

Badal, R. K. (1983). *Origins of the Underdevelopment of the Southern Sudan: British Administrative Neglect*. Monograph Series 16, Development Studies and Research Centre, Faculty of Economic and Social Studies, Khartoum.

Bailey, R. G. & Hickley, P. (1985). Nouvelles récoltes de *Nothobranchius virgatus* Chambers, un nouveau Killi du Sudan méridional. *Revue français d'Aquariologie* **12**, 77–8.

Baker, S. (1866). *Albert Nyanza, Great Basin of the Nile and Exploration of the Nile Sources*. London: Macmillan.

Baker, S. (1874). *Ismailia, A Narrative of the Exploration of Central Africa for the Suppression of the Slave Trade*. London: Macmillan

Baulny, H. L. de & Baker, D. (1970). *The Water Balance of Lake Victoria: Technical Note*. Entebbe: Uganda Water Development Department.

Beadle, L. C. (1974). *The Inland Waters of Tropical Africa. An Introduction to Tropical Limnology*. London: Longman.

Beauchamp, R. S. A. (1956). The electrical conductivity of the headwaters of the White Nile. *Nature, (London)* **178**, 616–19.

Beavan, J. (1930). *Notes on Northern District*. National Records Office, Khartoum [NRO] Civil Secretary 57/2/8.

Bedri, I. (1939). Dinka beliefs in their chiefs and rainmakers. *Sudan Notes and Records* **22**, 125–31.

Bedri, I. (1948). More notes on the Padang Dinka. *Sudan Notes and Records* **29**, 40–57.

Bell, R. H. V. (1969). *The Use of the Herb Layer by Grazing Ungulates in the Serengeti National Park, Tanzania*. PhD Thesis, Manchester University.

Berthelot, R. M. (1976). *Jonglei Canal: UNDP Fact-finding Mission, Sudan*. Khartoum: UNDP.

Beshir, M. O. (1974). *The Southern Sudan: From Conflict to Peace*. New York: Barnes & Noble.

Bews, J. W. (1917). The plant succession in the thorn veld. *South African Journal of Science* **15**, 153–72.

Bhalotra, Y. P. R. (1964). Wind energy for windmills in Sudan. *Sudan Meteorological Services Memo No. 7*. Khartoum.

Bishai, H. M. (1962). The water characteristics of the Nile in the Sudan with a note on the effects of *Eichhornia crassipes* on the hydrobiology of the Nile. *Hydrobiologia* **19**, 357–82.

Bishop, W. W. (1969). Pleistocene Stratigraphy in Uganda. *Geological Survey Uganda, Memoir 10*. Entebbe.

Boulenger, G. A. (1907). *The Fishes of the Nile*. London: Hugh Ross.

Brinkman, R. (1963). Ferrolysis, a hydromorphic soil-forming process. *Geoderma*, **3**, 199–206.

Brookman-Amissah, J., Hall, J. B., Swaine, M. D., & Attakorah, J. Y. (1980). A re-assessment of a fire protection experiment in north-eastern Ghana savanna. *Journal of Applied Ecology* **17**, 85–99.

Broun, A. F. (1905). Some notes on the *Sudd* formation of the Upper Nile. *Journal of the Linnaean Society (Botany)*, **37**, 51–8.

Brown, L. H., Urban, E. K. & Newman, K. (1982). *The Birds of Africa, Vol.1*. London & New York: Academic Press.

Brown, D. S., Fison, T., Southgate, V. R. & Wright, C. A. (1984). Aquatic snails of the

Jonglei region, Southern Sudan, and transmission of trematode parasites. *Hydrobiologia*, **110**, 247–71.

Bruton, M. N. (1979). The breeding biology and early development of *Clarias gariepinus* (Pisces: Clariidae) in Lake Sibaya, South Africa, with a review of breeding in species of the subgenus *Clarias (Clarias)*. *Transactions of the Zoological Society of London* **35**, 1–45.

Bruton, M. N. & Jackson, P. B. N. (1983). Fish and fisheries of wetlands. *Journal of the Limnological Society of South Africa* **9**, 123–33.

Bureau of Reclamation (1964). US Department of Interior. *Land and Water Resources of the Blue Nile Basin: Ethiopia*. 17 Vols. Washington, DC.

Burton, J. W. (1978). Ghost marriage and the cattle trade among the Atuot of the Southern Sudan. *Africa* **48**, 398–405.

Butcher, A. D. (1936). *The Jonglei Canal Scheme*. Cairo: Ministry of Public Works.

Butcher, A. D. (1936). *The Future of the Sudd Region*. Cairo: Ministry of Public Works.

Butcher, A. D. (1938). *Sadd Hydraulics*. Cairo: Ministry of Public Works.

Butcher, A. D. (1939). *Zeraf Hydraulics*. Cairo: Ministry of Public Works.

Chadwick, M. J. & Obeid, M. A. (1966). A comparative study on the growth of *Eichhornia crassipes* and *Pistia stratiotes* in water culture. *Journal of Ecology* **54**, 563–75.

Chan, Siu-on & Eagleson, P. S. (1980). *Water Balance Studies in the Bahr el Ghazal Swamp*. Department of Civil Engineering, Massachusetts Institute of Technology, Report No. 261.

Charter, J. R. & Keay, R. W. J. (1960). Assessment of the Olokemeji fire control experiment 28 years after institution. *Nigerian Forestry Information Bulletin* **3**, 1–32.

Chatterton, B. J. (1933). *Notes on the Quil Dinka*. Southern Records Office, Juba, Upper Nile Province 66.A.4.

Chatterton, B. J. (1934). *Ruweng Dinka*. Southern Records Office, Juba, Upper Nile Province 66.E.4.

Chatterton, B. J. (1954). *Reply to questionnaire* sent by P. P. Howell, Chairman SDIT, 28th June 1954.

Chenery, E. M. (1960). An Introduction to the soils of the Uganda Protectorate. *Memoirs of the Research Division of the Department of Agriculture, Uganda*: Series 1, No. 1.

Clayton, W. D. & Renvoize, S. A. (1982). Gramineae (Part 3). In *Flora of Tropical East Africa*, ed. R. M. Polhill. Rotterdam: A. A. Balkema.

Coe, M. J., Cumming, D. H. M. & Phillipson, J. (1976). Biomass and production of large African herbivores in relation to rainfall and primary production. *Oecologia* **22**, 341–54.

Collins, R. O. (1983). *Shadows on the Grass: Britain in the Southern Sudan 1918–1956*. New Haven & London: Yale University Press.

Collins, R. O. (1987). *The Jonglei Canal: the Past and Present of a future*. Sixth Trevelyan Lecture, 1986. University of Durham.

Coriat, P. (1923). *The Gaweir Nuer*. Khartoum.

Coriat, P. (1931). *Western Nuer District*. National Records Office, Khartoum; Civil Secretary 57/2/8.

Crazzolara, J. P. (1951). *The Lwoo. Part II: Lwoo Traditions*. Verona: Editrice Nigrizia.

Crazzolara, J. P. (1953). Zur Gesellschaft und Religion der Nueer. *Studia Instituti Anthropos, Vienna* **5**, 1–221.

Dahl, G. & Hjort, A. (1976). *Having Herds: Pastoral Herd Growth and Household Economy*. Department of Social Anthropology, University of Stockholm.

Daly, G. (1984). *In search of the 'New' Nuer: an Anthropological Study of New Fangak (a Nuer town) and its Hinterland*. Unpublished Interim Report, Department of Sociology, Trinity College, Dublin.

David, N. (1982*a*). The BIEA Southern Sudan expedition of 1979: interpretation of the ecological data. In *Culture History in the Southern Sudan*, ed. J. Mack & P. Robertshaw, pp. 49–57. Nairobi: British Institute in Eastern Africa.

David, N. (1982*b*). Prehistory and historical linguistics in Central Africa: points of contact. In *The Archaeological and Linguistic Reconstruction of African History*, ed. C. Ehret & M. Posnansky, pp. 78–103. Los Angeles: University of California Press.

David, N., Harvey, P. & Goudie, C. J. (1981). Excavations in the Southern Sudan, 1979. *Azania* **16**, 7–54.

Davis, J. W., Karstad, L. H. & Trainer, D. O. (eds) (1981). *Infectious Diseases of Wild Mammals*, 2nd edn. Iowa State University Press.

Debenham, F. (1947). The Bangweulu Swamp of Central Africa. *Geographical Review* **37**, 351–68.

Demiruren, A. S. (1974). *The Improvement of Nomadic and Transhumant Animal Production Systems*. Rome: FAO.

Deng, F. M. (1971). *Tradition and Modernization: a Challenge for Law among The Dinka of the Sudan*. New Haven: Yale University Press.

Deng, F. M. (1972). *The Dinka of the Sudan*. New York: Holt, Rinehart & Winston.

Deng, F. M. (1973). *Dynamics of Identification. A Basis for National Integration in The Sudan*. Khartoum University Press.

Denny, P. (1984). Permanent swamp vegetation of the Upper Nile. *Hydrobiologia* **110**, 79–90.

Diarra, A. (1978). Observations of wild rice and a study of control methods, Mopti 1977. In *3ᵉ Symposium sur le desherbage des cultures tropicales*, vol. 1, pp. 229–43. Dakar.

Doorenbos, J. & Pruitt, W. O. (1977). *Guidelines for Predicting Crop Water Requirements*. Rome: FAO.

DRS, MOA (Democratic Republic of Sudan, Ministry of Agriculture) (1979). *Current Agricultural Statistics*, vol. 1, no. 2, June 1979.

Duncan, P. (1975). *Topi and their Food Supply*. PhD Thesis, University of Nairobi.

Dyson-Hudson, N. & Dyson-Hudson, R. (1982). The structure of East African herds and the future of East African herders. *Development & Change* **13**, 213–38.

Dyson-Hudson, R. & Dyson-Hudson, N. (1970). The food production system of a semi-nomadic society. In *African Food Production Systems*, ed. P. F. M. McLoughlin, pp. 91–123. Baltimore: Johns Hopkins Press.

Eggeling, W. J. & Dale, I. R. (1952). *The Indigenous Trees of the Uganda Protectorate*. Entebbe: Government Printer.

Ehret, C. (1982). Population movement and culture contact in the Southern Sudan, *c*.3000BC to AD1000: a preliminary linguistic overview. In *Culture History in the Southern Sudan*, ed. J. Mack & P. Robertshaw, pp. 19–48. Nairobi: British Institute in Eastern Africa.

Ehret, C. (1983). Nilotic and the limits of Eastern Sudanic: classificatory and historical conclusions. In *Nilotic Studies. Proceedings of the International Symposium on Languages and History of the Nilotic Peoples, Cologne, Jan. 4–6, 1982*, ed. R. Vossen & M. Bechhaus-Gerst, pp. 377–421. Berlin: Dietrich Reiner Verlag.

Eid, M. T. *et al.* (1966). Preliminary estimated balance between irrigation requirements and river resources of the UAR. *Agricultural Research Review, Egypt*, **44**.

El Amin, M. & Ezeat, N. (1978). *Jonglei Canal Water Benefit (offtake at Bor)*. Khartoum: PJTC.

El Bashir, Mehdi (1981). *The Jonglei Canal*. PhD Thesis, University of Khartoum.

El Gendi, Saad & El Ghomry, O. (1977). *Re-use of Drainage Water for Irrigation Purposes*. MOI, Egypt; paper presented at UN Water Conference, Mar del Plata, Argentina, E/Conf. 70/TP 21; March 14–25.

El Hassen, M. A. K., Fadul, H. M., Ali, M. A. & El Tom, O. A. (1978). Exploratory soil survey and land suitability classification of Jonglei Projects area, Southern Region. *Soil Survey Report* **90**, Soil Survey Administration, Wad Medani, & Khartoum: JEO.

El Hemry, I. I. & Eagleson, P. S. (1980). *Water Balance Estimates of the Machar Marshes.* Department of Civil Engineering, Massachusets Institute of Technology, Report No. 260.

El Sammani, M. O. (1978) *The Democratic Characteristics of the Dinka of Kongor community.* Khartoum: JEO, Report No.7.

El Sammani, M. O. (1984). *Jonglei Canal: Dynamics of planned change in the Twic area.* Khartoum, Graduate College Publications, Monograph 8.

El Sammani, M. O. & Deng, P. L. (1978). *The Seasonal Migration of People and their Animals in Kongor and Bor Districts, Jonglei Province.* Khartoum: JEO.

El Sammani, M. O. & El Amin, F. M. (1978). *The Impact of the Extension of the Jonglei Canal on the Area from Kongor to Bor.* Khartoum, JEO.

El Sammani, M. O. & Hassan, A. (1978). *Agriculture in the Dinka and Nuer lands, Jonglei Province.* Khartoum: JEO. (Report No.10).

Eprile, C. (1974). *War and Peace in the Sudan, 1955–1972.* London, Davies & Charles.

Ethiopia (1977). *I. Water Resources Development; II. The Need for Co-operation among Co-Basin States.* UN Water Conference, Mar del Plata, March 1977, E/Conf.70/TP.

Euroconsult (1978). *Jonglei Environmental Aspects.* Arnhem, The Netherlands.

Euroconsult (1981). *Kongor Flood Protection Surveys.* Draft Final Report prepared for UN/FAO. Arnhem, The Netherlands.

Evans-Pritchard, E. E. (1934). The Nuer: Tribe and Clan. *Sudan Notes and Records,* **17**, 1–57.

Evans-Pritchard, E. E. (1936). The Nuer: Age Sets. *Sudan Notes and Records,* **19**, 233–71.

Evans-Pritchard, E. E. (1937). Economic Life of the Nuer. Cattle. *Sudan Notes and Records,* **20**, 209–45.

Evans-Pritchard, E. E. (1940a). *The Nuer.* Oxford: Clarendon Press.

Evans-Pritchard, E. E. (1940b). *The Political System of the Anuak of the Anglo-Egyptian Sudan.* London: Percy Lund, Humphries & Co., for the London School of Economics.

Evans-Pritchard, E. E. (1948) *The Divine Kingship of the Shilluk of the Nilotic Sudan.* The Frazer Lecture, 1948. Cambridge University Press.

Evans-Pritchard, E. E. (1951). *Kinship and Marriage among the Nuer.* Oxford: Clarendon Press.

Evans-Pritchard, E. E. (1956). *Nuer Religion.* Oxford: Clarendon Press.

EWMP (Egyptian Water Master Plan) (1981). *See* Arab Rep. of Egypt (1981).

Fitch, J. & Abdel Aziz, A. (1980). *Multiple Cropping Intensity in Egyptian Agriculture: a Study of its Determinants.* ARE; MOA, Microeconomic Study of the Egyptian Farm System, Project Paper No. 5, October 1980.

Freidel, J. W. (1979). Population dynamics of the water hyacinth *Eichhornia crassipes* (Mart.) Solms, with special reference to the Sudan. *Berichte aus dem Fachgebiet Herbologie der Universitat Hohenheim,* **17**, 1–132.

Fryxell, J. (1985). *Resource Limitation and Population Ecology of White-eared Kob.* PhD Thesis, University of British Columbia.

Gamachu, D. (1977). *Aspects of Climate and Water Budget in Ethiopia.* Addis Ababa: University Press.

Ganf, G. G. (1974). Incident solar irradiance and underwater light penetration as factors controlling the chlorophyll 'a' content of a shallow equatorial lake. *Journal of Ecology,* **62**, 593–609.

Garang, J. de M. (1981). *Identifying, Selecting and Implementing Rural Development Strategies for Socio-Economic Development in the Jonglei Projects Area, Southern Region, Sudan.* PhD Thesis, Iowa State University.

Garang, Mading de (1980) (Commissioner, JEO). *Address to Meeting for Co-ordination of JEO Works Plans for 1981.* Khartoum.

Garretson, A.H. (1967). *The Law of International Drainage Basins.* Dobbs Ferry.

Garstin, W. (1901). Report as to Irrigation Projects on the Upper Nile, etc. (In a Despatch from His Majesty's Agent and Consul-General at Cairo.) *Parliamentary Accounts and Papers 1901,* **91,** 1149 *et seq.*

Garstin, W. (1904). Report upon the basin of the Upper Nile (In a Despatch from His Majesty's Agent and Consul-General at Cairo). *Parliamentary Accounts and Papers 1904,* **111,** 315–735.

Garstin, W. (1909). Fifty years of Nile exploration and some of its results. *Geographical Journal,* **33,** 117–52.

Gaudet, J. J. (1979). Seasonal changes in nutrients in a tropical swamp: North Swamp, Lake Naivasha, Kenya. *Journal of Ecology,* **67,** 953–81.

Gay, P. A. (1958). *Eichhornia crassipes* in the Nile of the Sudan. *Nature, London,* **182,** 538–9.

Girgis, S. (1948). A list of common fish of the upper Nile with their Shilluk, Dinka and Nuer names. *Sudan Notes and Records,* **29,** 120–5.

Glickman, M. (1972). The Nuer and the Dinka: a further note. *Man, N. S.,* **7,** 586–94.

Glover, J., Dutchie, D. W. & French, M. H. (1957). The apparent digestibility of crude protein by the ruminant. I. A synthesis of the results of digestibility trials with herbage and mixed feeds. *Journal of Agricultural Science, Cambridge,* **48,** 373–8.

Goldschmidt, J. (1981). The failure of pastoral economic development programmes in Africa. In *The Future of Pastoral Peoples,* ed. J. Galaty, D. Anson, P. C. Salzman & A. Chouinard. Ottawa: IRDC.

Green, J. (1984). Zooplankton associations in the swamps of Southern Sudan. *Hydrobiologia* **113,** 93–8.

Greenway, P. J. (1973). A classification of the vegetation of East Africa. *Kirkia* **9,** 1–68.

Greenwood, P. H. (1976). Fish fauna of the Nile. In *The Nile: Biology of an Ancient River,* ed. J. Rzoska. The Hague: W. Junk.

Grimsdell, J. J. R. & Bell, R. H. V. (1975). *Ecology of the Black Lechwe in the Bangweulu Basin of Zambia.* Black Lechwe Research Project Final Report. Lusaka, Zambia: National Council for Scientific Research.

Grogan, E. S. (1901). *From the Cape to Cairo.* London: Nelson & Sons.

Grönblad, R. (1962). Sudanese Desmids II. *Acta Botanica Fennica* **63,** 1–19.

Grönblad, R., Prowse, G. A. & Scott, A. M. (1958). Sudanese Desmids I. *Acta Botanica Fennica* **58,** 48–82.

Grove, A. T. (1977). Desertification in the African Environment. In *Drought in Africa 2,* ed. D. Dalby, R. J. Harrison Church & F. Bezzaz, pp.54–64. London: International African Institute.

Gruenbaum, E. (1978). *Women's Labour in the Subsistence Sector, the Case of the Central Nuer of Jonglei Province.* Khartoum: Report No. 9., The Executive Organ for the Development Projects in the Jonglei area.

Guillet, A. (1978). Distribution and conservation of the Shoebill, *Balaeniceps rex,* in the southern Sudan. *Biological Conservation,* **13,** 39–50.

Gurdon, C. (1986). *Sudan in Transition: a Political Risk Analysis.* London: Special Report No. 226, Economist Intelligence Unit.

Harik, I. (1979). *Distribution of Land, Employment and Income in Rural Egypt.* Rural Development Committee, Cornell University.

Harvey, C. P. D. (1982). The archaeology of the Southern Sudan: environmental context. In *Culture History in the Southern Sudan,* ed. J. Mack & P. Robertshaw, pp. 7–18. Nairobi:

British Institute in East Africa.

Harvey, T. J. (1976). *The palaeolimnology of Lake Mobutu Sese Seko [Albert], Uganda – Zaire: the last 28,000 years*. PhD Thesis, Duke University, N.C.

Hasan, Y. F. (1967). *The Arabs and the Sudan*. Edinburgh: University Press.

Haynes, K. E. & Whittington, D. (1981a). *Coordination of the Operation of the Aswan High Dam with Upstream Developments*. Paper delivered to the International Conference on Water Researces Planning in Egypt, Cairo, Jan.10–13, 1981.

Haynes, K. E. & Whittington, D. (1981b). International management of the Nile – Stage Three? *Geographical Review*, **71**, 17–32.

Hefny, K. (1977). *Ground Water Potentialities in ARE (Arab Republic of Egypt)*. UN Water Conference, Mar Del Plata, Argentina.

Herring, R. S. (1979a). Hydrology and chronology: the Rodah nilometer as an aid to dating interlacustrine history. In *Chronology, Migration and Drought in Interlacustrine Africa*, ed. J. B. Webster, pp. 39–86. London: Longman & Dalhousie University Press.

Herring, R. S. (1979b). The influence of climate on the migrations of the central and southern Luo. In *Ecology and History in East Africa*, ed. B. A. Ogot, pp. 77–107. Nairobi: Kenya Literature Bureau.

Hickley, P. & Bailey, R. G. (1986). Fish communities in the perennial wetland of the *Sudd*, southern Sudan. *Freshwater Biology*, **16**, 695–709.

Hickley, P. & Bailey, R. G. (1987a). Fish communities in the eastern seasonal floodplain of the *Sudd*, southern Sudan. *Hydrobiologia*, **144**, 243–50.

Hickley, P. & Bailey, R. G. (1987b). Food and feeding relationships of fish in the *Sudd* swamps (River Nile, southern Sudan). *Journal of Fish Biology*, **30**, 147–59.

Hills, R. C. (1979). The structure of the Inter-Tropical Convergence Zone in Equatorial Africa and its relationship to East African rainfall. *Transactions of the Institute of British Geographers*, **4** (N.S.), 329–52.

Hoek, B. van der, Zanen, S. & Deng, P. L. (1978). *Social–Anthropological Aspects of the Jonglei Development Projects in South Sudan*. Leiden: Instituut voor Culturele Anthropologie en Sociologie der niet-Westerse Volken, University of Leiden.

Hofmayr, F. (1925). *Die Schilluk*. Vienna: Anthropos II.

Hope, C. M. (1902). The *Sudd* of the Upper Nile. *Annals of Botany* **16**, 495–516.

Hopson, A. J. (1969). *Seasonal Changes with Pattern of Salinity Distribution in the Northern Basin of Lake Chad*. Annual Report, 1966–67, Federal Fish Services, Republic of Nigeria, Appendix 2, 13–26.

Howard–Williams, C. & Walker, B. H. (1974). The vegetation of a tropical African lake: classification and ordination of the vegetation of Lake Chilwa (Malawi). *Journal of Ecology*, **62**, 831–54.

Howell, P. P. (1941). The Shilluk settlement. *Sudan Notes and Records* **24**, 47–67.

Howell, P. P. (1945a). A note on elephants and elephant hunting among the Nuer. *Sudan Notes and Records* **26**, 95–103.

Howell, P. P. (1945b). The Zeraf Hills. *Sudan Notes and Records* **26**, 319–28.

Howell, P. P. (1952). Observations on the Shilluk of the Upper Nile. *Africa*, **22**, 97–119.

Howell, P. P. (ed.) (1953). The Equatorial Nile Project and its effects in the Sudan. *Geographical Journal*, **119**, 33–48.

Howell, P. P. (1954). *A Manual of Nuer Law*. Oxford: University Press.

Howell, P. P. (1983). The impact of the Jonglei Canal in the Sudan. *Geographical Journal*, **149**, 286–300.

Howell, P. P. & Thompson, W. P. G. (1946). The death of a reth of the Shilluk and the installation of his successor. *Sudan Notes and Records*, **27**, 4–85.

Hunting Technical Services (1974). Development Plan: Southern Darfur Planning Survey.

Annex 3. *Animal Resources and Range Ecology*. Elstree.

Hrbek, I. (1977). Egypt, Nubia and the eastern deserts. In *The Cambridge History of Africa*, vol. 3, ed. R. Oliver, pp. 10–97. Cambridge University Press.

Hurst, H. E. (1944). *A Short Account of the Nile Basin*. Physical Department Paper No. 45, Cairo.

Hurst, H. E. (1952). *The Nile*. London: Constable.

Hurst, H. E. & Phillips, P. (1938). *The Nile Basin. Volume V. The Hydrology of the Lake Plateau and the Bahr el Jebel*. Cairo: Ministry of Works, Physical Department.

Hurst, H. E., Black, R. P. & Simaika, Y. M. (1947). *The Nile Basin, Volume VII. The Conservation of the Nile Waters*. Cairo: Ministry of Works, Physical Department.

Hutchinson, S. (1985). Changing concepts of incest among the Nuer. *American Ethnologist*, **12**, 625–41.

Hutchinson, S. (1988). *The Nuer in Crisis; Coping with Money, War and The State*, PhD Thesis, University of Chicago.

Hydromet (1974). *Hydrometeorological Survey of the Catchments of Lake Victoria, Kyoga and Albert*. Geneva: UNDP/WMO.

Hydromet (1981). *Hydrometeorological Survey of the Catchments of Lake Victoria, Kyoga, and Mobutu Sese Seko: Project Findings and Recommendations*. Geneva: UNDP/WMO.

Ibrahim, A. M. (1981). *The Development of the Nile River System*. Paper presented to the Nile Basin Conference, University of Khartoum.

Ibrahim, M. A. & El Amin, M. M. E. (1976). *Bahr el Jebel Discharge Losses as a Result of the Jonglei Canal*. Khartoum: PJTC.

Ibrahim, A. & Nur, M. A. (1981). *Increase of Nile Yield by Utilization of Lost Waters in Machar Marshes and Lost Waters in Ghazal Swamps*. Khartoum: PJTC.

IBRD (International Bank for Reconstruction and Development) (1979). *Sudan Agricultural Sector Survey*. Report No. 1836 a-Su, May 18.

IBRD (1981). *Egypt Irrigation Sub-Sector Report*.

Ilaco (1975). *Pengko Plain Pre-Feasibility Study. Vol. II*. Arnhem: Ilaco.

Ilaco (1977). *Sociological Investigations, Second Phase*. Arnhem: Ilaco.

Ilaco (1979a). *The Social and Economic Setting of Rural Bor Dinka, vol. 1*. Pengko Pilot Project Technical Note No. 9. Arnhem: Ilaco.

Ilaco (1979b). *Soils of the Pengko and Eastern Plains*. Pengko Pilot Project Technical Note No. 13. Arnhem: Ilaco.

Ilaco (1981a). *Agricultural Research under High and Low Input Levels, Wet Season, 1980*. Pengko Pilot Project Technical Note No. 18. Arnhem: Ilaco.

Ilaco (1981b). *Bor Dinka: Prospects for Development*. Pengko Pilot Project Technical Note No.20. Arnhem: Ilaco.

Ilaco (1981c). *Pengko Plain Development Study: Vol.1 – Evaluation and Conclusions; Vol. 2 – Technical Annexes* Arnhem: Ilaco.

Ilaco (1982a). *Bor Livestock Production System*. Pengko Pilot Project Technical Note No. 23. Arnhem: Ilaco.

Ilaco (1982b). *Grazing Trial – Pengko Plains*. Pengko Pilot Project Technical Note No. 24. Arnhem: Ilaco.

Ilaco (1982c). *Livestock Development in Bor Dinka*. Policy & Projects Part B: Veterinary Services. Arnhem: Ilaco.

Ilaco (1983). *Rangeland Productivity and Exploitation in Bor District*. Rural Development Activities, Sudan. Technical Note No. 2. Arnhem: Ilaco.

ILCA (International Livestock Commission for Africa) (1981a). *Systems research in the Arid Zones of Mali – initial results*. Addis Ababa: ILCA Systems Study No. 5.

ILCA (1981b). *Economic Trends – Small Ruminants*. Addis Ababa: ILCA Bulletin No. 7.

Irving, N. S. & Beshir, M. O. (1982). Introduction of some natural enemies of water hyacinth (*Eichhornia crassipes*) to the White Nile, Sudan. *Tropical Pest Management*, **28**, 20–6.

IUCN (International Union for the Conservation of Nature and Natural Resources) (1986). *Review of the Protected Areas System in The Afrotropical Realm*. Cambridge, UK & Gland, Switzerland: IUCN.

Jackson, H. C. (1923). The Nuer of the Upper Nile Province. *Sudan Notes and Records*, **6**, 59–107.

Jal, G. G. (1985). Letter to P. P. Howell.

Jal, G. G. (1987). *The History of the Jikany Nuer before 1920*. PhD Thesis, SOAS, University of London.

Jenness, J. (1982). *Planning for the Development of Land Use in the Jonglei Canal Area*. Paper presented to the Royal Geographical Society Conference on the Impact of the Jonglei Canal in the Sudan, London, Oct 1982 (*see also* Howell, 1983).

Jenness, J. (1985). Final Report of Project Manager/Land Use Planner to JEO. (Personal Communication).

JEO (Jonglei Executive Organ) (1974). *Statement to the Peoples' Assembly on the First Phase of the Jonglei Project*. Khartoum: JEO.

JEO (1975a). *Jonglei Project, Phase 1*. Khartoum: JEO.

JEO (1975b). *An Outline of the Proposed Socio-Economic Survey of the Jonglei Scheme*. Khartoum: JEO.

JEO (1976a). *Jonglei Socio-Economic Research Team: An Interim Report*. Khartoum: JEO.

JEO (1976b). *Labour Migration in the Jonglei Area*. Khartoum: JEO.

JEO (1976c). *A Request for Funding of Projects in the Jonglei Area*. Khartoum: JEO.

JEO (1978). *The Existing Services in Kongor and Bor Districts*. Khartoum: JEO & Economic and Social Research Council.

JEO (1979a). *Comparative Socio-Economic Benefits of the Eastern Alignment and the Direct Jonglei Canal Line*. Khartoum: JEO.

JEO (1979b). *Integrated Rural Development in Kongor District, Jonglei Province*. Khartoum: UNDP.

JEO (1979c). *Proposals for a Mid-term Program and a Crash Program for Development for Agriculture, Livestock and Socio-Economic Services in the Jonglei Area*. Khartoum: JEO.

JEO (1981–84). *Reports on Annual Meetings for Co-ordination of Work Plans*. Khartoum: JEO. [These reports were supported by working papers submitted by various agencies to each meeting. Copies of these are held by The Library, University of Durham]

JEO (1982). *The Jonglei Canal – Development Project in Sudan*. Khartoum: JEO.

JEO (1983). *A Comprehensive Plan for the Development of Community Services in the Jonglei Canal Area, the Sudan*. Khartoum: JEO.

JIT (1946). *First Interim Report of the Jonglei Investigation Team*. Sudan Government.

JIT (1947). *Second Interim Report of the Jonglei Investigation Team*. Sudan Government.

JIT (1948). *Third Interim Report of the Jonglei Investigation Team*. Sudan Government.

JIT (1949). *Progress Report of the Jonglei Investigation Team*. Sudan Government.

JIT (1954). *The Equatorial Nile Project and its Effects in the Anglo-Egyptian Sudan*. Report of the Jonglei Investigation Team. 4 Vols. Khartoum: Sudan Government.

Johnson, D. H. (1980). *History and Prophecy among the Nuer of the Southern Sudan*. PhD Thesis, UCLA.

Johnson, D. H. (1982). Tribal boundaries and border wars: Nuer–Dinka relations in the Sobat and Zaraf valleys c. 1860–1976. *Journal of African History*, **23**, 183–203.

Johnson, D. H. (1986a). Judicial regulation and administrative control: customary law and the Nuer, 1898–1954. *Journal of African History* **27**, 41–78.

Johnson, D. H. (1986b). On the nilotic frontier: Imperial Ethiopia in the Southern Sudan, 1898–1936. In *The Southern Marches of Imperial Ethiopia: Essays in History and Social Anthropology*, ed. D. Donham & W. James, pp. 219–45. Cambridge: University Press.

Johnson, D. H. (1988). Adaptation to floods in the Jonglei Area: a historical analysis. In *The Ecology of Survival. Case Studies from Northeast African History*, ed. D. H. Johnson & D. Anderson. London & Boulder, Colorado: Crook Academic Publishing & Westview Press.

Johnson, D. H. (in press). The twentieth century expansion of the common economy among pastoralists in the Upper Nile of the Sudan. In *Herders, Warriors and Traders: the Political Economy of African Pastoralism*, ed. J. Galaty & P. Bonte. Boulder, Colorado: Westview Press.

Jolly, G. (1969). Sampling methods for aerial censuses of wildlife populations. *East African Agriculture & Forestry Journal*, 33, 46–9.

Jurriëns, M. & Klaassen, G. J. (1979). *Evaluation of the Possibilities and the Effects of Bypassing Water along Marshy Areas*. Proc. Congress IAHR.

Kadouf, H. A. (1977). *An Outline of Dinka Customary Law in the Jonglei Area*. Khartoum: Customary Law Memorandum No. 2, Faculty of Law, University of Khartoum.

Kamal, A. M. (1982). *The Design and Construction of the Jonglei Project*. Paper presented to the Royal Geographical Society Conference on the Impact of the Jonglei Canal in the Sudan, London, Oct. 1982.

Kameir, El W. (1980). Nuer migrants in the building industry in Khartoum: a case of the concentration and circulation of labour. In *Urbanisation and Urban Life in the Sudan*, ed. V. Pons. Hull: Department of Sociology and Social Anthropology, University of Hull.

Karp, I. & Maynard, K. (1983). Reading *The Nuer. Current Anthropology*, 24, 481–503.

Kelly, R. (1985). *The Nuer Conquest: the Structure of an Expansionist System*. Ann Arbor: University of Michigan Press.

Kendall, R. L. (1969). An ecological history of the Lake Victoria Basin. *Ecological Monographs*, 39, 121–76.

Kinawy, I. Z. (1976). *The Efficiency of Water Use in Irrigation in Egypt*. Conference on Arid Lands Irrigation in developing countries. UNESCO/UNDP/Academy of Scientific Research.

Kite, G. W. (1981). Recent changes in level of Lake Victoria. *Hydrological Sciences Bulletin*, 26, 233–43.

Kite, G. W. (1982). Analysis of Lake Victoria levels. *Hydrological Sciences Journal*, 27, 99–110.

Kleppe, E. (1982). The *debbas* on the White Nile, Southern Sudan. In *Culture History in the Southern Sudan*, ed. J. Mack & P. Robertshaw, pp. 59–70. Nairobi: British Institute in East Africa.

Kurdin, V. P. (1968). Data on hydrological and hydrochemical observations on the White Nile. *Inf.Byull.Bio.Unutr.Vad.*, 2, 49–56 (in Russian).

Lako, G. T. (1985). The impact of the Jonglei scheme on the economy of the Dinka. *African Affairs*, 84, 15–38.

Lamb, H. H. (1966). Climate in the 1960s. *Geographical Journal*, 132, 183–212.

Lawry, S. (1982). *The Jonglei Canal and Endogenous Change: a New Framework for Policy Analysis*. Paper presented to the Royal Geographical Society Conference on the Impact of the Jonglei Canal in the Sudan, London, Oct 1982 (*see also* Howell 1983).

Laws, R. M., Parker, I. S. C. & Johnstone, R. C. B. (1975). *Elephants and their Habitats: the Ecology of Elephants in Northern Bunyoro, Uganda*. Oxford: Clarendon Press.

Leach, T. M. (1957). The occurrence of contagious bovine pleuro-pneumonia in species other than domestic cattle. *Bulletin of Epizootic Diseases of Africa*, 5, 325–8.

Letouzey, R. (1963). Balanitaceae. In *Flore du Cameroun*, ed. A. Aubréville, pt. 1, 159–67.

Paris: Muséum National d'Histoire Naturelle.

Lewis, B. A. (1951). Nuer spokesmen: a note on the institution of the *ruic*. *Sudan Notes and Records* **32**, 77–84.

Liddell, J. S. (1904). Report on march from Taufikia to Twi and visit to Twi by steamer. *Sudan Intelligence Report*, **119**, 3–9.

Lienhardt, G. (1958). The Western Dinka. In *Tribes without Rulers*, ed. J. Middleton & D. Tait, pp. 97–135. London: Routledge & Kegan Paul.

Lienhardt, G. (1961). *Divinity and Experience*. Oxford: University Press.

Livingstone, D. (1976). Palaeolimnology of headwaters. In *The Nile. Biology of an Ancient River*, ed. J. Rzoska, pp. 21–30. The Hague: Junk.

Livingstone, D. (1980). Environmental changes in the Nile headwaters. In *The Sahara and the Nile. Quaternary Environments and Prehistoric Occupation in Northern Africa*, ed. M. A. J. Williams & H. Faure, pp. 339–59. Rotterdam: A. A. Balkema.

Lock, J. M. (1972). The effects of hippopotamus grazing on grasslands. *Journal of Ecology*, **60**, 445–67.

Lock, J. M. (1977). The vegetation of Rwenzori National Park, Uganda. *Botanische Jahrbücher für Systematik, Pflanzengeschichte und Pflanzengeographie*, **98**, 372–448.

Lock, J. M. & Milburn, T. R. (1971). The seed biology of *Themeda triandra* in relation to fire. In *The Scientific Management of Animal and Plant Communities for Conservation*, ed. E. Duffey & A. S. Watt. Oxford: Blackwell Scientific Publications.

Lombardini, E. (1864). Saggio sull' idrologia dell' Africa centrale. *Milano Politecnico, 12. Mem.I, Lombardo 10.*

Longhurst, R., Chambers, R. & Swift, J. (1986). Seasonality and poverty: implications for policy and research. *Institute of Development Studies, Sussex, Bulletin*, **17**, 3.

Lowe-McConnell, R. H. (1975). *Fish Communities in Tropical Freshwaters*. London: Longman.

Lyons, H. G. (1905). The dimensions of the Nile and its basin. *Geographical Journal*, **26**, 198–201.

Lyons, H. G. (1906). *The Physiography of the River Nile and its Basin*. Cairo.

MacDonald, M. (1921). *Nile Control*. Cairo: Egyptian Government Press.

Maclaglan, T. A. (1931). *The Dunjol Dinka*. Southern Record Office, Juba; Upper Nile Province 66.a.4.

MacMichael, H. (1954). *The Sudan*. London: Benn.

Makec, J. W. (1986). *The Customary Law of the Dinka: a Comparative Analysis of an African Legal System*. Khartoum: St. George Printing Press.

Mann, E. (1977). *The Jonglei Canal: East Africa's Environmental Disaster*. Nairobi: Bild.

Mann, O. (1977). *The Jonglei Canal. Environmental and Social Aspects*. Nairobi: Environment Liaison Centre.

Mason, I. L. & Maule, J. P. (1960). *Indigenous Livestock of Eastern and Southern Africa*. Farnham Royal: Commonwealth Agricultural Bureaux, Technical Communication No. 14.

Mawson, A N. M. (1984). The Southern Sudan: a growing conflict. *The World Today*, **40**, 520–7.

McDermott, B. (1984). *Integrated Rural Development in Kongor District*. Rome: FAO.

McDermott, J. (1984). *Livestock Marketing in Kongor Rural Council*. Rome: FAO.

McDermott, J. & Ngor, M. D. (1984). *The Eastern Plain in Kongor Rural Council: Development Possibilities*. Rome: FAO.

Mefit (1977). *Southern Regional Development Plan*. Report to the Democratic Republic of the Sudan: Southern Region. Rome & Khartoum: Mefit spa.

Michel, C. (1901). *Mission de Bonchamps. Vers Fachoda, à la recontre de la Mission*

Marchand. Paris.

Migahid, A. M. (1947). An ecological study of the '*Sudd*' swamps of the Upper Nile. *Proceedings of the Egyptian Academy of Sciences*, **3**, 57–86.

Migahid, A. M. (1948). *Report on a Botanical Excursion to the Sudd Region*. Cairo: Fouad University Press.

Milford, R. & Minson, D. J. (1965). The relation between the crude protein content and the digestible crude protein content of tropical pasture grasses. *Journal of the British Grassland Society*, **20**, 177–90.

Ministry of Irrigation, DRS (1975). *Control and Use of the Nile Waters in Irrigation*. Khartoum.

Ministry of Irrigation, DRS (1978). *Blue Nile Waters Study*. Khartoum: Coyne & Bellier, Sir Alexander Gibb, Hunting Technical Services and Sir M. MacDonald.

Ministry of Irrigation, DRS (1979). *Nile Waters Study*. Khartoum.

Moghraby, A. I. & El Sammani, M. O. (1985). On the environmental and socio-economic impact of the Jonglei Canal Project, southern Sudan. *Environmental Conservation*, **12**, 41–8.

Mohammedein, M. A. (1982) *The Objective of the Jonglei Canal Project*. Paper presented to the Royal Geographical Society Conference on the Impact of the Jonglei canal in the Sudan, London, Oct. 1982 (*see also* Howell, 1983).

Monakov, A. V. (1969). The zooplankton and zoobenthos of the White Nile and adjoining waters in the Republic of Sudan. *Hydrobiologia* **33**, 161–85.

Morrice, H. A. W. & Allen, W. N. (1959). Planning for the ultimate hydraulic development in the Nile valley. *Proceedings of the Institution of Civil Engineers*, **14**, 101–56.

Newcomer, P. J. (1972). The Nuer are Dinka: an essay on origins and environmental determinism. *Man (N. S.)*, **7**, 5–11.

Newhouse, F. (1929). *The Problem of the Upper Nile*. Cairo: Unpublished Report.

Newhouse, F. (1939). *The Training of the Upper Nile*. London: Pitman.

Niamir, M. (1982). *A report on Animal Husbandry among the Ngok Dinka of the Sudan*. Abyei Rural Development Project; Harvard Institute for International Development, Harvard University.

Nicholson, S. E. (1980). Saharan climates in historic times. In *The Sahara and the Nile*, ed. M. A. J. Williams & H. Faure. Rotterdam: A. A. Balkema.

Nile Waters Agreement (1929). *Agreement between the British and Egyptian Governments on 7 May*. Cmd. 3348 of 1929.

Nile Waters Agreement (1959). *Agreement between the Republic of the Sudan and the United Arab Republic for full utilisation of the Nile Waters*. Signed at Cairo, 8 Nov. 1959.

Norton-Griffiths, M. (1981). *Counting Animals*, 2nd edn. Nairobi: African Wildlife Foundation.

O'Fahey, R. S. & Spaulding, J. (1974). *Kingdoms of the Sudan*. London: Methuen.

Obeid, M. O. (ed.) (1975). *Aquatic Weeds in the Sudan*. Khartoum: National Council for Research.

Ogilvy, S. (1977). *A Preliminary Study of the Food Habits and Nutritional Status of the Dinka of the Jonglei area*. Khartoum: UNDP.

Oliver, R. (1982). Reflections on the British Institute's expeditions to the Southern Sudan. In *Culture History in the Southern Sudan*, ed. J. Mack & P. Robertshaw, pp. 165–71. Nairobi: British Institute in East Africa.

Omar, M. H. & El Bakry, M. M. (1970). Estimation of evaporation from Lake Nasser. *Meteorological Research Bulletin*, **2**, 551–73.

Payne, W. J. A. (1970). *Cattle Production in the Tropics*. London: Longman.

Payne, W. J. A. (1976). *A Preliminary Report on the Livestock Industry in the Jonglei Canal area*. Khartoum: JEO.

Payne, W. J. A. (undated; *c*.1979). *Economic and Social Aspects of the Various Alignment Proposals for The Jonglei Canal*. Khartoum: UNDP.

Payne, W. J. A. & El Amin, F. M. (1977). *An interim report on the livestock industry in the Jonglei Canal area*. Khartoum: JEO.

Peacock, C. P. (1983). *Sheep and Goat Production at Elangata Wells Group Ranch in the Masai Pastoral System, Kenya*. ILCA Newsletter 2. Addis Ababa: ILCA.

Penman, H. L. (1948). Natural evaporation from open water, bare soil and grass. *Proceedings of the Royal Society of London*, **193**, 120–45.

Penman, H. L. (1963). *Vegetation and Hydrology*. Farnham Royal.

Penning de Vries, F. W. T. & Djiteye, M. A. (1982). *La Productivité des Pâturages Saheliennes*. Wageningen: PUDOC.

Pieterse, A. H. (1978). The water hyacinth (*Eichhornia crassipes*) – a review. *Abstracts on Tropical Agriculture*, **4**, 9–42.

Piper, B. S., Plinston, D. T. & Sutcliffe, J. V. (1986). The water balance of Lake Victoria. *Hydrological Sciences Journal* **31**, 25–37.

PJTC (Permanent Joint Technical Commission) (1981). *The Jonglei Canal Project. An Economic Evaluation*. Khartoum: PJTC.

Platenkamp, J. D. S. (1978). *The Jonglei Canal: its Impact on an Integrated System in the Southern Sudan*. Institut voor Culturele Anthropologie en Sociologie der niet-Westerse Volken, University of Leiden.

Plowright, W. (1981). Herpes Viruses. In *Infectious Diseases of Wild Animals*, 2nd edn., ed. J. W. Davis, L. H. Karstad, & D. O. Trainer, pp. 126–46. Iowa: State University Press.

Plowright, W. (1982). The effects of rinderpest and rinderpest control on wildlife in Africa. *Symposia of the Zoological Society of London*, **50**, 1–28.

Poulsen, G. (1983). *Community Forestry in the Jonglei Canal Area – Sudan. A Plan for Action*. Khartoum: JEO.

Pumphrey, M. E. C. (1941). The Shilluk tribe. *Sudan Notes and Records*, **24**, 1–45.

Rabie, J. W. (1964). Developmental studies on veld grasses. *South African Journal of Agricultural Sciences*, **7**, 583–8.

Redhead, K. (1984). Personal view. *British Medical Journal*, **289**, 760.

Rees, W. A. (1978). The ecology of the Kafue Lechwe: soils, water levels and vegetation. *Journal of Applied Ecology*, **15**, 163–76.

Renvoize, S. A., Lock, J. M. & Denny, P. (1984). A remarkable new grass from the southern Sudan. *Kew Bulletin* **39**, 455–61.

Republic of Kenya (1979). *National Master Water Plan: Stage 1, Vol. 1 – Water Resources and Demands*. New York & Nairobi: T.A.M.S. & Ministry of Water Development.

RES (Range Ecology Survey) (1980). (*As below*); Interim Report No.1. Glasgow, Khartoum & Rome: Mefit Babtie.

RES (Range Ecology Survey) (1983). *Development Studies in the Jonglei Canal Area. Technical Assistance Contract for Range Ecology Survey, Livestock Investigations and Water Supply. Final Report*. Glasgow, Khartoum & Rome: Mefit Babtie (often referred to elsewhere as Mefit Babtie, 1983).

Resource Management & Research (1977) *see* Watson *et al.* (1977).

Roberts, W. D. (1928). *Irrigation Projects in the Upper Nile and Their Effects on Tribal and Local Interests*. Cairo: Official report.

Robyns, W. (1947–55). *Flore des Spermatophytes du Parc National Albert, Vols I–III*. Brussels: Institut des Parcs Nationaux du Congo Belge.

Rölling, N. (1981). *Report on a Visit to the Rural Activities Component of Ilaco's Interventions in and around Bor, Republic of Sudan, Southern Region, with Special Reference to*

Extension, and Smallholder Agriculture Development. Wageningen: Netherlands International Agricultural Centre.

Rollinson, D. H. L. (1963). Reproductive habits and fertility of indigenous cattle to artificial insemination in Uganda. *Journal of Agricultural Science (Cambridge)*, **60**, 279–84.

Ryle, J. & Errington, S. (1982). *Warriors of the White Nile. The Dinka.* Amsterdam: Time-Life Books.

Rzóska, J. (1974). The Upper Nile swamps, a tropical wetland study. *Freshwater Biology*, **4**, 161–85.

Rzóska, J. (Ed.) (1976). *The Nile. Biology of an Ancient River.* The Hague: Junk. (Monogr. Biol. 29).

Sacks, K. (1979). Causality and change in the Upper Nile. *American Ethnologist*, **6**, 437–48.

Sahlins, M. (1961). The segmentary lineage, an organisation of predatory expansion. *American Anthropologist*, **63**, 322–44.

Sandon, H. (1950). *An Illustrated Guide to the Fishes of the Sudan.* Sudan Notes and Records Special Publication. London: McCorquodale & Co.

Sandon, H. (1951). The problems of fisheries in the area affected by the Equatorial Nile Project. *Sudan Notes and Records*, **32**, 5–36.

Sandon, H. & Tayib, A. (1953). The food of some common Nile fish. *Sudan Notes and Records*, **34**, 205–29.

Schove, D. J. (1977). African droughts and the spectrum of time. In *Drought in Africa 2*, ed. D. Dalby, R. J. Harrison, & F. Bezzaz, pp. 38–53. London: International African Institute.

Sculthorpe, C. D. (1967). *The Biology of Aquatic Vascular Plants.* London: Edward Arnold.

SDIT (Southern Development Investigation Team (1955)). *Natural Resources and Development Potential in the Southern Provinces of the Sudan.* Khartoum: Sudan Government.

SES (Swamp Ecology Survey) (1983). *Development Studies in the Jonglei Canal Area. Technical Assistance Contract for Swamp Ecology Survey. Final Report.* Glasgow, Khartoum & Rome: Mefit Babtie (often referred to elsewhere as Mefit Babtie 1983).

Sheppe, W. & Osborne, T. (1971). Patterns of use of a floodplain by Zambian mammals. *Ecological Monographs*, **41**, 179–205.

Shifrine, M. & Gourlay, R. N. (1965). Contagious bovine pleuro-pneumonia in wildlife. *Bulletin of Epizootic Diseases of Africa*, **15**, 319–21.

Sinclair, A. R. E. (1974). The natural regulation of buffalo populations in East Africa. Part 1. Introduction and resource requirements. *East African Wildlife Journal*, **12**, 135–54.

Sinclair, A. R. E. (1977). *The African Buffalo.* Chicago: University of Chicago Press.

Sinclair, A. R. E. & Norton-Griffiths, M. (Eds.) (1979). *Serengeti – Dynamics of an Ecosystem.* Chicago: University of Chicago Press.

Smith, A. (1980). Domesticated cattle in the Sahara and their introduction into West Africa. In *The Sahara and the Nile. Quaternary environments and prehistoric occupation in Northern Africa*, ed. M. A. J. Williams & H. Faure, pp. 503–26. Rotterdam: A. A. Balkema.

Smith, J. (1949). *Distribution of Tree Species in the Sudan in Relation to Rainfall and Soil Texture.* Khartoum: Sudan Ministry of Agriculture Bulletin No. 4.

Sokal, P. & Rohlf, F. (1973). *Biometrics.* New York: John Wiley.

South Dakota University (1978). *Remote Sensing Studies of the Jonglei Canal area.*

Southall, A. (1976). Nuer and Dinka are people: ecology, ethnicity, and logical possibility. *Man (N. S.)*, **11**, 463–91.

Spaulding, J. (1985). The end of the Nubian kingship in the Sudan, 1720–1762. In *Modernization in the Sudan*, ed. M. W. Daly, pp. 17–27. New York: Lilian Barber Press.

Stager, J. C. (1984). The diatom record of Lake Victoria (East Africa): the last 17,000 years. *Proceedings of the 7th Diatom Symposium, 1982*, 455–76.

Stemler, A. (1980). Origins of plant domestication in the Sahara and the Nile Valley. In The
    Sahara and the Nile. *Quaternary Environments and Prehistoric Occupation in Northern
    Africa*, ed. M. A. J. Williams & H. Faure, pp. 503–26. Rotterdam, A. A. Balkema.
Stigand, C. H. (1918). The Dabba of the Sudd area. *Sudan Notes and Records*, 1, 209–10.
Street, F. A. & Grove, A. T. (1976). Late quaternary lake level fluctuations in Africa:
    environmental and climatic implications. *Nature, London*, 261, 385–90.
Struvé, K. C. P. (1907). Report on Khor Atar District. *Sudan Intelligence Report*, 153, 8–11.
Stubbs, J. M. (1949). Freshwater fisheries in the northern Bahr el Ghazal waters. *Sudan
    Notes and Records*, 30, 245–51.
Sudan Government (1953). *Report of the Chief Conservator of Forests, Forestry Division of
    the Ministry of Agriculture, for 1952–53.*
Suliman, Abu el Gasim M. & Jackson, J. K. (1959). The heglig tree. *Sudan Sylva*, 9, 1–8.
Sutcliffe, J. V. (1957). *The Hydrology of the Sudd Region of the upper Nile*. PhD Thesis,
    University of Cambridge.
Sutcliffe, J. V. (1974). A hydrological study of the southern Sudd region of the upper Nile.
    Hydrological Sciences Bulletin, 19, 237–55.
Sutcliffe, J. V. (1979). Obituary: Harold Edwin Hurst. *Hydrological Sciences Bulletin*, 24,
    539–41.
Sutcliffe, J. V. & Parks, Y. P. (1982). *A Hydrological Estimate of the Effects of the Jonglei
    Canal on Areas of Flooding*. Wallingford, UK: Institute of Hydrology.
Sutcliffe, J. V. & Parks, Y. P. (1987). Hydrological Modelling of The Sudd and Jonglei
    Canal. *Hydrological Sciences Journal*, 32, 143–59.
Sutton, J. E. G. (1974). The aquatic civilisation in Middle Africa. *Journal of African History*,
    15, 527–46.
Tadros, T. M. (1940a). Structure and development of *Cyperus papyrus*. *Bulletin of the
    Faculty of Science, Fouad 1 University, Cairo*, 20, 1–28.
Tadros, T. M. (1940b). The daily changes in the concentration of O_2 and CO_2 in the internal
    atmosphere of *Cyperus papyrus* and the ventilation of submerged organs. *Bulletin of the
    Faculty of Science, Fouad 1 University, Cairo*, 20, 31–66.
Tahir, A. A. & El Sammani, M. O. (1978). *Environmental and Socio-Economic Impact of the
    Jonglei Canal Project*. Khartoum: JEO.
Talling, J. F. T. (1957a). The longitudinal succession of water characteristics in the White
    Nile. *Hydrobiologia*, 11, 73–89.
Talling, J. F. T. (1957b). The phytoplankton population as a compound photosynthetic
    system. *New Phytologist*, 56, 133–49.
Thompson, K., Shewry, P. R. & Woolhouse, H. W. (1979). Papyrus swamp development in
    the Upemba Basin, Zaire; studies of population structure in *Cyperus papyrus* stands.
    Botanical Journal of the Linnaean Society 78, 299–316.
Tosh, J. (1981). The economy of the Southern Sudan under the British, 1898–1955. *Journal
    of Imperial and Commonwealth History*, 9, 275–88.
Tothill, J. D. (Ed.) (1948). *Agriculture in the Sudan*. Oxford: University Press.
Trapnell, C. G. (1959). Ecological results of woodland burning experiments in Northern
    Rhodesia. *Journal of Ecology*, 47, 129–68.
UNDP (1977). *The Sudan: Multi-Temporal Landsat-Imagery Interpretation of the Flood
    Region Draining to the Sudd*. Rome: FAO.
UNDP/JEO. (1977). *Project proposals for Fisheries Development in the Jonglei Area.*
    Khartoum.
UNDP/JEO. (1978). *Sudd Fisheries Development Programme, Phase 1*. Khartoum.
USAID (1976). *Egypt: Major Constraints to Agricultural Production*. Washington, DC.
USDI (US Department of the Interior) (1964) *Land and Water Resources of the Blue Nile*

Basin: Ethiopia. 17 Vols. Washington, DC: Bureau of Reclamation.

Vesey-Fitzgerald, D. F. (1970). The origin and distribution of valley grasslands in East Africa. *Journal of Ecology*, **58**, 51–75.

Walker, D. (1970). Direction and rate in some British post-glacial hydroseres. In *Studies in the Vegetational History of the British Isles*, ed. D. Walker & R. G. West, pp. 117–39. Cambridge: Cambridge University Press.

Wall, L. L. (1976). Anuak politics, ecology, and the origins of Shilluk kingship. *Ethnology*, **15**, 151–62.

Waterbury, J. (1979). *Hydropolitics of the Nile Valley*. Syracuse University Press.

Waterbury, J. (1982). *Riverains and Lacustrines: Towards International Co-operation in the Nile Basin*. Discussion Paper No. 107, Research Program in Development Studies, Woodrow Wilson School, Princeton University.

Watson, C. M. (1876) Notes to accompany a traverse survey of the White Nile, from Khartoum to Rigaf. *Journal of the Royal Geographical Society*, **46**, 412–27.

Watson, R. M., Tippett, C. I., Rizk, J., Beckett, J. J. & Jolly, F. (1977). *Sudan National Livestock Census and Resource Inventory*. Nairobi: Resource Management & Research Ltd.
Vol. 18A. The results of an aerial census of Upper Nile Province in July/August 1976.
Vol. 18B. The results of an aerial census of Jonglei Province.
Vol. 20B. The results of an aerial census of El Buheyrat.
Vol. 26. Some of the ecological and wildlife implications.

Webster, J. B. (1979). Noi! Noi! Famines as an aid to interlacustrine chronology. In *Chronology, Migration and Drought in Interlacustrine Africa*, ed. J. B. Webster, pp. 1–37. London: Longman & Dalhousie University Press.

Welcomme, R. L. (1979). *Fisheries Ecology of Floodplain Rivers*. London: Longman.

Wendt, W. B. (1970). Responses of pasture species in eastern Uganda to phosphorus, sulphur and potassium. *East African Agriculture and Forestry Journal*, **36**, 211–19.

Werne, F. (1849). *Expedition to Discover the Sources of the Nile*, transl. C. W. O'Reilly. London: R. Bentley.

Westerman, D. (1912). *The Shilluk People: Their Language and Folklore*. Philadelphia, Pa.: Board of Foreign Missions of the United Presbyterian Church of North America; Berlin: Dietrich Reiner (Ernst Vohsen).

White, F. (1983). *The Vegetation of Africa*. Paris: UNESCO.

Whittington, D. & Haynes, K. E. (1985). Nile water for whom? In *Agricultural Development in the Middle East*, ed. P. Beaumont & K. McLachlan. Chichester: John Wiley & Sons.

Wilson, H. H. (1903). Report ... on the Dinkas of the White Nile. *Sudan Intelligence Report*, **104**, 11–18.

Wilson, H. H. (1905). Report ... on march from the Sobat (mouth of the Filus) to Bor. *Sudan Intelligence Report*, **128**, 5 9.

Wilson, R. J. (1976a). Studies on the livestock of Southern Darfur, Sudan. IV. Production traits in goats. *Tropical Animal Health and Production*, **8**, 221–32.

Wilson, R. J. (1976b). Studies on the livestock of Southern Darfur, Sudan. III. Production traits in sheep. *Tropical Animal Health and Production*, **8**, 103–14.

Wilson, R. J. (1983). Livestock production in Central Mali. The Macina wool sheep of the Niger inundation zone. *Tropical Animal Health and Production*, **15**, 17–31.

Wilson, R. J. & Clarke, S. E. (1976a). Studies on the livestock of Southern Darfur, Sudan. I. The ecology and livestock resources of the area. *Tropical Animal Health and Production*, **7**, 165–87.

Wilson, R. J. & Clarke, S. E. (1976b). Studies on the livestock of Southern Darfur, Sudan. II. Production traits in cattle. *Tropical Animal Health and Production*, **8**, 47–51.

Winder, J. (1939–40). *Unpublished reports on the Jonglei Scheme* (in Sudan Historical Archive, Durham University).

Wither, J. F. (1981). Brucellosis. In *Infectious Diseases of Wild Mammals*, 2nd edn, ed. J. W. Davis, L. H. Karstad & D. O. Trainer, pp. 280–7. Iowa State University Press.

Woodford, M. (1984). *Rinderpest in Wildlife in Sub-Sahelian Africa*. FAO Technical Analysis Report (AG:TCP/RAF/2323).

Wyld, J. G. (1930). *Report and Notes on Bor-Duk district*. National Record Office, Civil Secretary, 57/2/8.

Young, D. D. (1976). *The Development Aspects of the Jonglei Scheme*. Khartoum: JEO.

Zein, S. El (1975). The water resources of the Nile for agricultural development in the Sudan. In *Aquatic Weeds in the Sudan*, ed. M. Obeid. Khartoum: National Council for Research.

INDEX

Aak Nuer 204
Abbay river *see* Blue Nile
Abialang Dinka 207
Acacia
  drepanolobium 164
  effect of canal 392
  fistula 164
  polyacantha 164, 165, 179
Acacia seyal 164, 165, 166
  extent of woodland 179
  production 501
  regeneration of woodland 443
  use in construction 501
Acholi people 10
Addis Ababa Agreement (1972) 259, 436, 465
Addis Ababa Agreement, oil and mineral rights
  467
Adenota spp. 358
Adok 110
aerial surveys 7, 8, 486
Aeschynomene indica 157
Agreement for the Full Utilisation of the Nile
  Waters (1959) 47, 48, 62–3, 66
agriculture
  development 437–42
  large scale mechanised 438
Albert Nile 24
Alestes 331, 335, 407
alfisols 471
algae
  free-floating in seasonal pools 157
  and oxygen level of water 142
  planktonic 131, 137–8
Aliab
  channels 103
  Dinka 206
  river 113
Aliab Valley 113
  flooding 189–91, 193, 195–6
  hydrological changes 192–3
  land use 196

topography 189–91
vegetation distribution 191–2, 193–6
wild mammal population 195–6
alignment of canal 37, 56–8, 272
Alur 10
anabatids 340
anaplasmosis 299
Andropogon gayanus 161
Anglo-Egyptian government (1898–1955) 218–19
  see also Condominium Government
animal
  husbandry 234–5, 252
  management 312
  nutrition 301, 302
anthrax 300
Anuak people 10, 207, 214
aquaculture 407
aquifers 421
Arachis hypogea 318
Archimedean screw 30
army worm 323
Aswa torrent 25
Aswan High Dam 3, 5, 30, 33, 37
  displacement of inhabitants 33
  planning 43, 47
  preconditions to construction 66
  significance for southern Sudan 47
  storage of water 43
  surface evaporation rate 71
  water use by Egypt 66
Atbara
  development of storage sites 76, 77
  river 28
Atem river
  and alignment of Canal 56
  channel 7, 114
  chemical changes 130
  conductivity of water 131
  drying out 130
  flow-through 130
  stagnation 130